With Author's Compliments

Shiro Takaoshima

Electrical Properties of
Biopolymers and Membranes

Electrical Properties of Biopolymers and Membranes

Shiro Takashima

Department of Bioengineering
University of Pennsylvania

Adam Hilger, Bristol and Philadelphia

British Library Cataloguing in Publication Data

Takashima, Shiro
    Electrical properties of biopolymers and membranes
    1. Biopolymers. Electrical properties.
    I. Title
    574.19′24

    ISBN 0-85274-136-7

Library of Congress Cataloging-in-Publication Data

Takashima, Shiro, 1923–
    Electrical properties of biopolymers and membranes.

    Includes bibliographies and index.
    1. Biopolymers–Electric properties. 2. Biopolymers–Dipole moments.
    3. Membranes (Biology) I. Title.
    [DNLM: 1. Electrophysiology. 2. Membranes–physiology. 3. Polymers.
    QT34 T136e]
    QP517.B53T35 1989   574.19′127   88-34773
    ISBN 0-85274-136-7

The author and IOP Publishing Ltd have attempted to trace the copyright holder of all the figures and tables reproduced in this publication and apologize to copyright holders if permission to publish in this form has not been obtained.

Published under the Adam Hilger imprint by IOP Publishing Ltd
Techno House, Redcliffe Way, Bristol BS1 6NX, England
242 Cherry Street, Philadelphia, PA 19106, USA

Typeset by KEYTEC, Bridport, Dorset
Printed in Great Britain by Billing and Sons Ltd, Worcester

Contents

Preface

The origin of dipole moment of most biological materials is different from that of small polar molecules and, therefore, different theories are required to explain its extraordinary dielectric behaviour. Namely, the concept of interfacial polarization plays as important a role as the theory of polar molecules for a discussion of the electric polarization of biological polyions. Within the limited space of this book it is virtually impossible to review all the work done in this field. My intention is to outline the basic principles and discuss how these theories have been used to explain the dielectric properties of biopolymers and membranes.

I do not intend to present complete proofs of the theories. Since step by step derivations are not given in this book, the discussion on some theories may be difficult to follow. References are given for the convenience of the reader so that more detail can be found in the original paper or in more specialized books.

Chapters 1–3 discuss the definition of a dipole moment, the mechanism of the polarization of a group of dipoles and their frequency dependence. These basic discussions may be redundant to some readers who are familiar with dielectric theories, yet this book is aimed at either the biochemist or biophysicist who has little or no experience in this field, so these chapters are needed. Chapter 4 discusses the electric polarization of synthetic polymers. The early development of the theories and some experimental results are presented as briefly as possible. Some of the theories discussed in this chapter are excessively complex; however, an understanding of them is not an essential prerequisite for the subsequent chapters. Moreover, I did not attempt to review recent progress in this field. After all, the material covered in this chapter is not directly related to biophysical problems.

Chapter 5 discusses the dipole moment and relaxation of protein molecules. Although classical polar theory still plays a dominant role for the protein molecules, the concept of interfacial polarization begins to emerge in this chapter. Protein molecules are highly charged and the distribution of fixed and mobile charges is the major determinant of the extraordinarily large dipole moment. Chapter 6 is devoted to a discussion on the interfacial polarization of inhomogeneous systems. The dielectric properties of many biological systems cannot be explained by the theory of polar molecules and it was necessary to introduce a different concept. Interfacial polarization has undergone a series of evolutions. It all started with the so-called Maxwell–Wagner theory many years ago. Progress was accelerated when the concept of mobile ion fluctuation was introduced and applied to polyelectrolyte molecules. We cannot discuss the dielectric behaviour of polyelectrolyte without some knowledge of these theories.

Chapter 7 discusses the dielectric behaviour of DNA molecules at low and high frequencies. The low-frequency relaxation of DNA is due to counterion fluctuations and this topic can be considered as an application of the theories discussed in Chapter 6. In addition, the polarization of DNA at very high frequencies due to resonance absorption is discussed. The structure of liquid water and ice and their dielectric properties are presented in Chapter 8. The first few sections discuss the properties of pure water and ice and the rest of the chapter is devoted to a discussion of the hydration water of proteins and DNA.

Usually, dielectric measurements are carried out using small electrical fields. Under these circumstances the measured dielectric constant is independent of input signals. These are called linear properties. When the intensity of the applied fields increases, the systems begin to exhibit non-linear dielectric behaviour. Chapter 9 discusses the mechanism of non-linear polarizations. Finally the electrical impedance problems of nerve and muscle membranes are discussed in Chapter 10. This chapter begins with the structure and transport properties of excitable membranes. Ion transport has a profound effect on membrane impedance or admittance. The chapter then focuses on the linear and non-linear membrane impedances of nerve and muscle fibres.

Before I conclude this preface, mention has to be made of the units used in this book. The early dielectric theories were developed using cgs esu units. However, in recent years, more people are inclined to use rationalized MKSA units. Therefore, the logical choice is MKSA units rather than cgs units. However, I found it a difficult task to convert all the early theories from cgs units to MKSA units. This is the reason why I adhered to cgs units in spite of the possible objection from the reader.

The author is deeply indebted to the late Dr K S Cole from whom I learned so much about excitable membranes. The memories of several summers I spent at Woods Hole, Mass. and the numerous discussions I had with Dr Cole will remain with me for ever. I am privileged to have had Professor H P Schwan as my mentor and friend for so many years. I have benefited a great deal from his vast knowledge on dielectric theories and techniques and related subjects. There are many other scientists to whom I am indebted; Professors H Froehlich, E H Grant, A Minakata, I Tasaki, H Pauly, G Schwarz, H M Fishman and K R Foster, to name but a few. I also would like to thank Professor J Bateman who induced me to the writing of this book. Finally, I would like to express my gratitude to our able secretaries. Without their assistance, this book would not have been possible. The last of my acknowledgments goes to my patient wife Yuki to whom this book is dedicated with my affection and appreciation.

Shiro Takashima
June 1988

1 The Origin of Dipole Moment

1.1 Introduction

Whereas dielectric phenomena in general can be treated using classical theories, the origin of dipole moment is by nature quantum mechanical. In this chapter, the concept of dipole moment and polarizability will be discussed using simple examples.

1.2 Wave Functions and the Dipole Moment Operator

Roughly, molecules can be divided into two categories: non-polar and polar. Non-polar molecules are those which do not have a dipole moment and polar molecules are those having a non-zero dipole moment. A dipole moment is produced when the centres of positive and negative charges are displaced by a certain distance r. If the centre of positive charges due to nuclei of diatomic molecules coincides with the centre of negative charges due to electrons, then the molecule has no dipole moment. The hydrogen molecule, H–H, is one of these examples. On the other hand, in hetero-polar diatomic molecules such as H–Cl, the distribution of electrons is asymmetric because Cl is more electronegative than hydrogen. This means that the centre of negative charges is pulled toward Cl causing separation of the charge centres. Hence a non-zero dipole moment.

The above qualitative statement can be formalized using the following equations. The centre of positive charges due to nuclei, D_n, is defined by

$$D_n = \sum_k R_k Q_k \left(\sum_k Q_k \right)^{-1} \tag{1.1}$$

where R_k is the position vector of the kth nucleus and Q_k is the charge. On the other hand, the centre of negative charges due to electrons is given by the following equation:

$$D_e = \frac{1}{eN} \sum_k \int \psi_k (er) \psi_k \, d\tau \qquad (1.2)$$

where ψ_k is the wave function of valence electrons, r is the displacement vector of the ith electron and the term in brackets (er) is the dipole moment operator. 'e' is electronic charge and N is the number of electrons. The wave function ψ_k usually represents valence electrons only and non-valance electrons are 'collapsed' into the nucleus. If the molecule is neutral, the total positive charges due to nuclei must be equal to the total negative charges of electrons

$$\left| \sum_k Q_k \right| = |-eN| \qquad (1.3)$$

and the dipole moment of the molecule is given by

$$\mu = eN(D_n - D_e). \qquad (1.4)$$

Substituting equations (1.1), (1.2) and (1.3) into (1.4), we obtain

$$\mu = \sum_k R_k Q_k - \sum_k \int \psi_k (er) \psi_k \, d\tau. \qquad (1.5)$$

Evaluation of the integrals in (1.2) or (1.5) is discussed by McGlynn *et al* (1972) for various wave functions.

In equation (1.5), the integration is performed only for valence electrons and non-bonding electrons are usually considered as part of the nucleus. In spite of this simplification, the integration is difficult to perform even for simple diatomic molecules. Above all, careful choice of wave functions is very important to attain sufficient accuracy. Thus, it is more expedient to use a simplistic approach, treating the charge separation r as an empirical quantity. With this simplification, dipole moment μ is defined as

$$\mu = er. \qquad (1.6)$$

The magnitude of elementary charge 'e' is 4.8×10^{-10} esu in cgs units. Assuming the charge separation to be about 1 Å (10^{-8} cm), the dipole moment will be

$$\mu = 4.8 \times 10^{-18} \text{ esu cm.}$$

Usually dipole moment is expressed in Debye units (D), i.e. 10^{-18} esu cm. Using this unit, the dipole moment calculated above becomes 4.8 D, a much easier number to handle. The dipole moments of some chemical bonds and simple diatomic molecules are listed in tables 1.1 and 1.2. As shown in these tables, the dipole moments of asymmetric hetero-polar diatomic molecules are non-zero and their magnitudes seem to depend on the difference in electronegativities of the nuclei. For example, the dipole moments of hydrogen halides decrease in the following order:

$$HF > HCl > HI.$$

Electronegativity is a measure of the electron attracting power of elements and there is a scale, x, associated with it. Table 1.3 shows some examples. The important observation is that the dipole moment of a bond A–B is nearly equal to the difference $x_A - x_B$ as shown in table 1.4. Since electronegativity difference is a measure of the ionic character of chemical bonds, the magnitude of the dipole moment can also be used as a measure of the extent of ionic nature of the bond. Namely, the dipole moments of chemical bonds increase as the ionic nature increases.

Table 1.1[†] The dipole moments of chemical bonds in Debye units.

H–C	H–N	H–C	C–Cl
0.4	1.31	1.51	1.46, 1.7 [‡]
C–N	C=N	C–O	C=O
0.22	0.90	0.74	2.3, 2.7[‡]

[†] Taken from Smyth (1955).
[‡] Taken from Davies (1965).

Table 1.2[†] The dipole moments of diatomic molecules in Debye units.

H–H	Br–Br	N=N
0	0	0
H–F	H–Cl	H–I
1.75, 1.9[‡]	1.03	0.38

[†] Taken from Liberles (1968).
[‡] Taken from Coulson (1952).

Table 1.3[†] The electronegativity scale for elements.

Elements	H	F	Cl	Br	I
x_i	2.1	4.0	3.0	2.8	2.5

[†] Taken from Coulson (1952).

Table 1.4[†] Dipole moments and electronegativity differences.

	HF	HCl	HBr	HI
$x_A - x_B$	1.9	0.9	0.8	0.4
μ (D)	1.91	1.03	0.78	0.38

[†] Taken from Coulson (1952).

So far, we have discussed the dipole moments of diatomic molecules and single chemical bonds. The dipole moments of polyatomic molecules can be calculated if the dipole moments of chemical bonds are known. In order to add bond moments vectorially, we use the following equation:

$$\mu_r^2 = \mu_1^2 + \mu_2^2 + 2\mu_1\mu_2 \cos \theta \qquad (1.7)$$

where μ_r is the net dipole moment of a polyatomic molecule, μ_1 and μ_2 are the dipole moments of bonds 1 and 2 (see figure 1.1) and θ is the angle between them.

Figure 1.2 illustrates two examples of linear triatomic molecules. As shown Cl–Be–Cl and O=C=O are symmetric and therefore

$$\mu_1 = \mu_2 \qquad \text{and} \qquad \theta = 180°.$$

Using equation (1.7), we can find readily that the net dipole moment of these molecules must be zero. On the other hand the angle between two O–H bonds in water is known to be 104.5°, as shown in figure 1.2. Therefore, the dipole moment of water is calculated to be 1.839 D and the dipole vector must be in the direction of the axis of symmetry which bisects the bond angle.

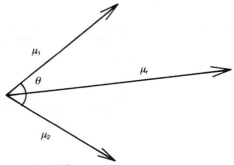

Figure 1.1 The vectorial sum of two bond moments in order to calculate the net dipole moment of triatomic molecules.

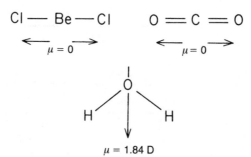

Figure 1.2 The dipole moments of linear and non-linear triatomic molecules.

Likewise, the dipole moment of polyatomic molecules such as carbon tetrachloride can be calculated using bond moments and angles. As shown in figure 1.3, C–Cl$_4$ has four polar bonds C–Cl. However, these bonds have a tetrahederal configuration and the vectorial sum of these bond moments will be exactly zero. If some of these chlorines are replaced by hydrogen, introducing asymmetry, the net dipole moments become non-zero as shown in figure 1.3.

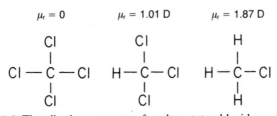

Figure 1.3 The dipole moments of carbon tetrachloride and chloromethanes.

In the above, we calculated the dipole moment of water by adding the dipole moments of two O–H bonds vectorially. In this calculation, we used 104.5° as the angle between the two O–H bonds instead of 90°. The reason for this is as follows. In the oxygen atom, two electrons in the 2s orbital and four electrons in 2p orbitals form sp^3 hybridized orbitals. These four bonds, therefore, form a tetrahedron with bond angles of 104.5°. Two O–H bonds are formed using two of these hybridized bonds and the 1s electrons of the H atoms. The remaining two orbitals are filled with two electrons and they extend away from the hydrogen atoms without participating in bond formation. These are called 'lone pair' electrons. The negative charges of lone pair electrons will cause further separation of the negative and positive charge centres of the O–H bond which entails a larger dipole moment. This means that a part of the dipole moment of an O–H bond arises from lone pair electrons as shown in figure 1.4 (two small arrows). As discussed before, the charge centre of electrons was calculated using equation (1.2).

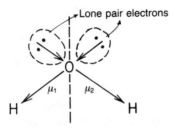

Figure 1.4 Lone pair electrons of sp^3 hybridized O–H bonds.

However, the presence of lone pair electrons shifts the negative charge centre further toward oxygen, thereby increasing the magnitude of the dipole moment. This again shows the difficulty of calculating the dipole moment of the chemical bonds non-empirically using the quantum mechanical principle.

1.3 The Calculation of π-bond Moments

The calculation of dipole moments of polyatomic molecules as a vectorial sum of bond moments has limited applicability to organic molecules which have conjugated double bonds. It is well known that double bonds are made up of a σ-bond as well as a π-bond. σ-electrons are localized around the nucleus. However, π-electrons are delocalized and diffusely distributed over the entire molecule. Therefore, the concept of bond moment does not apply to π-electrons and we cannot calculate the dipole moment of conjugated molecules as the vectorial sum of bond moments. Instead, we have to use molecular orbital theory as explained in this section.

If the structure of an organic molecule is symmetric and consists of conjugated hydrocarbon chains, the distribution of π-electrons will be uniform all over the molecule. Under these circumstances, the contribution of π-electrons to the dipole moment of the molecule reduces to zero. For example, the distribution of π-electrons in a benzene molecule is symmetrical within the molecular plane and this entails a zero π-electron moment. If one of the carbon atoms in benzene is methylated, this will introduce asymmetry in the distribution of π-electrons and a non-zero dipole moment will result. In order to calculate the moment due to π-electrons, the distribution of π-electrons over the entire molecule must be calculated using appropriate molecular orbital theories.

In general, molecular orbitals of conjugated molecules can be written as a linear combination of atomic orbitals

$$\Psi_r = C_1\phi_1 + C_2\phi_2 + \ldots + C_n\phi_n \tag{1.8}$$

were ϕ_js are atomic wave functions. C_j is an orthonormalized coefficient and the square of C_j represents the density of π-electrons around the jth element. The energy levels of these molecular orbitals and the coefficients C_j can be calculated using various molecular orbital methods. A brief discussion of molecular orbital calculations is given in Appendix A.

Let us apply molecular orbital calculations to a simple molecule methylene cyclopropanol (see figure 1.5) and use the result to calculate the π-bond moment. Using the simplest molecular orbital method HMO (the Hückel molecular orbital method), the energy levels of molecular orbitals and the coefficients C_{ij} are calculated and are shown in table 1.5. (See Appendices A and B for the method of calculation.)

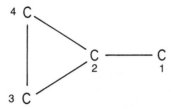

Figure 1.5 The molecular structure of methylene cyclopropanol.

In order to calculate dipole moment using this table, we still have to compute π-electron densities q_r as defined by

$$q_r = \sum_j n_j C_{rj}^2 \tag{1.9}$$

where n_j is the number of electrons occupying the jth orbital. For example, as discussed in Appendix A, we fill low lying molecular orbitals with two π-electrons (see figure 1.6). Since there are four π-electrons altogether in this molecule, the orbitals ϕ_1 and ϕ_2 are occupied by two π-electrons each while the orbitals ϕ_3 and ϕ_4 are vacant. Therefore, the numerical values of n_1, n_2 are 2 and n_3, n_4 are zero. The values of q_r are also listed in table 1.5. The last line in this table shows the net charge densities. The actual charge at these elements is the difference between the positive charge due to the nucleus and the negative charge due to electrons. Therefore, the net charge density is defined as

$$\zeta_r = 1 - q_r. \tag{1.10}$$

Table 1.5† The results of the molecular orbital calculation for methylene cyclopropanol.

	x	C_1	C_2	C_3	C_4
1	-2.170	0.278	0.612	0.524	0.524
2	-0.311	0.814	0.253	-0.368	-0.368
3	1.000	0	0	0.707	-0.707
4	1.418	0.520	-0.736	-0.304	-0.304
q_r		1.4778	0.882	0.802	0.820
ζ_r		-0.478	0.118	0.180	0.180

† Taken from Streitweiser (1961).

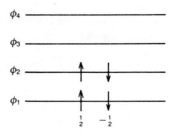

Figure 1.7 shows the charge densities in methylene cyclopropanol graphically. Choosing atom 2 as the origin of coordinates, we can calculate the dipole moment in the x direction as follows. Assuming the bond length to be 1.4 Å,

$$\mu_\pi = [2(0.18)(1.40)(\cos 30°) + (0.478)(1.40)]4.8$$

$$= 5.27 \text{ (D)}. \tag{1.11}$$

Because of symmetry, the moment in the y direction reduces to zero. Also because this molecule is planar, the moment in the z direction must vanish. Thus, the dipole moment of methylene cyclopropanol lies along the axis of symmetry as shown in the same figure.

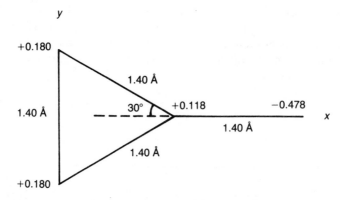

Figure 1.7 The charge densities ζ_r in methylene cyclopropanol. (From Streitwieser 1961.) Reproduced by permission of John Wiley & Sons, Inc. © 1961.

The discussion presented above is based on the result of the HMO calculation. It is known that the dipole moments calculated by HMO tend to be too large. Refinement of the calculation using more advanced molecular orbital theories would yield considerably smaller values.

1.4 Polarizablity

When molecules are placed in an electromagnetic field, the distribution of electrons will undergo a small change. These changes are treated as perturbation. Usually, the first-order perturbation term vanishes and polarizability can be defined as a second-order perturbation (Cohan *et al* 1957)

$$\alpha_{xx} = 4e^2 \sum_p \sum_q \frac{\langle \phi_p | x | \phi_q \rangle^2}{E_q - E_p} \qquad (1.12)$$

where α_{xx} is the polarizability along the x coordinate and e is the electronic charge. $\langle \phi_p | x | \phi_q \rangle$† is the element of the polarizability tensor, ϕ_p and ϕ_q are wavefunctions with energies E_p and E_q. Here, E_q is the energy of the excited state and E_p that of the ground state. The small perturbation of electron distribution by the applied external fields produces an induced dipole moment in the direction of the E-vector. Induced dipole moment is different from the permanent dipole moment discussed above in that the former appears only in the presence of electromagnetic fields whereas permanent dipole moment exists in the molecule even in the absence of the field. It is important to point out at this juncture that polarizability is closely related to refractive index in the optical region. Furthermore refractive index is related to the dielectric constant by Maxwell's relation:

$$n^2 = \varepsilon. \qquad (1.13)$$

Let us use equation (1.12) to calculate the polarizability due to π-electrons. Using molecular orbitals, $\psi_r = \sum C_{pj}\phi_j$, the matirx elements can be rewritten as

$$\langle \phi_p | x | \phi_q \rangle = \sum C_{pj} C_{qj} x_j \qquad (1.14)$$

where x_j is the x coordinate of the jth atom. Using this, polarizability can be expressed as follows:

$$\alpha_{xx} = -e^2 \sum_{p=1}^{m} \sum_{q=m+1}^{n} \pi_{jk} x_j x_k \qquad (1.15)$$

where the quantity π_{jk} is defined by the following equation (Coulson and Longuet-Higgins 1947a, b):

$$-\pi_{jk} = 4 \sum_{j}^{m} \sum_{k}^{n} \frac{C_{pj} C_{qj} C_{pk} C_{qk}}{E_q - E_p}. \qquad (1.16)$$

Similar expressions can be obtained for the y and z directions. Clearly,

† $\langle \phi_p | x | \phi_q \rangle$ can also be written as $\int \phi_p x \phi_q \, d\tau$.

the calculation of polarizability α_{xx} reduces to computing molecular orbitals using appropriate methods such as HMO or more advanced theories. As an example, the calculation of polarizabilites of uracil and adenine is given in table 1.6. It is well known that double helical DNA has a large optical polarizability in the transverse direction, i.e. along the planes of base pairs, while the polarizability along the helical axis is very small. As shown in this table, both uracil and adenine have large polarizablities in the x–y plane while that in the z direction is zero. Therefore, the π-electron polarizabilities of planar base molecules are markedly anisotropic giving rise to a strong optical birefringence in DNA and synthetic polynucleotides.

Table 1.6† The polarizabilities of uracil and adenine in units of 10^{-25} cm³.

	α_x	α_y	α_z
Uracil	80.44	43.61	0
Adenine	100.8	131.6	0

† Taken from Takashima (1969).

References and Suggested Reading

Cohan N V, Coulson C A and Jamieson J B 1957 π-electron polarizablities of some condensed aromatic molecules *Trans. Faraday Soc.* **53** 582–8

Coulson C A 1952 *Valence* (Oxford: Clarendon)

Coulson C A and Longuet-Higgins H C 1947a The electronic structure of conjugated systems, I. General theory *Proc. R. Soc.* A **191** 39–60

—— 1947b The electronic structure of conjugated systems. II. Unsaturated hydrocarbons and their hetero derivatives *Proc. R. Soc.* A **192** 16–32

Davies M 1965 *Some Electrical and Optical Aspects of Molecular Behavior* (Oxford: Pergamon)

Liberles A 1968 *Introduction to Theoretical Organic Chemistry* (New York: MacMillan)

McGlynn S P, Vanquickenborne L G, Kinosita M and Caroll D G 1972 *Introduction to Applied Quantum Chemistry* (New York: Rinehart and Winston)

Smyth C P 1955 *Dielectric Behavior and Structure* (New York: McGraw-Hill)

Streitwieser A Jr 1961 *Molecular Orbital Theory for Organic Chemists* (New York: Wiley)

Takashima S 1969 Optical anisotropy of synthetic polynucleotides *Biopolymers* **8** 199–216

2   The Polarization of Groups of Polar Molecules

2.1 Introduction

We have discussed the origin of dipole moment in polar molecules in the previous chapter. As a next step, we will discuss the behaviour of a group of polar molecules in the gaseous, liquid and solid states. Various theories of electric polarization of dipoles in static fields will be presented.

2.2 The Behaviour of a Polar Molecule in an Electric Field

Polar molecules can exist as gases, liquids or solids. As is well known, these molecules undergo rapid Brownian motion in gases, and their distribution can be described as completely random if there is no external electric field. In liquids, particularly those in which constituent molecules are hydrogen bonded one to another, intermolecular forces become significant, restricting thermal motion considerably. Under these circumstances, a liquid can no longer be considered as a random state. Due to the considerable order which may exist in the liquid state, the orientation of polar molecules becomes different from those in the gaseous state. Let us start the discussion about simple gaseous molecules with an assumption that there are no intermolecular interactions.

When we apply an electric field to a group of molecules, the vector of the E-field will not, in general, be parallel to the vector of the dipoles because of their nearly random distribution. For the sake of simplicity, let us consider the orientational behaviour of one polar molecule in a static electrical field. Since we are assuming that there is no interaction between these molecules, the information we obtain from this analysis can only be extended to a group of dipoles if the orientation of a dipole does not influence the behaviour of other molecules. As shown in figure 2.1, the dipole vector makes an angle θ with the vector of the field. The negative charge at one end of the dipole tends to migrate in the direction of applied field. However, this movement will be counteracted

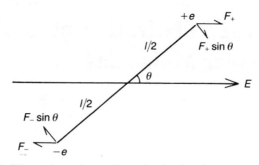

Figure 2.1 The orientation of a dipole in the presence of an electrical field E. e^+ and e^- are the charges at both ends of the dipole. F_+ and F_- are the forces due to the interaction of the field with these charges. l is the distance between charges. (After Gabler 1978.)

by the opposite movement of the positive charge at the other end and, consequently, the net translational movement will be zero. However, the torque consists of two components, namely $F_+ \sin \theta$ and $F_- \sin \theta$, acting on the dipole as shown in the figure. These forces tend to reorient the molecule along the direction of electrical field. As is well known, torque is a quantity which is defined as a cross product of force and length vectors (E and L in this case), i.e.

$$\boldsymbol{\Gamma} = \boldsymbol{L} \times \boldsymbol{E}. \tag{2.1}$$

Using scalar notation, the total torque is given by the following equation:

$$\Gamma = F_+(l/2) \sin \theta + F_-(l/2) \sin \theta = Fl \sin \theta. \tag{2.2}$$

Since the force F is given by qE, the torque is rewritten as

$$\Gamma = qEl \sin \theta \tag{2.3}$$

where q is the charge. Using the relation $\mu = ql$, we can write the torque as

$$\Gamma = \mu E \sin \theta. \tag{2.4}$$

The incremental work dw which must be done to reorient the dipole by an angle dθ is

$$\mathrm{d}w = \Gamma \, \mathrm{d}\theta. \tag{2.5}$$

Integrating this equation, we obtain

$$w = \int \Gamma \, \mathrm{d}\theta. \tag{2.6}$$

Inserting equation (2.4) into equation (2.6):

$$w = \mu E \int \sin \theta \, d\theta = -\mu E \cos \theta. \qquad (2.7)$$

Equation (2.6) shows the amount of work or energy which is needed to reorient a dipole by an angle θ when an electrical field E is applied. However, we must remember that even in the presence of an electrical field, dipoles are still under the influence of thermal motion which amounts to an energy kT where k is the Boltzmann constant $(1.38 \times 10^{-16} \, \text{erg K}^{-1})$ and T is the temperature. Therefore, the torque which tends to orient the dipole along the field is always counteracted by thermal energy. The function $\exp(\mu E/kT)$ plays an essential role in calculating the average orientation of a group of dipoles in an applied electrical field as will be discussed presently.

2.3 An Ensemble of Dipoles and Polarization

In the previous section we discussed the behaviour of one dipole which is placed in an electrical field. Now we will apply the same principle to a group of dipoles consisting of a very large number of the same polar molecules. Again we assume that the molecules are in the gaseous state. We assume a hypothetical sphere of radius r surrounding a group of molecules, as shown by figure 2.2. Before we apply a field, the orientation of these molecules is completely random, that is to say the probability of finding these molecules pointing in some direction is uniformly equal to $1/4\pi r^2$. By inspection of figure 2.2, we can calculate the probability of finding molecules pointing in a direction between θ

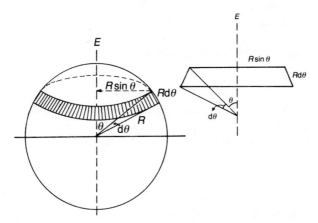

Figure 2.2 The orientation of a group of dipoles which are placed in a spherical cavity. The inset shows the trapezoid which represents the narrow strip on the surface of the sphere. E is electrical field.

and $\theta + d\theta$ using the following equation:

$$\frac{dN}{N} = \frac{2\pi r \sin \theta r d\theta}{4\pi r^2} \tag{2.8}$$

namely, the probability is given as the ratio between the area of the narrow strip on the surface and the entire surface area of the sphere. Simplifying this equation, we obtain

$$\frac{dN}{N} = \tfrac{1}{2}\sin \theta \, d\theta. \tag{2.8a}$$

If a field E is applied, the probability is given as the product of equation (2.8a) and the Boltzmann factor $\exp(-w/kT)$, i.e.

$$\exp(-w/kT)\tfrac{1}{2}\sin \theta \, d\theta.$$

Unlike equation (2.8), this function is not normalized, therefore a normalization factor must be introduced:

$$\frac{dN}{N} = \frac{1}{2}\exp\left(\frac{-w}{kT}\right)\sin \theta \, d\theta \left(\int_0^\pi \frac{1}{2}\exp\left(\frac{-w}{kT}\right)\sin \theta \, d\theta \right)^{-1}. \tag{2.9}$$

Using equation (2.7) we can rewrite (2.9) more explicitly:

$$\frac{dN}{N} = \exp\left(\frac{\mu E \cos \theta}{kT}\right)\sin \theta \, d\theta \left(\int_0^\pi \exp\left(\frac{\mu E \cos \theta}{kT}\right)\sin \theta \, d\theta \right)^{-1}. \tag{2.10}$$

Equation (2.10) represents the probability of dipoles pointing toward the same narrow strip in the presence of a field E. The next step is to calculate the mean direction cosine $\langle \cos \theta \rangle$. In the absence of a field, needless to say, the mean direction cosine will be zero. However, in the presence of a field, the probability of finding dipoles pointing along the field vector increases and the mean direction cosine becomes non-zero. Using a well known principle, we can calculate the mean of an unknown variable X using the following equation:

$$\langle x \rangle = \int X p(x) \, dX \tag{2.11}$$

where $p(x)$ is the probability density function that a random variable X has a numerical value x. At present, the random variable is $\cos \theta$ and the probability density function is dN/N. Therefore, the mean of $\cos \theta$ can be calculated using (2.11) and $X = \cos \theta$:

$$\langle \cos \theta \rangle = \int_0^\pi \cos \theta p(\theta) \, d\theta \tag{2.12}$$

where $p(\theta)$ is given by (2.10). Once the value of the mean direction cosine $\langle \cos \theta \rangle$ is found, the mean dipole moment $\langle m \rangle$ of the group of dipoles can be calculated by

$$\langle m \rangle = \mu \langle \cos \theta \rangle \tag{2.13}$$

where μ is the dipole moment of the molecule. Using equations (2.10), (2.11) and (2.12), we obtain

$$\langle m \rangle = \int_0^\pi \mu \cos \theta \exp\left(\frac{\mu E}{kT}\right) \sin \theta \, d\theta \left(\int_0^\pi \exp\left(\frac{\mu E}{kT}\right) \sin \theta \, d\theta\right)^{-1}. \tag{2.14}$$

In the absence of the field, the right-hand side (RHS) of (2.13) reduces to zero and the mean dipole moment vanishes. The remaining task is to integrate (2.14). Using the following substitutions, we can perform this integration fairly easily:

$$y = \mu E \cos \theta / kT = a \cos \theta \tag{2.15}$$

where

$$a = \mu E / kT. \tag{2.16}$$

Using these and changing the integration variable from θ to y, we can rewrite (2.14) in the following simple form:

$$\langle \cos \theta \rangle = \int_0^a y e^y \, dy \left(\int_0^a e^y \, dy\right)^{-1}. \tag{2.17}$$

Note the change in the limits of the integrals. Integration of (2.17) gives rise to the following result:

$$\langle \cos \theta \rangle = \frac{e^a + e^{-a}}{e^a - e^{-a}} - \frac{1}{a} \tag{2.18}$$

or

$$\langle \cos \theta \rangle = \coth a - 1/a. \tag{2.19}$$

The RHS of (2.18) is called the Langevin function and is denoted as

$$L(a) = \coth a - 1/a. \tag{2.20}$$

The behaviour of the Langevin function is shown schematically in figure 2.3.

As noted, the Langevin function is described as a 'saturation function' and can never exceed the value of unity no matter how large the value of 'a' may be. Physically, this means that the mean moment of a group of polar molecules can never exceed the dipole moment of each molecule. Obviously, the value $\langle m \rangle$ reaches μ only when all of the polar molecules are aligned parallel to the applied field. However, the behaviour of $L(a)$ with weak electrical field is more important from the experimental point of view unless one is interested in the non-linear dielectric behaviour. (This topic is dealt with in Chapter 9.) The Langevin function can be expanded in a power series:

$$L(a) = a/3 - a^3/45 + \dots$$

or (2.21)

$$L(\mu E/kT) = \frac{1}{3}\left(\frac{\mu E}{kT}\right) - \frac{1}{45}\left(\frac{\mu E}{kT}\right)^3 + \dots$$

Therefore, if the value of a is sufficiently small, we can ignore higher-order terms. Namely, the Langevin function $L(a)$ becomes a linear function of 'a' with a slope of $\frac{1}{3}$ as shown in figure 2.3. As illustrated, this approximate formula is a fairly accurate representation of $L(a)$ if the value of 'a' is small. However, as 'a' increases, the deviation of $a/3$ from $L(a)$ becomes apparent. While $L(a)$ approaches a value of 1 as 'a' increases, $a/3$ continues to increase indefinitely. Numerical values of $L(a)$ and $a/3$ are tabulated in table 2.1 for various values of 'a'.

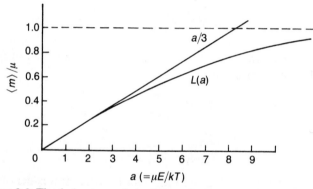

Figure 2.3 The behaviour of the Langevin function. The ordinate is $\langle m \rangle/\mu$ and the abscissa is 'a' ($= \mu E/kT$). $L(a)$ is the full Langevin function and $a/3$ is the linear approximation.

Table 2.1 Numerical values of the Langevin function $L(a)$.

$a = \mu E/kT$	$L(a)$	$a/3$
0.1	0.0333	0.0333
0.2	0.0665	0.0666
0.5	0.1640	0.1667
1.0	0.3130	0.3333
2.0	0.5373	0.6666

This result implies that if the intensity of applied field E is sufficiently small, the higher-order terms can be truncated and the mean moment $\langle m \rangle$ can be written simply as

$$\langle m \rangle = \frac{\mu^2 E}{3kT} \qquad \text{if } \mu E \ll kT. \qquad (2.22)$$

The linear approximation discussed above applies only when the field intensity is sufficiently small. Let us calculate the intensity of applied field E which is large enough to make $\mu E / kT$ unity (E_c). If we assume the value of dipole moment to be $10\,D$ and $T = 300\,K$, the field intensity E_c will be

$$E_c = kT/\mu$$
$$= \frac{(1.38 \times 10^{-16})(300)}{10 \times 10^{-18}}$$
$$= 4.1 \times 10^3 \text{ esu.}$$

Since 1 esu $= 300$ V cm^{-1}, the field intensity E is

$$E_c = 12.3 \times 10^5 \text{ V cm}^{-1}.$$

This is indeed a very large value. Usually, field intensities of 1–10 V cm^{-1} are used for most experiments. Therefore, linearization of the function $L(a)$ is a valid approximation under normal circumstances. However, if the dipole moment of a molecule is very large, the value of E_c decreases markedly and the validity of the linear approximation becomes questionable. For example, as will be discussed in Chapter 9, biological polymers are known to have exceedingly large dipole moments. In particular, the dipole moment of DNA molecules reaches a value of $10^5\,D$ easily. Under these circumstances, the value of E_c decreases markedly and becomes of the order of 100–300 V cm^{-1}. Table 2.2 shows the values of E_c for various biological molecules.

Table 2.2 The values of E_c and E_s for biological molecules. E_c is defined as the field intensity which makes the value of $\mu E / kT$ unity. E_s is the field strength which causes the polarization near saturation.

Molecule	Dipole moment (D)	E_c (V cm^{-1})	E_s (V cm^{-1})
Water	1.84	18.58×10^5	9.29×10^6
Glycine	15.5	8.01×10^5	4.0×10^6
Di-glycine	25	4.96×10^5	2.48×10^6
Myoglobin	170	7.30×10^4	3.65×10^5
Haemoglobin	≈ 480	2.58×10^4	1.29×10^5
DNA	$\approx 100\,000$	1.24×10^2	6.2×10^2

As shown in this table, the value of E_c, for some biopolymers, approaches the range which is used for actual measurements. Even with $a = 1.0$ the value of the Langevin function is still only 0.2, i.e. the average orientation of dipoles is far from complete. In order to attain $L(a) = 1$, i.e. a complete orientation parallel to applied fields, a much higher field strength is required. However, because of thermal motion which opposes the torque due to the field, complete orientation can never be achieved. The last column of table 2.2 shows the field intensity which is needed to reach a value of $L(a) = 0.8$, instead of 1.0. Based on the calculations shown in table 2.2 it is clear that the orientation of polar molecules is usually minutely small unless either the field strength or dipole moment is exceedingly large.

When the intensity of applied fields is very high, the mean moment $\langle m \rangle$ can no longer be represented by (2.21) and higher-order terms become necessary to describe accurately the mean moment. Since these include E^2, E^3, . . . terms, the mean moment becomes a non-linear function of electrical field E. This phenomenon is commonly called dielectric saturation. Under these conditions, the Langevin function (equation (2.20)) can be approximated as

$$L(a) \simeq 1 - kT/\mu E \tag{2.23}$$

for large value of 'a'. We can see from this equation that even under this condition, the value of the second term on the RHS is likely to be 0.1–0.15 and this means that the value of $L(a)$ is actually considerably smaller than 1. In other words, it is virtually impossible to orient dipoles parallel to the field vector no matter how large the intensity of the applied field may be.

The discussion presented above is the manifestation of macroscopic orientation effects of externally applied electrical fields. The symbol E is meant to be the intensity of the external field. However, this casual assumption turns out to be quite inaccurate. We need to consider the microscopic mechanism carefully, taking into account the actual field in the proximity of the molecules. If an electrical field is applied to a group of dipoles, the field E causes orientation of the dipoles. The extent of orientation can be represented by 'polarization' as defined by

$$P = N\langle m \rangle \tag{2.24}$$

where N is the number of dipoles per unit volume. In other words, the polarization P is the dipole moment per unit volume. The orientation of dipoles will create an additional electrical field. Therefore, the actual field in the proximity of dipolar molecules is larger than the applied field. This is generally called the internal field and its magnitude is given by the following equation:

$$F = E + 4\pi P/3. \tag{2.25}$$

This is the field which is felt by individual molecules and the torque is created by F rather than E. Therefore, the mean dipole moment $\langle m \rangle$ should be proportional to the total field F, namely

$$\langle m \rangle = \alpha F \qquad (2.26)$$

where α is the proportionality constant which, in the case of dipole orientation, should be equal to $\mu^2/3kT$. The quantity α is called collectively the polarizability. In addition to the polarizability due to dipole orientation, there are other polarizabilities. One of them is due to the distortion of electron distribution. As discussed in the previous chapter, electron clouds surrounding nuclei undergo a slight distortion in the direction of electric field. The extent of distortion is proportional to the field F and the electric moment thus produced can also be given by equation (2.26). Although the mathematical expression for these polarizabilities is the same, their physical meanings are quite different. Furthermore, the bending and stretching of chemical bonds also produce polarization. Therefore, the polarizability used in (2.26) actually consists of three components:

$$\alpha = \alpha_o + \alpha_e + \alpha_a \qquad (2.27)$$

where α_o is the orientation polarizability and α_e and α_a are the electronic and atomic polarizabilities. So far we have defined (i) the polarization P as the dipole moment per unit volume, (ii) the total field F by (2.25) and (iii) α as the polarizability consisting of three components. However, dipole moment is a quantity which cannot be measured experimentally. The quantity which is measurable is the dielectric constant or permittivity ε. Therefore, in order to determine the value of dipole moment experimentally, we need to find the connection between dielectric constant and dipole moment. In general, we can relate the dielectric constant ε to polarization P by the following equation:

$$P = \frac{(\varepsilon - 1)E}{4\pi}. \qquad (2.28)$$

The RHS of this equation is known as the electric susceptibility. This equation is written more often as

$$D = E + 4\pi P \qquad (2.29)$$

where D is the electric displacement and is defined as

$$D = \varepsilon E. \qquad (2.30)$$

The dielectric constant ε is one of the fundamental quantities and is accessible experimentally. ε is usually a scalar quantity. Although tensor notation may have to be used for some anisotropic substances, ε is very rarely treated as a tensor, particularly for biological materials. The dielectric constant of free space ε_0 is $8.64 \times 10^{-12} \, \text{F m}^{-1}$ in the MKS system and those of other substances are defined as multiples of this value, i.e. $\varepsilon\varepsilon_0$ where ε is the relative permittivity or dielectric constant. However, in the cgs system, the relative dielectric constant is used simply as a dimensionless quantity. The dielectric constants of various substances are tabulated in table 2.3. Now, using equations (2.24)–(2.26), we can eliminate m, E and P and derive the following equation:

$$\frac{\varepsilon - 1}{\varepsilon + 2} = \frac{4\pi N}{3}\alpha. \tag{2.31}$$

This equation was derived many years ago by Clausius and Mosotti, and even now is called the Clausius–Mosotti equation. However, it was Debye (1929) who pointed out that the RHS of equation (2.31) consists of three terms: α_o, α_e and α_a, as shown by equation (2.27). It is customary to combine α_e and α_a and simply call the combination the 'polarizability'. As discussed previously, α_o is equal to $\mu^2/3kT$. Also, converting equation (2.31) into molar basis, we can finally write an equation which has played a crucial role in the development of dielectric theories:

$$[p] = \frac{\varepsilon - 1}{\varepsilon + 2}\frac{M}{d} = \frac{4\pi N_A}{3}\left(\alpha + \frac{\mu^2}{3kT}\right) \tag{2.32}$$

where N_A is Avogadro's number, and M and d are the molecular weight and density respectively. $[p]$ is called the molar polarization.

Table 2.3 The relative dielectric constant of various substances.

Substance	Dielectric constant	Temperature (°C)
Water	78.5	25
Methyl alcohol	56.6	−80
Ethyl ether	8.14	−100
Benzene	2.32	−80
Dioxane	2.0	—
Carbon tetrachloride	2.24	20

Obviously, the Debye equation consists of two parts: a temperature-dependent term $(4\pi N_A\mu^2/9kT)$ and a temperature-independent term $(4\pi N_A\alpha/3)$. Therefore, if we measure the dielectric constant ε experimentally at various temperatures, calculate the left-hand side (LHS) of equation (2.32) and plot the value against $1/T$, we should obtain a straight line, as shown schematically in figure 2.4. The slope of this plot yields $4\pi N_A\mu^2/9k$, from which we can calculate the value of the dipole moment μ. Also, the plot intercepts the ordinate at $4\pi N_A\alpha/3$ and we

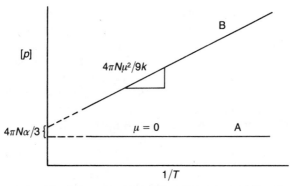

Figure 2.4 A plot of molar polarization against $1/T$. Curve A represents the behaviour of non-polar molecules where $\mu = 0$. Curve B represents the behaviour of polar molecules.

can then find the value of the polarizability. Note that the molar polarization $[p]$ of polar molecules makes a positive slope against $1/T$ and those of non-polar molecules are parallel to the abscissa, i.e. $\mu = 0$. The dipole moments and polarizabilities of some gas molecules obtained by this method are tabulated in table 2.4.

Table 2.4 The dipole moments and polarizabilities of gases.

Molecule	Dipole moment (D)	Polarizability (cc mol^{-1})
HCl	1.03	2.63×10^{-24}
HBr	0.79	3.58×10^{-24}
H$_2$	0	0.79×10^{-24}
H$_2$O	1.84	1.68×10^{-24}
CH$_4$	0	2.61×10^{-24}
CH$_3$Cl	1.86	4.56×10^{-24}

The Debye equation was originally derived for dilute gases. For the cases described here, the dielectric constants are not much larger than that of free space. However, many polar molecules have dielectric constants which are considerably higher than 1. It is well known that Debye's equation breaks down for highly polar molecules having a large dielectric constant. In the derivation of the Debye equation, the internal field F was defined by equation (2.25). Using equation (2.28) with this equation, we obtain the following expression:

$$F = \frac{\varepsilon + 2}{3}E. \tag{2.33}$$

This equation was derived by Lorentz and the field F is known as the Lorentz internal field. The use of Lorentz field is the essential part of the Debye theory and led him to equation (2.32). However, the main reason why the Debye equation is inadequate for polar molecules arises from the use of Lorentz field. The inadequacy of Debye's theory was remedied by Onsager by the introduction of a then new concept of reaction field. Although Onsager's theory is rather complex, nevertheless the detail of this theory is discussed in the following section because of its importance.

2.4 The Concept of Reaction Field (Onsager Theory, 1936)

Onsager introduced a small spherical cavity of molecular dimension, at the centre of which a point dipole m is placed (see figure 2.5). Inside the cavity the dielectric constant is assumed to be unity but polarizable. So, under the influence of a field, it produces an induced moment $\alpha_o F$, where F is the total field which, according to Onsager's calculation, turns out to be substantially different from the one used by Debye. The calculation of the total field F is, in effect, the most important part of Onsager's theory. First of all, it was assumed that the total field consists of two components: (a) internal field G which is induced in the cavity by applied field E; and (b) reaction field R which is induced by the presence of a dipole m.

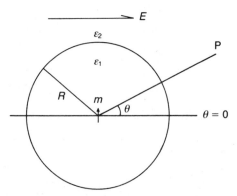

Figure 2.5 A spherical cavity surrounding a dipole m. ε_1 and ε_2 are the dielectric constants of the cavity and the external medium respectively. E is applied electrical field. R is the radius of the spherical cavity and P is an arbitrary point outside the cavity.

The determination of internal field G is based on the solution of a boundary value problem for the spherical cavity in the presence of an electrical field. This calculation reduces to solving the Laplace equation $\nabla^2\phi = 0$ for a sphere with appropriate boundary conditions (see the suggested reading at the end of this chapter, e.g. Stratton (1941) and Plonsey and Collin (1961)). The electrical potential ϕ inside the cavity is given, as the solution of the Laplace equation, by the following expressions:

$$\phi = Er\cos\theta + \left(\frac{\varepsilon - 1}{2\varepsilon + 1}\right)Er\cos\theta$$

$$= \left(\frac{3\varepsilon}{2\varepsilon + 1}\right)Er\cos\theta. \tag{2.34}$$

Since electrical field E is defined as

$$E = -(\partial\phi/\partial r) \tag{2.35}$$

we can calculate the internal field by differentiating (2.34) assuming $\cos\theta = 1$:

$$G = \frac{3\varepsilon}{2\varepsilon + 1}E. \tag{2.36}$$

Obviously, G is a vector which is parallel to the applied field.

In order to calculate reaction field R, we determine the potential due to the point dipole m alone. Namely, the potentials inside and outside the cavity are given by the following equations:

$$\phi_{in} = \frac{m\cos\theta}{r^2} + Ar\cos\theta \tag{2.37}$$

$$\phi_{out} = \frac{m\cos\theta}{r^2} + \frac{B}{r^2}\cos\theta. \tag{2.38}$$

In these equations, the term $m\cos\theta/r^2$ is the potential due to the dipole m in free space, and the second term is the potential modified by the presence of the boundary. A and B are constants which can be determined using two boundary conditions, namely the continuity of potential at the boundary

$$\phi_{in} = \phi_{out} \qquad \text{at } r = a \tag{2.39}$$

and the continuity of the normal component of displacement

$$\varepsilon_{in}\frac{\partial\phi_{in}}{\partial r} = \varepsilon_{out}\frac{\partial\phi_{out}}{\partial r} \qquad \text{at } r = a. \tag{2.40}$$

Using these equations, the potential inside the cavity is determined:

$$\phi_{in} = \frac{m\cos\theta}{r^2} - \frac{2m}{a^3}\frac{\varepsilon - 1}{2\varepsilon + 1}r\cos\theta. \tag{2.41}$$

Of the terms on the RHS, $m\cos\theta/r^2$ is the potential that arises from the point dipole and continues to decrease regardless of the presence of the boundary. Therefore, the component which changes the magnitude of the dipole in the cavity is only the second term. Differentiating this term with respect to r, we obtain an expression for the reaction field R setting $\cos \theta = 1$:

$$R = \frac{2m}{a^3}\left(\frac{\varepsilon - 1}{2\varepsilon + 1}\right). \tag{2.42}$$

It should be noted that R is parallel to the dipole vector m. Combining G and R, the total field is $F = G + R$:

$$F = \frac{3\varepsilon}{2\varepsilon + 1}E + \frac{2(\varepsilon - 1)}{a^3(\varepsilon + 1)}m. \tag{2.43}$$

As has been discussed, the dipole moment m consists of a permanent dipole term $\mu\mu_0$, where μ is a unit dipole, and also an induced dipole term $\alpha_o F$, thus

$$m = \mu_0\mu + \alpha_o F. \tag{2.44}$$

Using the Clausius–Mosotti equation (equation (2.31)) and equating ε to n^2 (see equation (1.13), where n is the refractive index), the polarizability α_o can be written as

$$\alpha_o = \left(\frac{n^2 - 1}{n^2 + 2}\right)a^3. \tag{2.45}$$

Substituting this equation into equation (2.44) and using equation (2.43), we obtain

$$m = \mu\bar{\mu} + \frac{\varepsilon(n^2 - 1)}{2\varepsilon + n^2}a^3 E \tag{2.46}$$

where $\bar{\mu}$ is defined as follows:

$$\bar{\mu} = \frac{(n^2 + 2)(2\varepsilon + 1)}{3(2\varepsilon + n^2)}\mu_0. \tag{2.47}$$

The remaining task is to find the average value of m using the same method as for the Debye theory. However, before we calculate the mean moment, we have to find the torque acting on the dipoles. As discussed, the total field F consists of two components: G and R. Figure 2.6 illustrates the mutual relation between the dipole and field vectors. Note in this figure that the reaction field R is parallel to the dipole vector m. Therefore, the reaction field cannot orient the dipole m because the cross product $m \times R$ is zero. With this consideration it is clear that internal field G alone would cause the orientation of the dipole m. The role of R is to change the magnitude of the dipole moment from μ_0 to $\bar{\mu}$. Thus, our task is to calculate the following term:

$$T = m \times G + E \times E. \tag{2.48}$$

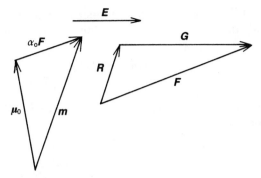

Figure 2.6 Vectorial relationships between the dipole moment μ_0, the polarizability αF and the total moment m with internal and reaction fields. E is applied field and F is total field. Note that m is parallel to R and that the polarizability αF is parallel to F. Internal field G is parallel to applied field E. (After Edsall and Wyman 1958.)

Note that the term $E \times E = 0$. Therefore, we end up with the following equation for torque T:

$$T = m \times E\left(\frac{3\varepsilon}{2\varepsilon + 1}\right)\bar{\mu}. \tag{2.49}$$

In scalar notation

$$T = \left(\frac{3\varepsilon}{2\varepsilon + 1}\right)\bar{\mu}E \sin \theta$$

$$= \mu^* E \sin \theta \tag{2.50}$$

where

$$\mu^* = \frac{3\varepsilon}{2\varepsilon + 1}\bar{\mu}. \tag{2.51}$$

The energy of interaction between the field and dipole is, as before

$$w = -\mu^* E \cos \theta.$$

It is now clear that for the calculation of mean moment it is necessary to use (2.14) with μ^* instead of $\bar{\mu}$. As before, assuming that the magnitude of applied field E is small, we can ignore the non-linear terms and obtain

$$\langle m \rangle = \left(\frac{\bar{\mu}\mu^*}{3kT} + \frac{\varepsilon(n^2 - 1)a^3}{2\varepsilon + n^2}\right)E. \tag{2.52}$$

It should be noted that although the first term in the brackets resembles that of the Debye equation, the physical meaning of $\bar{\mu}\mu^*$ is quite different from μ^2. Remembering equation (2.26) and $P = N\langle m \rangle$, and that the size of the spherical cavity is comparable to that of molecules, then $4\pi a^3/3 = 1/N$, where N is the number of dipoles in the molecular assembly. With this, we finally obtain, after rearranging, a well known equation:

$$\varepsilon - 1 = \frac{4\pi N\bar{\mu}\mu^*}{3kT} + \frac{3\varepsilon(n^2 - 1)}{2\varepsilon + n^2}. \tag{2.53}$$

This equation is called Onsager's equation, which can be rewritten alternatively in the following form:

$$\frac{(\varepsilon - n^2)(2\varepsilon + n^2)}{\varepsilon(n^2 + 2)^2} = \frac{4\pi N\mu_0^2}{9kT}. \tag{2.54}$$

This equation groups the refractive index and dielectric constant on the LHS and the dipole moment on the other. If the dielectric constant is very large, as for water, equation (2.54) reduces to

$$\varepsilon = \frac{(n^2 + 2)^2}{2} \frac{4\pi N\mu_0^2}{9kT} \tag{2.55}$$

i.e. the dielectric constant becomes proportional to the square of dipole moment as refractive indices do not vary much from one material to another. On the other hand, if ε approaches unity, Onsager's equation should reduce to the Debye equation.

In Debye's theory, the interaction between polar molecules is ignored. This assumption does not cause serious errors so long as only dilute gases are dealt with. However, as the distance between molecules decreases as in dense gases and liquids, the neglect of intermolecular interaction causes serious errors. It is well known that the Debye equation breaks down for highly polar substances where the dielectric constant is much larger than unity. The introduction of reaction field in Onsager's theory is a remedy for the inadequacy of Debye's theory. The effect of reaction field is equivalent to saying that the dipole moment of a polar molecule is modified by the presence of neighbouring molecules. Onsager's analyses formally incorporated intermolecular interaction in the theory. Although Onsager's theory is applicable to many polar substances which have large dielectric constants, its applicability to polar liquids is still limited. We will now turn to another theory derived by Kirkwood.

2.5 The Dielectric Constant of Polar Liquids (Kirkwood Theory)

In the derivation of Debye's theory it was assumed that the distance between molecules is so large that there is hardly any interaction between them. Due to this assumption, the angular distribution of these molecules is completely random if electric field is absent. This situation will alter drastically for highly polar molecules, in particular those in the liquid state. For the analysis of polar liquids one can select a particular molecule in the group and look at its influence on neighbouring dipoles. This molecule may enhance the orientation of other molecules because of attractive or repulsive interactions, or it may interfere with the orientation of others by steric hindrance. Thus, the average moment of a group of dipoles may be different from those in free space. In other words, the dipole moment immersed in a large number of molecules will be different from the moment of a single molecule. In dilute gases we define the orientation of each molecule with respect to one reference vector, i.e. the electric field. In liquids we have to define the orientation of dipoles with respect to at least two references: one, the dipole vector of a central molecule, and the other, the vector of applied field.

First, we define a spherical coordinate which is shown in figure 2.7. Let us denote the dipole moment of a molecule by μ_i and the vector of applied field by $E_0 = E_0 e$, where e is a unit vector. The projection of μ_i on the field vector is given by $\mu_i \cdot e$ or $\mu_i \cos \theta'$, where θ' is the angle between E_0 and μ_i. We define a central molecule as the vector sum of the dipole moments of all the neighbouring molecules, i.e. $M = \Sigma \mu_k$. The angle between M and μ_i is defined as θ_μ, and that between the field vector E and M as θ_e. Also, the field vector is so directed as to make the azimuthal angle ϕ_e. Note that these angles are defined relative to the vector of the central dipole rather than to the vector of applied field with the exception of the angle θ'.

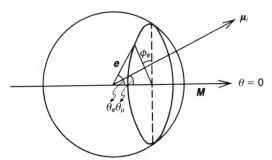

Figure 2.7 The coordinate system used to describe Kirkwood's theory of dipole moments and dielectric constants. μ_i is the permanent dipole of the ith molecule, M is the vectorial sum of all the dipoles. e is a unit vector in the direction of applied field. (After Edsall and Wyman 1958.)

The mean value of μ_i is given by the following equation:

$$\langle \mu_i \cdot e \rangle = \int_0^{2\pi} \int_0^{\pi} \mu_i \cos \theta' \, p(\theta_e) \, \mathrm{d}\theta_e \, \mathrm{d}\phi_e \qquad (2.56)$$

where $p(\theta')$ is the probability density function, i.e.

$$p(\theta_e) = \exp\left(\frac{ME_0 \cos \theta_e}{kT}\right) \sin \theta_e \left(\int_0^{2\pi} \int_0^{\pi} \exp\left(\frac{ME_0 \cos \theta_e}{kT}\right) \sin \theta_e \, d\theta_e \, d\phi_e\right)^{-1}.$$

(2.57)

The interaction energy between E_0 and dipole molecules is defined by $-E_0 M \cos \theta_e$. Obviously, the integration of (2.56) is not possible unless the angle θ' is expressed explicitly in terms of θ_e. Fortunately, such an expression can be found as shown by the following:

$$\cos \theta' = \cos \theta_\mu \cos \theta_e + \sin \theta_\mu \sin \theta_e \cos \phi_e.$$

(2.58)

Substituting this equation into (2.56) and (2.57), and retaining only non-zero terms, we obtain the following equation:

$$\langle \boldsymbol{\mu}_i \cdot \boldsymbol{e} \rangle = \mu_i \cos \theta_\mu \int \int \cos \theta_e p(\theta_e) \, d\theta_e \, d\phi_e$$

(2.59)

where $p(\theta_e)$ is defined by (2.57).

Equation (2.59) is similar to (2.14) and the solution can be obtained analogously:

$$\langle \boldsymbol{\mu}_i \cdot \boldsymbol{e} \rangle = \mu_i \cos \theta_\mu L(a)$$

(2.60)

where $L(a)$ is the Langevin function and $a = ME_0/kT$. For a weak electrical field, we can approximate (2.60) by

$$\langle \boldsymbol{\mu}_i \cdot \boldsymbol{e} \rangle = \mu_i \cos \theta_\mu \left(\frac{ME_0}{3kT}\right) \qquad \text{or} \qquad = \boldsymbol{M} \cdot \boldsymbol{\mu}_i \frac{E_0}{3kT}.$$

(2.61)

The quantity $\langle \boldsymbol{\mu}_i \cdot \boldsymbol{e} \rangle$ is the mean value for the moment of the ith molecule in the direction of the field, in a fixed configuration, with $\boldsymbol{M}$ as the central moment. What we need, however, is the mean moment of $\boldsymbol{M}$ for all the configurations in the sample and so $\boldsymbol{M}$ must be replaced by its mean $\bar{\boldsymbol{M}}$. Thus, (2.61) should be written as follows:

$$\langle \boldsymbol{\mu}_i \cdot \boldsymbol{e} \rangle = (\bar{\boldsymbol{M}} \cdot \boldsymbol{\mu}) \frac{E_0}{3kT}.$$

(2.62)

In addition, note that E_0 is the intensity of the applied field outside the cavity. Therefore, using (2.33) discussed before, we have to convert E_0 into internal field. This can be done easily so long as we use a spherical cavity:

$$E_i = \frac{3E_0}{\varepsilon + 2}.$$

(2.63)

Using this equation, we obtain a general expression

$$\langle \boldsymbol{\mu} \cdot \boldsymbol{e} \rangle = \frac{\varepsilon + 2}{9kT} (\boldsymbol{\mu} \cdot \bar{\boldsymbol{M}}) E_i.$$

(2.64)

The term $\bar{M}$ consists of two parts. (i) The mean moment due to reaction field, i.e. the moment modified by the polarization of molecules outside the cavity. This may be considered a long-range interaction. (ii) The mean moment around the molecule. This is designated as $\bar{\mu}$. This may be termed a short-range interaction. Kirkwood considered only the short-range interaction in his derivation (Kirkwood 1939). He proved that $\bar{M}$ is related to $\bar{\mu}$ by the following equation:

$$\bar{M} = \frac{9\varepsilon}{(\varepsilon + 2)(2\varepsilon + 1)}\bar{\mu}.$$ (2.65)

Substituting this equation into $(\varepsilon - 1)E = 4\pi N\bar{M}$ and (2.64), we obtain

$$\frac{(\varepsilon - 1)(2\varepsilon + 1)}{9\varepsilon} = \frac{4\pi N}{3}\left(\frac{\mu\cdot\bar{\mu}}{3kT} + \alpha\right).$$ (2.66)

Note that a polarizability term was added in this equation. In his derivation, Kirkwood assumed that the long-range interaction term due to reaction field can be ignored. This may appear to be an arbitrary assumption. However, Froehlich (1948), using a more general treatment, proved that equation (2.65) can be derived as a special case from the general equation without arbitrary assumptions.

In order to utilize this equation, we must calculate $\bar{\mu}$. Often, the term $\mu\cdot\bar{\mu}$ is written as $g\mu^2$, where g is Kirkwood's correlation parameter. The approximate form proposed for the g-factor is given by the following equation:

$$g = 1 + z\langle\cos\gamma\rangle$$ (2.67)

where z is the number of nearest neighbours surrounding the molecule. $\langle\cos\gamma\rangle$ is the mean value of the cosine of the angle between the dipole moments of adjacent molecules. Equation (2.67) contains two structural parameters: z and $\langle\cos\gamma\rangle$. An estimate of these parameters must be based on detailed information on the configuration of molecules in liquids. For example, water has a tetrahedral configuration in which four molecules surround one central dipole. Using this basic coordination, we can calculate the value of the g-factor. This problem will be discussed in detail in Chapter 8.

Froehlich's general theory

As mentioned briefly earlier, a general theory which holds widely for a variety of dielectric substances was developed by Froehlich in 1948. This theory was derived in a very general way and, therefore, is applicable as long as the substance is not permanently polarized as for some solid dielectric materials. The equation derived by Froehlich is

$$\varepsilon_s - n^2 = \frac{3\varepsilon_s}{2\varepsilon_s + n^2}\frac{4\pi N}{3}\frac{\overline{mm}^*}{kT}$$ (2.68)

where m^* represents the moment of a large spherical region polarized by one of its unit whose moment is kept at a value of m. Froehlich pointed out that the term $\overline{mm^*}$ is equivalent to Kirkwood's $\mu \cdot \bar{\mu}$ for polar liquids. The derivation of this elegant theory is discussed in detail in his book (1949).

Empirical theory

As already discussed, dielectric theories have been developed for dilute gases and have reached an extreme sophistication for polar liquids. On the other hand, a simple empirical formula which fits experimental results nicely was developed by Wyman (see Edsall and Wyman 1958). This equation was derived based on many experimental data for known values of refractive indices, density and dipole moment. Wyman's equation reads

$$\frac{\varepsilon + 1}{8.5} = \frac{4\pi N}{3}\left(\alpha + \frac{\mu^2}{3kT}\right). \tag{2.69}$$

It must be pointed out that this equation is similar to Kirkwood's equation in its limiting form for large dielectric constant, i.e.

$$\frac{\varepsilon}{4.5} = \frac{4\pi N}{3}\left(\alpha + \frac{\mu \cdot \bar{\mu}}{3kT}\right). \tag{2.70}$$

We have already discussed the theories which relate the dipole moment of polar substances to the dielectric constant. They were derived with the assumption that electric field is static. The dynamic behaviour of polar molecules in periodic fields will be discussed in the next chapter.

2.6 Dipoles in Solids

Although our interests in biological molecules lie almost exclusively with the liquid state, it became known that many biologically important macromolecules function in the matrix of cell membranes. As will be discussed later, proteins are embedded in a highly ordered lipid moiety of the membranes. The configuration of these molecules is dictated by a strong field existing in the membrane and attempts to rotate these protein molecules must overcome the force due to the field which tends to keep the molecules in the equilibrium position. Also, the ordered structure of the membrane imposes a very high viscous drag which makes the free rotation of dipoles virtually impossible. In other words, the dipole molecules which are embedded in the membrane matrix are, in some way, similar to the dipoles found in crystalline solids.

Although there may be a strong interaction between neighbouring molecules in liquid states, molecules in the liquid state do not possess a distinct preferential direction for orientation as a whole. However, in crystalline solids the interaction between molecules is much stronger than in the liquid state, and their preferential orientation depends upon the direction of the crystal field. The potential barrier between various equilibrium orientations is very high and virtually no free rotation of molecules is allowed unless the temperature is near or higher than the melting point and/or critical temperature T_0 (see figure 2.8(a)). At T_0, an order–disorder transition will take place and the potential energy which separates one preferential orientation from another will be removed, thereby enabling dipoles to reorient themselves.

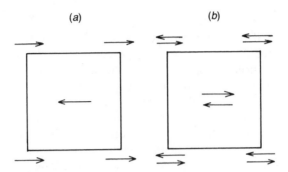

Figure 2.8 Equilibrium positions of dipoles for a simple crystalline solid model. (a) an ordered state with permanent polarization and (b) a disordered state where both directions have equal probabilities. (From Froehlich 1949.) Reproduced by permission of Oxford University Press.

For the analysis of dipolar orientation in solids, a simple face centred cubic lattice is considered. As shown in figure 2.8, the dipole at the centre is directed to the left and those at the corners are directed to the right. For the centre dipole, the direction ← is the favourable orientation with a minimum energy. Likewise, the direction → is the favourable orientation for corner dipoles. The energy difference between opposite orientations is very high, and the reorientation ← to → or vice versa will not occur if the temperature is much lower than the critical temperature. However, if temperatures rise, the average energy difference between these equilibrium orientations will gradually diminish and finally vanish at T_0. Now let w be the probability of finding a dipole in an unfavourable orientation, then $(1 - w)$ is the probability of finding it in a favourable position. Thus the ratio of these probabilities can be given by

$$\frac{w}{1 - w} = \exp(-V(T)/kT) \qquad (2.71)$$

where $V(T)$ is the energy difference between two orientations.

From this equation, we can derive

$$w = \frac{\exp(-V(T)/kT)}{1 + \exp(-V(T)/kT)} \tag{2.72}$$

and

$$1 - w = \frac{1}{1 + \exp(-V(T)/kT)}. \tag{2.73}$$

According to these equations

$w \ll 1$ and $1 - w \simeq 1$ if $V(T) \gg kT$

$w = 1 - w = \frac{1}{2}$ if $V(T) = 0$ at $T \geq T_0$.

In other words, if the temperature T is very low, then the reorientation of dipoles, either at the centre or at the corners, is virtually prohibited and the contribution of dipoles to the dielectric constant of crystals is due only to minor elastic oscillation caused by applied fields. Therefore, the dielectric constant at these temperatures should be only slightly higher than the square of the refractive index, n^2. As temperatures rise, however, the reorientation of the centre as well as corner dipoles becomes possible. Thus, the mean moment of a unit cell, consisting of centre and corner dipoles, is determined statistically by the relative orientations of these dipoles. Table 2.5 shows these relative orientations, and the probabilities associated with them. Let us recall that a quantity m^* which was defined in (2.68) is the mean moment of a region surrounding a unit cell. Using this definition, $\overline{mm}^*$ can be rewritten as

$$\overline{mm}^* = \sum_i m_i m_i^* p_i \tag{2.74}$$

where p_i is the probability and is either w or $(1 - w)$.

Table 2.5 The relative orientation of dipoles in a unit cell.

Orientation		m_i	p_i	$V(T)$
←	→	0	$(1 - w)^2$	0
←	←	-2μ	$w(1 - w)$	$V(T)$
→	→	2μ	$w(1 - w)$	$V(T)$
→	←	0	w^2	$2V(T)$

Using table 2.5 and equation (2.73), we can derive

$$\sum_i m_i m_i^* p_i = 8w(1 - w)\boldsymbol{\mu} \cdot \bar{\boldsymbol{\mu}}. \tag{2.75}$$

Substituting this into (2.68) we obtain

$$\frac{(\varepsilon_s - \varepsilon_\infty)(2\varepsilon_s + n^2)}{3\varepsilon_s} = \frac{4\pi N \boldsymbol{\mu} \cdot \bar{\boldsymbol{\mu}}}{3kT} 4w(1 - w). \tag{2.76}$$

Clearly, the RHS of this equation depends on the probability w and $(1 - w)$. At very low temperatures, where $w \ll 1$, ε_s is only equal to ε_∞; usually ε_∞ is slightly larger than n^2. However, at temperatures near or above the critical temperature, w becomes $\frac{1}{2}$ and (2.76) reduces to the Kirkwood equation for polar liquids.

As a conclusion to this chapter, table 2.6 shows a summary of the theories discussed. These are simplified equations for a large dielectric constant, i.e. for $\varepsilon \gg n^2$. Thus, the polarizability term α is ignored except for the Debye equation.

Table 2.6 A summary of dielectric theories in static fields.

Debye† (1929)	$\dfrac{\varepsilon - 1}{\varepsilon + 2} V = \dfrac{4\pi N}{3}\left(\alpha + \dfrac{\mu^2}{3kT}\right)$
Onsager (1936)	$\dfrac{2\varepsilon}{(n^2 + 2)^2} V \simeq \dfrac{4\pi N \mu_0^2}{9kT}$
Kirkwood (1939)	$\dfrac{2\varepsilon}{9} V = 4\pi N \dfrac{\boldsymbol{\mu} \cdot \bar{\boldsymbol{\mu}}}{3kT}$
Froehlich (1948)	$\dfrac{2\varepsilon}{3} V = 4\pi N \dfrac{\overline{mm^*}}{3kT}$
Wyman (1958)	$\dfrac{\varepsilon}{8.5} V = \dfrac{4\pi N}{3}\left(\dfrac{\mu^2}{3kT}\right)$

† Debye's theory was derived for a small dielectric constant. Due to this limitation, the approximate equation for a large dielectric constant can not be shown.

References and Suggested Reading

Boettcher C J F and Bordewijk P 1987 *Theory of Electric Polarisation* 2nd edn (Amsterdam: Elsevier)

Debye P 1929 *Polar Molecules* (New York: Dover)

Edsall J T and Wyman J Jr 1958 in *Biophysical Chemistry* vol 1 (New York: Academic) ch 6

Froehlich H 1948 General theory of static dielectric constant *Trans. Faraday Soc.* **44** 238

—— 1949 *Theory of Dielectrics* (Oxford: Clarendon)

Gabler R 1978 *Electrical Interactions in Molecular Biophysics* (New York: Academic)

Hill N E, Vaughan W E, Price A H and Davies M 1969 *Dielectric Properties and Molecular Behavior* (New York: Van Nostrand)

Kirkwood J G 1939 The dielectric polarization of polar liquids *J. Chem. Phys.* **7** 911

Onsager L 1936 Electric moments of molecules in liquids *J. Am. Chem. Soc.* **58** 1486

Plonsey R and Collin R E 1961 *Principles and Applications of Electromagnetic Fields* (New York: McGraw-Hill)

Rushbrooke G S 1949 *Introduction to Statistical Mechanics* (Oxford: Clarendon)

Smyth C P 1955 *Dielectric Behavior and Structure* (New York: McGraw-Hill)

Stratton A 1941 *Electromagnetic Theory* (New York: McGraw-Hill)

3

Dynamic Aspects of Electric Polarization

3.1 Introduction

We discussed, in the last chapter, the mathematical formulation for the electric polarization of polar molecules in static fields. In this chapter, the dynamic aspects of electric polarization in alternating fields will be discussed. It is well known that the dielectric constant of polar molecules decreases markedly if the frequency of applied AC field increases above a certain critical value. Therefore, 'electrical permittivity' or simply 'permittivity' are preferred terms used by some investigators instead of dielectric constant.

In the treatment of electric polarization in a static field, we ignored the transient state which follows immediately after an electrical field is turned on or off. Our previous analyses were focused on the polarization of polar molecules only after the stationary state had been reached. The main topic of this chapter is how the molecules behave during the transient period following the application of a step pulse. This analysis will be extended to the polarization behaviour of these dipoles in the presence of alternating fields. Information on the molecular structure can be extracted from the analysis of dielectric relaxation curves.

3.2 Electric Polarization with a Pulsed Field

If an electrical field is applied as a step function to a group of dipoles, these molecules begin to orient in the direction of the field. However, orientational polarization is slow and the increase of the polarization lags behind the rise of electrical field as shown in figure 3.1. In contrast to the orientation of the entire molecule, atomic and electronic polarizations are much faster and rise almost instantaneously as depicted in the same figure. The polarization P consists of roughly two parts:

$$P_t = P_o + P_\infty \tag{3.1}$$

where P_t is the total polarization, P_o the orientational polarization and

P_∞ the sum of the atomic and electronic polarizations. P_t and P_∞ satisfy the following equations as pointed out in the last chapter:

$$P_t = \frac{\varepsilon_s - 1}{4\pi} E \tag{3.2}$$

$$P_\infty = \frac{\varepsilon_\infty - 1}{4\pi} E. \tag{3.3}$$

The time required for the dipole to reach the stationary state varies widely and ranges from 10^{-3} to 10^{-13} s. As will be discussed later, the time constant (or relaxation time) depends on many factors such as the dimension and shape of the molecule. In addition, it depends on the temperature and the viscosity of the solvent.

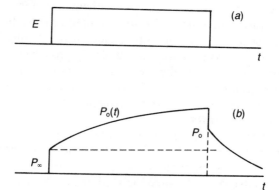

Figure 3.1 The response of a dipole to a square pulse. (*a*) An applied pulse. (*b*) The polarization of dipoles. The initial rise is due to electronic and atomic polarizations and the exponential phase is due to the orientational polarization of permanent dipoles.

Using the diagram shown in figure 3.1, we can calculate the polarization as a function of time after a field E is applied suddenly. If we define $P_o(t)$ as the orientational polarization during the transient period, its time derivative implies the rate of increase in the polarization with time. The rate should be proportional to the number of unpolarized dipoles, i.e. $P_o - P_o(t)$. Therefore

$$\frac{dP_o(t)}{dt} = \frac{1}{\tau}(P_o - P_o(t)) \tag{3.4}$$

where τ is the relaxation time and $1/\tau$ represents the rate constant. The solution of this equation is easily obtained assuming the initial value of $P_o(t)$ is zero:

$$P_o(t) = P_o(1 - e^{-t/\tau}) \tag{3.5}$$

namely, the polarization of dipoles $P_o(t)$ in response to a square pulse will increase exponentially as shown schematically by figure 3.1. Likewise, the decay of polarization is also an exponential function:

$$P_o(t) = P_o e^{-t/\tau'}. \tag{3.6}$$

3.3 The Polarization of Dipoles in Alternating Fields

The discussion presented above applies also to the polarization of dipoles in alternating fields. However, before we discuss this problem, we have to introduce the concept of a complex dielectric constant. So far, the dielectric constant has been treated as a real number. However, when an alternating field instead of a static field is applied to a group of dipoles, there will be a phase shift between the field E and the polarization P, as shown by figure 3.1. The dissipation of electrical energy will occur under such conditions, and the dielectric constant must be treated as a complex quantity.

Alternating fields are expressed using complex notation, as shown below:

$$E^* = E_0 e^{j\omega t} \tag{3.7}$$

where ω is the angular frequency and is given by $\omega = 2\pi f$ (where f is the frequency) and E_0 is the amplitude. Using this equation, the dielectric displacement D can be defined as follows:

$$D^* = D_0 e^{j(\omega t - \delta)} \tag{3.8}$$

where δ is introduced to account for the phase shift between the applied field and the displacement. As defined previously, the dielectric constant is the ratio between electric field E and displacement D. Likewise, the complex dielectric constant is the ratio between D^* and E^*, including the phase shift. Thus

$$\varepsilon^* = |\varepsilon| e^{-j\delta} \tag{3.9}$$

where $|\varepsilon| = D_0/E_0$. We can rewrite this equation as follows:

$$\varepsilon^* = |\varepsilon| \cos \delta - j|\varepsilon| \sin \delta \tag{3.10}$$

namely, the complex dielectric constant consists of a real part $|\varepsilon| \cos \delta$ and an imaginary part $|\varepsilon| \sin \delta$. An alternate expression for the complex dielectric constant can be used:

$$\varepsilon^* = \varepsilon' - j\varepsilon''. \tag{3.11}$$

Usually 'dielectric constant' means the real part ε' and ε'' is often called the dielectric loss indicating that this quantity represents the dissipation of electrical energy into the system. Using the definition of real and imaginary parts of the complex dielectric constant, we can define the 'loss tangent' as

$$\tan \delta = \frac{\varepsilon''}{\varepsilon'} \tag{3.12}$$

where δ is the phase angle.

Using these definitions and equations (3.2) and (3.3), we can express the polarization P_0^* in alternating fields E^* as

$$P_0^* = \frac{\varepsilon_s - \varepsilon_\infty}{4\pi} E^* \tag{3.13}$$

where E^* is defined by equation (3.7). Substituting (3.13) into (3.4)

$$\frac{dP_0^*(t)}{dt} = \frac{1}{\tau}\left(\frac{\varepsilon_s - \varepsilon_\infty}{4\pi} E_0 e^{j\omega t} - P_0^*(t) \right). \tag{3.14}$$

The solution of this equation is

$$P_0^*(t) = Ce^{-t/\tau} + \frac{1}{4\pi} \frac{\varepsilon_s - \varepsilon_\infty}{1 + j\omega\tau} E_0 e^{j\omega t}. \tag{3.15}$$

The first term on the RHS shows the transient response following the onset of alternating field. This term, however, will diminish to zero as time increases. Therefore, we can ignore this term for the steady state solution. Using only the second term on the RHS of (3.15), we can write the equation for P_0^* as the sum of two terms, i.e.

$$P_0^* = P_\infty^0 + P_0^*(t)$$
$$= \frac{1}{4\pi}\left((\varepsilon_\infty - 1) + \frac{\varepsilon_s - \varepsilon_\infty}{1 + j\omega\tau} \right) E_0 e^{j\omega t}. \tag{3.16}$$

Now recalling that the displacement is given by $D^* = E^* + 4\pi P^*$ and using the steady state solution discussed above, we obtain the following equation:

$$D^* = \left(\varepsilon_\infty + \frac{\varepsilon_s - \varepsilon_\infty}{1 + j\omega\tau} \right) E_0 e^{j\omega t}. \tag{3.17}$$

Since $D^* = \varepsilon^* E^*$, we can finally obtain an equation for the complex dielectric constant:

$$\varepsilon^* = \varepsilon_\infty + \frac{\varepsilon_s - \varepsilon_\infty}{1 + j\omega\tau}. \tag{3.18}$$

This equation was first derived by Debye and is still called the 'Debye equation'. Separating the real and imaginary parts of equation (3.18), we obtain the following two equations:

$$\varepsilon' = \varepsilon_\infty + \frac{\varepsilon_s - \varepsilon_\infty}{1 + (\omega\tau)^2} \tag{3.19}$$

$$\varepsilon'' = \frac{(\varepsilon_s - \varepsilon_\infty)\omega\tau}{1 + (\omega\tau)^2}. \qquad (3.20)$$

In these expressions, ε_s and ε_∞ are the low- and high-frequency limiting dielectric constants and τ is the time constant or relaxation time. Equation (3.19) indicates that the real part of the complex dielectric constant decreases from ε_s to ε_∞ in the approximate frequency range $10 > \omega\tau > 0.1$. It is worth noting at this stage that the derivation of (3.18) is based on an assumption that the rise and/or decay of polarization follows a single exponential function with one time constant. This means that the Debye equation holds only for the case in which molecules are uniform and isotropic. It will be shown later that if the size of the molecules is not uniform or if the shape is not spherical, the Debye equation is no longer applicable.

Normalizing the amplitude of the dielectric dispersion and using a normalized frequency f/f_r (where f_r is the centre relaxation frequency of dispersion), we can plot the real and imaginary parts at various frequencies. The result of this calculation is illustrated in figure 3.2. As expected, the real part ε' decreases from ε_s to ε_∞ over four decades of frequency. At the centre of the dispersion curve, i.e. where $f = f_r$, the following equality holds:

$$\tau = 1/2\pi f_r. \qquad (3.21)$$

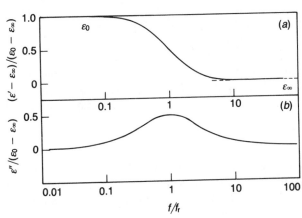

Figure 3.2 (a) The frequency dependence of the real part of the complex dielectric constant. The ordinate is the normalized dielectric increment and the abscissa is the frequency on a logarithmic scale. f_r is the relaxation frequency and ε_s and ε_∞ are the low- and high-frequency limiting values of the dielectric constant. (b) The frequency profile of the imaginary part of the complex dielectric constant. The maximum value of ε'' is half of the amplitude of the dielectric increment shown in part (a). Note that the peak frequency corresponds to f_r.

On the other hand, equation (3.20) indicates that the plot of ε'' against frequency must be a bell shaped curve with a maximum value ε''_m at $f = f_r$, as shown by figure 3.2(b). At the relaxation frequency, the magnitudes of ε' and ε'' are given by the following equations:

$$\varepsilon' = (\varepsilon_s + \varepsilon_\infty)/2 \tag{3.22}$$

$$\varepsilon''_m \leqslant (\varepsilon_s - \varepsilon_\infty)/2. \tag{3.23}$$

Equation (3.23) is particularly important in that the magnitude of ε''_m is equal to one half of the amplitude of dielectric dispersion. As will be discussed later, the value of ε''_m becomes smaller than $(\varepsilon_s - \varepsilon_\infty)/2$ if there is more than one relaxation time. This problem will be discussed later. Equation (3.23) represents the limiting case which means that ε''_m can never exceed the value of $(\varepsilon_s - \varepsilon_\infty)/2$.

3.4 Intrinsic Relaxation Time

We have derived equations (3.19) and (3.20) which describe mathematically the frequency dependence of the dielectric constant of polar molecules. These equations are widely applicable to gases and liquids so long as the molecules are spherical and there are no intermolecular interactions. At present, we are interested in how the relaxation time is related to the dimension and geometry of polar molecules. The model used in this discussion is a spherical particle with a radius a. The relaxation time is calculated by studying how the polarized state of a group of dipoles will decay when an applied external field is turned off. The transition from polarized state to random distribution by thermally driven rotary diffusion is hindered by the viscosity η. As before, we calculate the number of dipoles which are pointing in the directions θ and $\theta + d\theta$. This number is given by

$$dN = fN \sin \theta \, d\theta/2 \tag{3.24}$$

where N is the total number of molecules and f has a value of 1 when no external field is present. When an electrical field is applied, the function f is given by the Boltzmann factor as discussed before:

$$f = A \exp(\mu E \cos \theta/kT) \tag{3.25}$$

where A is a constant. Assuming that the intensity of applied field is very small, the exponential term can be expanded in a power series. Ignoring higher-order terms in the series, we obtain

$$f = A\left(1 + \frac{\mu E}{kT} \cos \theta\right). \tag{3.26}$$

When the electrical field is removed, the polarization decreases exponentially and the system transforms into a random state. Debye accounted for the gradual decrease of polarization by multiplying the second term on the RHS of (3.26) by an exponential decay function:

$$f(t) = A\left[1 + \frac{\mu E}{kT}\exp\left(-\frac{2kTt}{\zeta}\right)\cos\theta\right] \tag{3.27}$$

where ζ is the constant of Stokes's law for rotary diffusion of spheres and is given by

$$\zeta = 8\pi\eta a^3 \tag{3.28}$$

where η is the viscosity around the molecule and a is the radius. Note that f is now a function of time and reduces to 1 after a long time, leading the system to a random distribution. For gaseous molecules, the friction coefficient ζ is very small causing rapid decay of the exponential term. However, the decrease of polarization is a slow process for viscous fluids.

The mean value of the dipole moment during the decay is given by the following equation:

$$\langle m_t \rangle = \int_0^\pi \mu\cos(\theta)p(t)\,d\theta \tag{3.29}$$

where the distribution function $p(t)$ is given by

$$p(t) = f(t)N\sin\theta\,d\theta\left(\int_0^\pi f(t)N\sin\theta\,d\theta\right)^{-1}. \tag{3.30}$$

Inserting (3.30) into (3.29)

$$\langle m_t \rangle = \frac{\mu\displaystyle\int_0^\pi \cos\theta\sin\theta + \frac{\mu^2 E}{kT}\int_0^\pi \exp(-2kTt/\zeta)\cos^2\theta\sin\theta\,d\theta}{\displaystyle\int_0^\pi \sin\theta\,d\theta + \frac{\mu E}{kT}\int_0^\pi \exp(-2kTt/\zeta)\cos\theta\sin\theta\,d\theta}. \tag{3.31}$$

Of these integrals, the terms containing $\cos\theta\sin\theta$ disappear and we can rearrange the remaining terms in the following form:

$$\langle m_t \rangle = \frac{\mu^2 E}{kT}\exp(-2kTt/\zeta). \tag{3.32}$$

The term $\zeta/2kT$ has a dimension of time and is called the intrinsic relaxation time, i.e.

$$\tau^* = \zeta/2kT. \tag{3.33}$$

Combining this equation and (3.28), we obtain the following formula:

$$\tau^* = 4\pi a^3\eta/kT. \tag{3.34}$$

According to this equation, the intrinsic relaxation time is proportional to the cube of the radius of the spheres. Also, τ^* is proportional to viscosity as well as $1/T$.

The intrinsic relaxtion time τ^* is calculated as the time constant of decay of the polarization P after the field is suddenly removed. This process is controlled by rotary diffusion of the molecule, and, therefore, is affected strongly by the conditions of its environment, in particular, the viscosity. The relaxation time τ dictates the time course of polarization in the presence of electrical fields without due consideration of viscous drag. Therefore, these two relaxation times are not the same quantities. It has been proven that the intrinsic relaxation time τ^* is related to the dielectric relaxation time τ by the following equation (Debye 1929):

$$\tau^* = \tau \, \frac{\varepsilon_\infty + 2}{\varepsilon_s + 2}. \tag{3.35}$$

Debye's theory for one relaxation time holds when the following conditions are satisfied.

(i) An absence of interaction between dipoles. In the presence of interaction between neighbouring molecules, the dipoles tend to behave non-uniformly. In order to have only one relaxation time, it is necessary that all of the dipoles are equivalent, i.e. all behave in an identical way.

(ii) Only one process leading to equilibrium. This requires that molecules are all uniform in size and shape. Moreover, these molecules must be spherical. If they are not spherical, rotation along the long axis produces a longer relaxation time than that along the short axis.

The viscous medium surrounding the dipole is considered as a continuum. This assumption is particularly appropriate for large dipolar molecules which are immersed in a solvent. Therefore, Debye's equation (3.34) is a good approximation for globular proteins of spherical shape. Debye's equation can be generalized to ellipsoidal molecules as will be discussed later.

3.5 The Relaxation of Solid Dielectric Material

In contrast to gases and polar liquids, the interaction between neighbouring molecules is very strong in the solid state. Under these circumstances, these molecules are allowed to undergo only an elastic displacement around the equilibrium position. The dipolar orientations which occur in gases and liquids are, therefore, unlikely in solids. However, if the temperature approaches the melting and/or critical point, there will be several equilibrium positions, and dipoles are allowed to jump from one position to another when an electrical field is applied. Under these conditions, even solid materials are able to exhibit dielectric dispersion due to discrete transitions from one equilibrium position to another.

For the sake of simplicity, we assume that molecules in solids can jump between only two equilibrium positions (see figure 2.8). Namely, the dipoles can have only two orientations, i.e. $+z$ and $-z$. If we apply a field along the $+z$ direction, certain numbers of molecules which are in the $-z$ direction would flip over to the $+z$ direction. However, due to the randomizing effect of thermal motion, some molecules will change their orientation from the $+z$ to the $-z$ direction.

We assume that the number of molecules in the $+z$ and $-z$ directions is n_1 and n_2 respectively. Also, we define the probability of direction change per unit time from $+z$ to $-z$ as w_{12} and the probability of transition from $-z$ to $+z$ as w_{21}. Using these notations, we can set up the following equations:

$$\frac{dn_1}{dt} = -w_{12}n_1 + w_{21}n_2 \tag{3.36}$$

$$\frac{dn_2}{dt} = w_{12}n_1 - w_{21}n_2. \tag{3.37}$$

At equilibrium

$$\frac{dn_1}{dt} = \frac{dn_2}{dt} = 0. \tag{3.38}$$

Therefore, using either (3.36) or (3.37), we obtain

$$\frac{w_{12}}{n_2} = \frac{w_{21}}{n_1}. \tag{3.39}$$

Now let us equate these terms to $1/2\tau$:

$$\frac{w_{12}}{n_2} = \frac{1}{2\tau} \qquad \frac{w_{21}}{n_1} = \frac{1}{2\tau} \tag{3.40}$$

where τ is a quantity having dimensions of time, as will be discussed further. At equilibrium, the number of molecules in the $+z$ and $-z$ directions is given by the following two equations, with the assumption that dipoles in solids still follow the Boltzmann distribution:

$$n_1 = A \exp(\mu E/kT) \tag{3.41}$$

$$n_2 = A \exp(-\mu E/kT). \tag{3.42}$$

Substituting these into (3.40), we obtain expressions for the probabilities w_{12} and w_{21}:

$$w_{12} = \frac{A}{2\tau} \exp(-\mu E/kT) \tag{3.43}$$

$$w_{21} = \frac{A}{2\tau} \exp(\mu E/kT). \tag{3.44}$$

Substituting these into (3.36) and (3.37), we obtain the following equations, with the aid of the assumption that the field is very small:

$$2\tau \frac{dn_1}{dt} = -(n_1 - n_2) + \frac{\mu E}{kT}(n_1 + n_2) \tag{3.45}$$

$$2\tau \frac{dn_2}{dt} = (n_1 - n_2) - \frac{\mu E}{kT}(n_1 + n_2). \tag{3.46}$$

The above equations were derived assuming the field E is static. However, we can easily extend the discussion to periodic fields using $E = E_0 e^{j\omega t}$. We assume that the number of molecules which change direction with applied field is very small. The molecules which do not respond to the field can be expressed as n_0; therefore, we can define n_1 and n_2 as follows:

$$n_1 = n_0 + v_0 e^{j\omega t} \tag{3.47}$$

$$n_2 = n_0 - v_0 e^{j\omega t}. \tag{3.48}$$

Substituting these into (3.45) or (3.46), we obtain

$$v_0 = \frac{n_0}{1 + j\omega\tau} \frac{\mu E_0}{kT}. \tag{3.49}$$

It is clear now that the parameter τ has the meaning of dielectric relaxation time. Even at equilibrium, the number of molecules pointing in the direction of the field is not much greater than those pointing in the opposite direction. Therefore, the net moment can be calculated as the difference between $n_1\mu$ and $n_2\mu$ as shown below:

$$n_1\mu - n_2\mu = 2v_0 e^{j\omega t} = \frac{2n_0}{1 + j\omega\tau} \frac{\mu^2 E_0}{kT}. \tag{3.50}$$

This is the net moment of a group of molecules of total number $2n_0$. Therefore, the net moment per molecule is obtained by dividing the RHS of (3.50) by $2n_0$:

$$\bar{\mu} = \frac{1}{1 + j\omega\tau} \frac{\mu^2 E_0}{kT} e^{j\omega t} \quad \text{or} \quad = \frac{1}{1 + j\omega\tau} \frac{\mu^2 E}{kT}. \tag{3.51}$$

This equation is similar to those derived for gases and liquids with the exception of the numerical factor in the denominator of the RHS, i.e. kT instead of $3kT$.

Although the theory for solid dielectrics discussed above does not pertain to biological materials directly, the conditions existing in some biological systems resemble those of solid material. For example, protein molecules embedded in biological membranes are not free to rotate because of the strong interaction between protein and lipid molecules. In addition, the extremely high viscosity of membranes prohibits the orientation of membrane proteins below a critical temperature. Under these conditions, we can only assume that dipolar protein molecules can undergo a limited orientation between certain allowed equilibrium positions. Therefore, the orientation of membrane proteins does not resemble those of dipolar liquids. Thus, solid state analogy rather than the liquid state approach may be more appropriate for the study of the orientation behaviour of membrane proteins.

3.6 A Generalization of Relaxation Theory to Multiple Relaxation Time

As pointed out previously, equation (3.18) was derived with an assumption that the relaxation process is characterized by only one time constant. This assumption led us to (3.19) and (3.20), and the behaviour of these functions is graphically presented in figure 3.2. However, it is well known that the dispersion curves observed with polar molecules often deviate from Debye's equation. Actually, Debye-type behaviour is exceptional, and the majority of polar molecules which have been investigated display a certain deviation from the Debye theory. The dispersion curves are usually broader than the Debye curve and dielectric losses are lower than that predicted by the theory. The deviations from the Debye equation take place when the relaxation process involves more than one relaxation time. For example, if the molecule under investigation has the shape of a general ellipsoid and the dipole axis does not coincide with the molecular axes, then the polarization of this molecule will involve three rotations around the major as well as minor axes. Therefore, these rotations are characterized by three relaxation times instead of one. If the lengths of the major and minor axes are very different, the RHS of (3.18) consists of three dispersion terms having different relaxation times and dielectric increments. In general, the dispersion equation must be rewritten if there are more than one discretely different relaxation process:

$$\varepsilon' = \varepsilon_\infty + \sum_{i=1}^{n} \frac{\Delta\varepsilon_i}{1 + j\omega\tau_i}. \tag{3.52}$$

A further generalization can be made by introducing a distribution function $G(\tau)$ in equation (3.52), assuming that the distribution of relaxation time is continuous:

$$\varepsilon' = \varepsilon_\infty + \int_0^\infty \frac{G(\tau)\Delta\varepsilon\,d\tau}{1 + j\omega\tau}. \tag{3.53}$$

For the sake of generality, the type of distribution function will not be specified at this stage. In any case, $G(\tau)$ must satisfy the normalization condition such that

$$\int_0^\infty G(\tau)\,d\tau = 1. \tag{3.54}$$

Separating the real and imaginary parts, we obtain the following two equations:

$$\varepsilon' = \varepsilon_\infty + \int_0^\infty \frac{G(\tau)\Delta\varepsilon\,d\tau}{1 + (\omega\tau)^2} \tag{3.55}$$

$$\varepsilon'' = \int_0^\infty \frac{G(\tau)\Delta\varepsilon(\omega\tau)\,d\tau}{1 + (\omega\tau)^2}. \tag{3.56}$$

The assumption underlying these equations is the continuous distribution of relaxation times, that is to say, the relaxation process is characterized by a continuous spectrum of time constants. The question is what type of distribution function is most reasonable and realistic for physical and/or chemical systems. For example, if we are dealing with a synthetic polymer solution, the length of these polymer molecules will not be uniform and the molecular weight will vary continuously. It is known that the distribution of length of synthetic polymers follows the logarithmic–normal distribution function. Therefore, we have reason to believe that the distribution of dielectric relaxation time arises from the length distribution and also follows the logarithmic–normal distribution function. In the following, the possible causes of the distribution of relaxation time are listed.

(i) Size distribution. As mentioned above, the size or length distribution (of polymers) is one of the major causes of the distribution of relaxation time. Even if the molecule is spherical, a mixture of spheres with different radii will display non-Debye-type dispersion.

(ii) Non-spherical molecules. In contrast to spherical molecules, ellipsoidal molecules, for example, have three axes of rotation. The frictional coefficients along these axes are different and this, in turn, produces the distribution of relaxation time.

(iii) Flexible chain polymer molecules. For some polymer molecules such as polyvinyl chloride, the C–C axes are relatively free to rotate and bend. Therefore, free-rotating segments of different lengths cause the distribution of relaxation time.

(iv) Intermolecular interactions. If there are electrostatic interactions between molecules, they will create non-uniform local fields. This is also one of the causes of the distribution of time constants.

3.7 Use of the Gaussian Distribution Function

As discussed, the distribution of relaxation times can be produced by size and shape non-uniformity. Therefore the choice of distribution function depends upon the cause of multiple relaxation times. As an example, let us use a logarithmic Gaussian function, as shown below (Wagner 1914):

$$G(\tau)\,d\tau = \frac{b}{\sqrt{\pi}}\exp(-b^2z^2)\,dz \qquad (3.57)$$

where $z = \log(\tau/\tau_0)$. Accordingly, the distribution function has a bell shaped symmetric curve with a centre value of τ_0 when plotted on a logarithmic scale. The parameter b is the standard deviation and determines the spread of distribution curves as shown in figure 3.3.

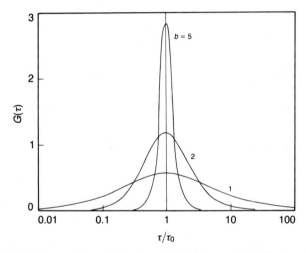

Figure 3.3 The profile of the logarithmic–normal distribution function plotted against $\ln(\tau/\tau_0)$, where τ_0 is the median value of the relaxation time. The value of the parameter b is varied from 0 to ∞. Note that the shape of the function becomes broader as the value of b decreases.

If one substitutes $z = \log(\tau/\tau_0)$ into (3.57), we obtain

$$G(\tau) = \frac{b}{\sqrt{\pi}}\frac{1}{\tau}\exp\{-b^2[\log(\tau/\tau_0)]^2\}. \qquad (3.58)$$

The variation of this Gaussian function is shown in figure 3.3 for various values of the parameter b. The abscissa is the normalized relaxation time (τ/τ_0). As shown, the shape of the Gaussian function depends markedly on the value of b; it reduces to the delta function when b approaches infinity (see Appendix C). On the other hand, if b decreases, the curve becomes very broad. However, the normalization condition requires that the area under these curves must be the same regardless of the value of b.

Having discussed the behaviour of the Gaussian function, let us use this function with (3.55) and (3.56). When (3.58) is substituted into (3.55), we obtain

$$\frac{\varepsilon' - \varepsilon_\infty}{\varepsilon_0 - \varepsilon_\infty} = \frac{b}{\sqrt{\pi}} \int_{-\infty}^{\infty} \frac{\exp(-b^2 z^2)}{1 + \omega^2 \tau_0^2 e^{2z}} \, dz. \tag{3.59}$$

Changing the integration variable from z to u such that

$$u = z + z_0$$

where $z_0 = \log \omega\tau_0$, the integral can be simplified slightly:

$$\frac{\varepsilon' - \varepsilon_\infty}{\varepsilon_0 - \varepsilon_\infty} = \frac{b}{\sqrt{\pi}} \int_{-\infty}^{\infty} \frac{\exp[-b^2(u - z_0)^2]}{1 + e^{2u}} \, du \tag{3.60}$$

$$= \frac{b}{\sqrt{\pi}} \exp(-b^2 z_0^2) \int_{-\infty}^{\infty} \exp(-b^2 u^2 + 2b^2 z_0 u) \frac{du}{e^u + e^{-u}} \tag{3.61}$$

$$= \frac{b}{\sqrt{\pi}} \exp(-b^2 z_0^2) \int_{-\infty}^{\infty} \exp(-b^2 u^2) \frac{\exp[(2b^2 z_0 - 1)u]}{2 \cosh u} \, du. \tag{3.62}$$

The last equation can be converted to a form which is more convenient for numerical integration:

$$\frac{\varepsilon' - \varepsilon_\infty}{\varepsilon_0 - \varepsilon_\infty} = \frac{b}{\sqrt{\pi}} \exp(-b^2 z_0^2) \int_0^{\infty} \exp(-b^2 u^2) \frac{\cosh(2b^2 z_0 - 1)u}{\cosh u} \, du. \tag{3.63}$$

Although this integral cannot be solved analytically, the calculation can be performed by numerical integration with various values of the parameters b and z_0. Likewise, the imaginary part ε'' can be computed using the following equation:

$$\frac{\varepsilon''}{\varepsilon_0 - \varepsilon_\infty} = \frac{b}{\sqrt{\pi}} \exp(-b^2 z_0^2) \int_0^{\infty} \exp(-b^2 u^2) \frac{\cosh(2b^2 z_0)u}{\cosh u} \, du. \tag{3.64}$$

Numerical computations of ε' and ε'' using these equations were performed by Yeager in 1936 and the results are shown in figures 3.4 and 3.5. Based on these figures, we are able to find the following.

(i) The relationship between b and ε' and ε''. The value of b determines the slope of the curve of ε' versus frequency. Obviously, when b is infinity, (3.60) reduces to the Debye equation with a single relaxation time. The curve when $b = \infty$ shows the Debye limiting slope. Likewise, (3.64) must reduce to the Debye equation (3.20) and the curve when $b = \infty$ in figure 3.5 is identical to that predicted by Debye's theory. However, as the value of b decreases, the slope of ε' versus frequency becomes broader and the height of ε'' becomes lower. As the parameter b reaches zero, both the real and imaginary parts reduce to a straight horizontal line.

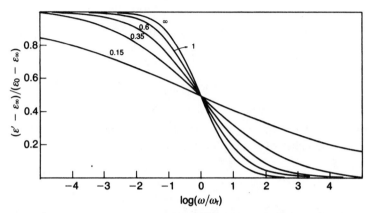

Figure 3.4 The normalized dielectric constant plotted against frequency. The curves are calculated using equation (3.63). Note that the dispersion curves become broader as the value of b (the number indicated by each curve) decreases. The abscissa is the normalized angular frequency on a logarithmic scale.

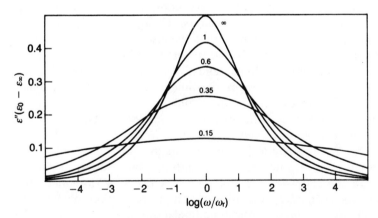

Figure 3.5 The imaginary part of the complex dielectric constant computed using equation (3.64). The number indicated by each curve is the value of the parameter b. The abscissa is the normalized angular frequency on a logarithmic scale.

(ii) The relationship between ε', ε'' and ω.

(a) $\omega = 0$. In this case, normalized ε' and ε'' become

$$\frac{\varepsilon' - \varepsilon_\infty}{\varepsilon_0 - \varepsilon_\infty} = 1 \qquad\qquad (3.65)$$

$$\frac{\varepsilon''}{\varepsilon_0 - \varepsilon_\infty} = 0. \qquad\qquad (3.66)$$

(b) $\omega = \omega_c$ where ω_c is the centre frequency and ε'' is a maximum:

$$\frac{\varepsilon' - \varepsilon_\infty}{\varepsilon_0 - \varepsilon_\infty} = 0.5 \qquad\qquad (3.67)$$

$$\frac{\varepsilon''}{\varepsilon_0 - \varepsilon_\infty} = \frac{b}{\sqrt{\pi}} \int_0^\infty \frac{\exp(-b^2 u^2)}{\cosh u} du. \qquad\qquad (3.68)$$

In this case, the value of ε' is constant regardless of the value of b. However, the value of ε'' is still a function of b.

(c) $\omega = \infty$:

$$\frac{\varepsilon' - \varepsilon_\infty}{\varepsilon_0 - \varepsilon_\infty} = 0 \qquad\qquad (3.69)$$

$$\frac{\varepsilon''}{\varepsilon_0 - \varepsilon_\infty} = 0. \qquad\qquad (3.70)$$

In this case, both the real and imaginary parts reduce to zero.

As already discussed, use of the Gaussian distribution function in equations (3.55) and (3.56) gives reasonable results which are quite consistent with experimental observations. The slope of the dispersion curve and the spread of ε'' can be changed by adjusting only one parameter b. The problem is that the use of this function makes the mathematical treatment of dispersion equations exceedingly cumbersome. The integrals in (3.63) and (3.64) cannot be solved analytically and we have to resort to numerical integrations. In order to simplify the computation, we can use simpler functions such as triangular or even rectangular functions with appropriate spread as shown in figure 3.6. However, the physical meaning of these functions is not clear and may not represent the physical reality of the system. Therefore, non empirical approaches are not used frequently for the analysis of experimental data. In the following we will discuss a few empirical methods which enable us to analyze the distribution of relaxation time quite easily without complex mathematical manipulations.

3.8 Cole–Cole's Empirical Equation (1941)

In view of the mathematical difficulties we encountered with the analytical approach discussed above, it became apparent that there may be some empirical methods able to analyze the distribution of relaxation

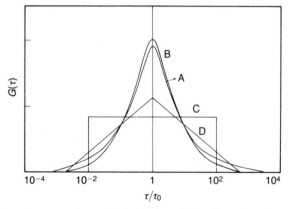

Figure 3.6 A comparison of the shape of the Gaussian distribution function with $b = 2.0$ (curve B) and the Cole–Cole distribution function with $\alpha = 0.3$ (curve A). Note that the Cole–Cole distribution function decreases more rapidly than the Gaussian function. Curves C and D are schematic representations of square and triangular functions used as approximate distribution functions of relaxation time.

time. We will discuss two such empirical techniques proposed by Cole and Cole (1941) and also by Fuoss and Kirkwood (1941). As the starting point, Cole and Cole rewrote Debye's equation by introducing an empirical parameter, as shown below:

$$\varepsilon^* = \varepsilon_\infty + \frac{\varepsilon_s - \varepsilon_\infty}{1 + (j\omega\tau)^{1-\alpha}} \tag{3.71}$$

where α is a parameter which ranges from 0 to 1. In order to separate the real and imaginary parts of the complex dielectric constant, we have to use the following equation for j ($= \sqrt{-1}$):

$$j = \cos(n\pi/2) + j\sin(n\pi/2) \tag{3.72}$$

where $n = 1 - \alpha$. Substituting this equation into (3.71), we obtain

$$\varepsilon^* = \varepsilon_\infty + \frac{\varepsilon_0 - \varepsilon_\infty}{1 + (\omega\tau)^n[\cos(n\pi/2) + j\sin(n\pi/2)]}. \tag{3.73}$$

Separating the real and imaginary parts, the following two equations are obtained:

$$\varepsilon' - \varepsilon_\infty = (\varepsilon_s - \varepsilon_\infty)\frac{1 + (\omega\tau)^n\cos(n\pi/2)}{1 + 2(\omega\tau)^n\cos(n\pi/2) + (\omega\tau)^{2n}} \tag{3.74}$$

$$\varepsilon'' = (\varepsilon_s - \varepsilon_\infty)\frac{(\omega\tau)^n\sin(n\pi/2)}{1 + 2(\omega\tau)^n\cos(n\pi/2) + (\omega\tau)^{2n}}. \tag{3.75}$$

The calculation of ε' and ε'' using these two equations does not require complex mathematical manipulations and is much easier than when using (3.63) and (3.64). In spite of its simplicity, the results obtained with these empirical equations are quite similar to those shown in figures 3.4 and 3.5. Careful examination reveals, however, that the dispersion curves calculated by Cole–Cole's equation are slightly different from those obtained with (3.63) and (3.64). This means that the Cole–Cole empirical formalism is equivalent to introducing a distribution function which is different from the Gaussian distribution function. It is pointed out by Cole and Cole that the distribution function must be of the following form (see Appendix D for the derivation):

$$G(s) = \frac{1}{2\pi} \frac{\sin n\pi}{\cosh ns + \cos n\pi} \tag{3.76}$$

where $s = \log(\tau/\tau_0)$. This distribution function is shown graphically in figure 3.6 and is compared with the Gaussian distribution function. Curve A in this figure is calculated assuming $\alpha = 0.3$ and curve B was computed assuming $b = 2.0$. Since the functional relationship between the parameters b and α is not clearly defined, it is sufficient to detect the difference in the shape of these two curves qualitatively. According to this figure, the Cole–Cole distribution function decreases more rapidly for small values of s than the Gaussian function. For large values of s, equation (3.76) reduces to

$$G(s) = (1/\pi)e^{-ns} \sin \alpha\pi.$$

As shown, the decrease in $G(s)$ for large s is slower than for the Gaussian function. When s is near zero, equation (3.76) reduces to

$$G(s) = \frac{\sin n\pi}{2\pi(1 - \cos n\pi)}.$$

If $\alpha = 0$ or $n = 1$, $G(s)$ is infinitely large, becoming the delta function.

An important finding made by Cole and Cole is that when ε' is plotted against ε'', a circular arc is obtained. When $\alpha = 0$, the arc is a complete semicircle. However, as the value of α increases, the circular arc becomes depressed with its centre below the real axis. In the following, we will discuss the mathematical basis of the circular plot ε' versus ε''. Let us rearrange the Debye equations (3.19) and (3.20) as follows:

$$\frac{\varepsilon' - \varepsilon_\infty}{\varepsilon_s - \varepsilon_\infty} = \frac{1}{1 + (\omega\tau)^2} \tag{3.77}$$

$$\frac{\varepsilon''}{\varepsilon_s - \varepsilon_\infty} = \frac{\omega\tau}{1 + (\omega\tau)^2}. \tag{3.78}$$

Dividing equation (3.78) by equation (3.77):

$$\frac{\varepsilon''}{\varepsilon' - \varepsilon_\infty} = \omega\tau \tag{3.79}$$

and substituting this into either (3.77) or (3.78), we can obtain

$$\left(\varepsilon' - \frac{\varepsilon_s + \varepsilon_\infty}{2}\right)^2 + \varepsilon''^2 = \left(\frac{\varepsilon_s - \varepsilon_\infty}{2}\right)^2. \tag{3.80}$$

This is the equation of a circle having a radius of $(\varepsilon_s - \varepsilon_\infty)/2$. Thus the plot of ε' against ε'' would produce a circle with its centre on the real axis. The circle is shown in figure 3.7. We now repeat the same derivation starting with equations (3.74) and (3.75). The derivation is more complex this time. Nevertheless, we can derive the following equation without much difficulty using the same procedure:

$$\left(\varepsilon' - \frac{\varepsilon_s + \varepsilon_\infty}{2}\right)^2 + \left(\varepsilon'' + \frac{\varepsilon_s - \varepsilon_\infty}{2}\cot\frac{n\pi}{2}\right)^2 = \left(\frac{\varepsilon_s - \varepsilon_\infty}{2}\operatorname{cosec}\frac{n\pi}{2}\right)^2. \tag{3.81}$$

This equation, although it looks more complex, still represents a circle with its centre below the real axis. Namely, the centre of this circle is located at the coordinates

$$\frac{\varepsilon_s + \varepsilon_\infty}{2} \qquad \text{and} \qquad -\frac{\varepsilon_s - \varepsilon_\infty}{2}\cot\frac{n\pi}{2}$$

and the radius is

$$\frac{\varepsilon_s - \varepsilon_\infty}{2}\operatorname{cosec}\frac{n\pi}{2}.$$

An example of a Cole–Cole plot is given in figure 3.7. As shown, the circle is depressed with its centre below the abscissa. The line connecting the centre 0 and the intercept between the arc and the real axis

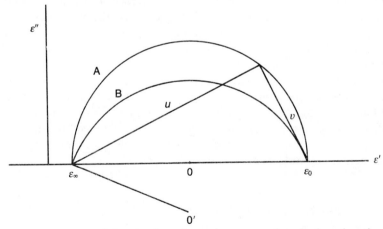

Figure 3.7 Cole–Cole circular plots: A, an arc for single relaxation time; B, a depressed circle for multiple relaxation times. The angle between the abscissa and the line connecting the centre of the circle and the intercepts ε_0 and ε_∞ is equal to $\alpha\pi/2$, from which we can calculate the value of α. See the text for the explanation of u and v.

makes an angle of $n\pi/2$. Therefore, the value of n $(= 1 - \alpha)$ can be determined graphically by measuring the angle θ and dividing it by 90°. So long as the Cole–Cole plot is symmetric, the distribution parameter can be determined unequivocally using this technique.

The line segments u and v are related to the relaxation time and frequency by the following equation:

$$v/u = (\omega\tau)^{1-\alpha}. \tag{3.82}$$

Taking the log of both sides, we obtain

$$\log(v/u) = (1 - \alpha)\log\omega + (1 - \alpha)\log\tau_0. \tag{3.83}$$

Therefore, if the log of v/u is plotted against $\log\omega$, a straight line should be found. The slope of this plot, according to equation (3.83), is equal to the value of n or $1 - \alpha$, and the intercept of this plot with the vertical axis at $\log\omega = 0$ should give rise to the value of $(1 - \alpha)\log\tau_0$. Thus we can determine both the distribution parameter α and the relaxation time τ_0 using either a circular arc or linear plot.

Not infrequently, we encounter dielectric material which exhibits an asymmetric dispersion curve. The Cole–Cole plots of these samples are skewed. Equation (3.71), which was derived to analyze symmetrical circles, cannot be used for this case. Davidson and Cole (1951) modified (3.71) as follows:

$$\varepsilon^* = \varepsilon_\infty + \frac{\varepsilon_s - \varepsilon_\infty}{(1 + j\omega\tau)^{1-\alpha}}. \tag{3.84}$$

Asymmetrical dispersion curves are the results of a skewed distribution of relaxation times. This is particularly important for synthetic polymers which are usually non-uniform in size. At times, the molecular weight distribution of some polymers is asymmetrical and thus gives rise to a skewed distribution function. Equation (3.84) is useful to analyze the relaxation of these cases.

The significance of the parameter α or its relevance to the physical reality of the system is not clearly understood. However, the Cole–Cole distribution parameter is treated as a mathematical method which enables us to find the extent of the distribution of relaxation time. This technique has been used widely and has been one of the most important experimental methods in dielectric measurements.

3.9 The Fuoss–Kirkwood Method (1941)

Another empirical approach was proposed by Fuoss and Kirkwood for the analysis of distribution of relaxation time. In contrast to the Cole–Cole approach, this method utilizes only the imaginary part of the complex dielectric constant. The derivation starts with equation (3.20) (the equation is reproduced once again for convenience):

$$\varepsilon'' = (\varepsilon_s - \varepsilon_\infty)\frac{\omega\tau}{1 + (\omega\tau)^2}. \tag{3.20}$$

In addition, we use the following relations:

$$\omega_m = 1/\tau \tag{3.85}$$

$$\varepsilon'' = \frac{\varepsilon_s - \varepsilon_\infty}{2} \tag{3.24}$$

where ω_m is the radian frequency at which the imaginary part ε'' is a maximum. Combining these three equations, we obtain

$$\varepsilon'' = \varepsilon_m'' \frac{2}{z + 1/z} \tag{3.86}$$

where $z = \omega/\omega_m$. Since $z = \exp(\ln z)$, we can rewrite (3.86) as follows:

$$\varepsilon'' = \varepsilon_m'' \operatorname{sech} \ln(\omega/\omega_m). \tag{3.87}$$

This equation gives rise to a bell shaped curve when ε'' is plotted against frequency with its peak at ω_m. Namely, when $\omega = \omega_m$, $\ln(\omega/\omega_m)$ reduces to zero and $\operatorname{sech}(\ln(\omega/\omega_m))$ has a maximum value of 1. If ω is either smaller or greater than ω_m, the value of ε'' is always smaller than ε_m'' and the curve is symmetric on both sides of ω_m. Equation (3.87) was derived starting with the Debye theory where there is only one relaxation time. In order to account for the distribution of relaxation time, an empirical parameter β was introduced, as shown below:

$$\varepsilon'' = \varepsilon_m'' \operatorname{sech}(\beta \ln (\omega/\omega_m)). \tag{3.88}$$

The value of β ranges from 1 to 0. If the value of β is smaller than 1, the shape of the ε'' versus frequency plot becomes broader. However, the value of ε_m'' is always 1. The results of numerical calculations using (3.88) for various values of β are shown in figure 3.8. The procedure to obtain the value of β is to plot normalized ε'' against frequencies and curve-fit the experimental data with the calculated curves by trial and error until a reasonable agreement is obtained.

   Another method is to transform (3.88) into the following form in order to generate a linear plot, namely

$$\cosh^{-1}(\varepsilon_m''/\varepsilon'') = \beta \ln(\omega/\omega_m). \tag{3.89}$$

If the values of ε_m'' and ω_m are known, we can plot the value of $\cosh^{-1}(\varepsilon_m''/\varepsilon'')$ against normalized frequency (ω/ω_m) on a logarithmic scale. Theoretically, this plot is a straight line with a slope of β. Even if the value of ω_m is not known, we can still generate a linear plot by plotting $\cosh^{-1}(\varepsilon_m''/\varepsilon'')$ against $\ln \omega$. An example of this plot for dilute NaCl

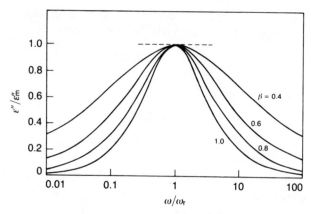

Figure 3.8 Calculated dielectric loss curves for various values of the parameter β. Note that the peak value of $\varepsilon''/\varepsilon''_m$ is always 1 and that the ε''–frequency plot becomes broader as the value of β decreases.

solution is shown in figure 3.9. From the slope of this plot, we find the value of β to be 1.0. This value is equivalent to $\alpha = 0$ as discussed below.

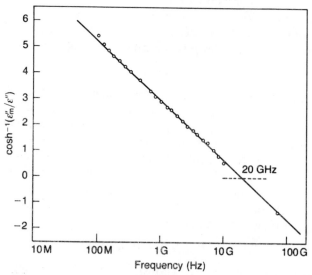

Figure 3.9 An example of a plot using equation (3.89) for dilute NaCl solution at 20 °C. The intercept between this plot and the zero line gives rise to a relaxation frequency of 18 GHz and a slope of approximately 1. (Takashima, unpublished.)

The parameter β is related to the Cole–Cole parameter α or n by the following equation:

$$\beta = \frac{n}{\sqrt{2}(\cos n\pi/4)}. \tag{3.90}$$

The values of β and α calculated using (3.90) are shown in table 3.1.

Table 3.1 Numerical values of the parameters α and β. Data are taken from Boettcher and Bordewijk (1978).

n	α	β
0.00	1.00	0.00
0.25	0.75	0.181
0.50	0.50	0.383
0.75	0.25	0.638
1.00	0.00	1.00

Analogous to the Cole–Cole parameter, the physical significance of the β-parameter is not understood. However, the distribution function for the Fuoss–Kirkwood empirical parameter is given by the following equation (see Appendix D for the derivation):

$$F(s) = \frac{\beta}{\pi} \frac{\cos(\beta\pi/2)\cosh(\beta s)}{\sinh^2(\beta s) + \cos^2(\beta\pi/2)} \tag{3.91}$$

where $s = \log(\tau/\tau_0)$. For large s, this equation reduces to

$$F(s) = \frac{\beta}{\pi} e^{-\beta s} \cos(\beta\pi/2). \tag{3.92}$$

This equation is similar to the limiting form of the Cole–Cole distribution function. This function also decreases more slowly than the Gaussian function for large s. When s approaches zero, equation (3.91) reduces to

$$F(s) = \frac{2\beta}{\pi} \frac{1}{\cos(\beta\pi/2)}. \tag{3.93}$$

When $\beta = 1$, the function $F(s)$ is infinitely high, becoming identical to the delta function. This behaviour is quite similar to the Cole–Cole function.

As has been discussed, simple empirical methods are available for the analysis of the distribution of relaxation time. In this sense, the use of these empirical parameters provides us with a very effective means of calculation, which is more expedient than the analytical approach discussed earlier.

3.10 A Treatment of Dielectric Relaxation Using Elementary Rate Theory (Glasstone *et al* 1941)

Before we discuss the application of the absolute rate theory to electric polarization, the basic principle of Eyring's treatment must be explained. The scheme of a chemical reaction, envisioned by Eyring, is illustrated in figure 3.10. According to this diagram, reactants form an activated complex and the complex will decompose into products at a certain rate. The abscissa is the reaction coordinate which represents the progress of chemical reactions. The rate of the reaction is determined by the average velocity of decomposition of the activation complex.

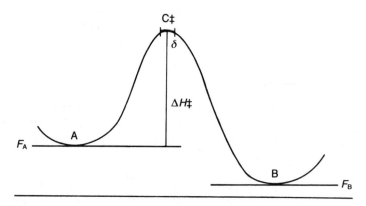

Figure 3.10 A schematic representation of reaction coordinates. The abscissa represents the progress of a chemical reaction and the ordinate is free energy. Reactants form an activation complex C‡, with a free energy which is higher than those of reactants and products. δ indicates a short path for which the activation complex is stable.

According to Eyring's calculation, this velocity is given by

$$v = C\ddagger \frac{kT}{h} \tag{3.94}$$

where $C\ddagger$ is the concentration of activation complex and k and T are Boltzmann's constant and absolute temperature respectively. h is Planck's constant. The velocity v is also given by the rate of formation of activation complex, as shown by

$$v = k_c C_A C_B C_C \ldots \tag{3.95}$$

where k_c is the rate constant, and C_A, C_B and C_C are the concentrations of reactants. Equating (3.94) and (3.95), we obtain

$$k_c = \frac{kT}{h} \frac{C\ddagger}{C_A C_B C_C \dots}. \tag{3.96}$$

If we represent the chemical reaction by the following equation:

$$A + B + C + \dots = C\ddagger \tag{3.97}$$

and assume that the reactants and activation complex $C\ddagger$ are in equilibrium, then the equilibrium constant is given by

$$K\ddagger = \frac{C\ddagger}{C_A C_B C_C \dots}. \tag{3.98}$$

Substituting (3.98) into (3.96):

$$k_c = \frac{kT}{h} K\ddagger. \tag{3.99}$$

Inserting the relation $\Delta F\ddagger = RT \ln K\ddagger$ ($\Delta F\ddagger$ is the free energy of activation) in (3.99):

$$k_c = \frac{kT}{h} \exp(-\Delta F\ddagger/RT). \tag{3.100}$$

This equation relates the rate constant to the free energy of activation. We can further modify this equation using $\Delta F\ddagger = \Delta H\ddagger - T\Delta S\ddagger$:

$$k_c = \frac{kT}{h} \exp(-\Delta H\ddagger/RT) \exp(\Delta S\ddagger/R) \tag{3.101}$$

where $\Delta H\ddagger$ and $\Delta S\ddagger$ are standard enthalpy and entropy changes of the activation process. Equation (3.100) shows that the rate of a chemical reaction is determined by the free energy of activation whereas, according to (3.101), $\Delta H\ddagger$ alone does not determine the rate of reaction. Even if $\Delta H\ddagger$ is unfavourable for the reaction, a large value of entropy $\Delta S\ddagger$ can compensate for a large heat of activation. It is well known, for example, that the $\Delta H\ddagger$ of protein denaturation is very large (approximately $100\,\text{kcal mol}^{-1}$) but usually a large entropy increase facilitates the denaturation reaction. This is a typical example of enthalpy–entropy compensation.

Eyring applied the elementary rate theory to many other molecular processes such as viscosity, diffusion and electrochemical processes. For example, diffussion was considered as a rate process which must overcome a potential barrier before the flow can start. Eyring also applied this theory to dielectric polarization. As depicted in figure 3.11, the process of dipole orientation from position A to another position B can be considered a rate process. The equilibrium positions A and B are separated by a potential barrier and the dipole must overcome the barrier in order to move over to another equilibrium position B. The rate of transition is determined by the inverse of relaxation time τ.

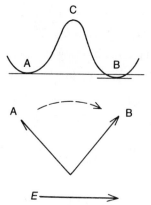

Figure 3.11 A schematic representation of the orientation of a dipole between two equilibrium positions. A potential barrier is assumed between two equilibrium positions.

Therefore, we can insert $k_c = 1/\tau$ in either (3.100) or (3.101) to obtain

$$\tau = \frac{h}{kT}\exp(\Delta F\ddagger/RT) \tag{3.102}$$

or

$$= \frac{h}{kT}\exp(\Delta H\ddagger/RT)\exp(-\Delta S\ddagger/R). \tag{3.103}$$

Multiplying both sides of (3.103) by T and taking the log of these terms, we obtain

$$\ln(\tau T) = \ln(h/K) - \Delta S\ddagger/R + \Delta H\ddagger/RT. \tag{3.104}$$

Differentiating both sides with respect to $1/T$:

$$\frac{\partial \ln(\tau T)}{\partial(1/T)} = \frac{\Delta H\ddagger}{R}. \tag{3.105}$$

This equation indicates that a plot of $\ln(\tau T)$ against $1/T$ gives a straight line with a slope of $\Delta H\ddagger/R$. Experimentally, it was found that this plot is indeed a straight line for many organic substances including biological polymers. Using this method, we can calculate the enthalpy of activation accompanying electric polarization. Equation (3.102) indicates that the free energy of activation is directly related to relaxation time τ. Therefore, $\Delta F\ddagger$ is readily calculated using the value of τ. Once both the free energy and enthalpy are determined, we can compute the entropy of activation. Table 3.2 shows the result of this type of calculation based on the experimental data for ice. The value of the enthalpy of activation is 13.2 kcal mol^{-1}. This is about three times the energy of a hydrogen bond (see chapter 8). The magnitude of the potential barrier depends upon the number of bonds disrupted by rotating a dipolar molecule. This indicates that the orientation of a water molecule in ice requires the breaking of three hydrogen bonds. Also, the sign of the entropy change is positive; this is particularly interesting because it indicates the disordering of the system accompanying electric polarization. In ice, therefore, the structure is slightly disordered by the application of electrical fields, indicated by entropy increases. This result gives us considerable insights into the dynamics of water molecules in ice in polarized and unpolarized states and clearly shows that extensive ordering does not occur upon polarization in ice. This finding may be generalized to a variety of other molecules.

Table 3.2 The changes in free energy, enthalpy and entropy accompanying the electric polarization of ice. Data are taken form Stearn and Eyring (1937).

Temperature (K)	τ (s)	$\Delta F\ddagger$ (kcal mol^{-1})	$\Delta H\ddagger$ (kcal mol^{-1})	$\Delta S\ddagger$ (cal mol^{-1})
227.3	10×10^{-5}	9.1	13.2	18.1
240.6	2.7×10^{-5}	9.0	13.2	17.5
261.3	2.4×10^{-5}	8.5	13.2	18.0
269.3	1.0×10^{-5}	8.4	13.2	17.9
272.3	8.6×10^{-4}	8.4	13.2	17.6

Suggested Reading

Boettcher C J F and Bordewijk P 1978 *Theory of Electric Polarisation* (Amsterdam: Elsevier)
Debye P 1929 *Polar Molecules* (New York: Dover)
Hill N E, Vaughan W E, Price A H and Davies M 1969 *Dielectric Properties and Molecular Behavior* (London: Van Nostrand–Reinhold)
Smyth C P 1955 *Dielectric Behavior and Structure* (New York: McGraw-Hill)

References

Cole K S and Cole R H 1941 Dispersion and absorption in dielectrics *J. Chem. Phys.* **9** 341–51
Davidson D W and Cole R H 1951 Dielectric relaxation of glycerol, propylene glycol and n-propanol *J. Chem. Phys.* **18** 1417–24
Fuoss R M and Kirkwood J G 1941 Electrical properties of solids. VIII. Dipole moments in polyvinyl chloride-diphenyl systems *J. Am. Chem. Soc.* **63** 385–94

Glasstone S, Laidler K J and Eyring H 1941 *The Theory of Rate Processes* (New York: McGraw-Hill)

Stearn E and Eyring H 1937 The deduction of reacting mechanisms from the theory of absolute rates *J. Chem. Phys.* **5** 113–24

Wagner K W Erklarung der dielecktrischen Nachwirkungsvorgange auf grund Maxwellscher Forstellungen 1914 *Ann. Phys, Lpz* **40** 817

Yeager W A 1936 *Physica* **7** 434

4   Dielectric Properties of Polymer Molecules

4.1 Introduction

This chapter will discuss the basic theories of electric polarization of polymer molecules and some exemplary results of measurement. The investigation of dipole moment of polymer molecules began in the late 1940s. The basic theories of electric polarization of polymers were formulated and massive experimental work began at the same time.

This chapter discusses only the early development of dielectric theories for synthetic polymers. Biopolymers will be discussed separately in later chapters. Dipole moment is closely related to the conformation of polymer molecules. The structure and/or conformation of biological polymers is much more complex than synthetic polymers and, moreover, the origin of the dipole moment of biopolymers may be different from that of synthetic polymers. Therefore, theories which were developed for synthetic polymers may not necessarily apply to biopolymers such as proteins and nucleic acids.

4.2 The Structure of Polymer Molecules

Polymer molecules can be synthesized by either addition polymerization or by the condensation of monomer units. The simplest of all polymers are linear polymers. The basic structure of these polymers is shown below schematically:

$$A'-A-A-A- \ldots -A-A''$$

or

$$A'-(-A-)_{x-1}- \ldots -A''$$

where A is a monomer unit and x is the degree of polymerization. In addition, A' and A'' are terminal groups. Examples of linear polymers are polyisobutylene (A) and polyethylene (B) and their repeating units are shown below:

$$
\begin{array}{ccc}
& \text{H} & \text{CH}_3 \\
& | & | \\
- & \text{C} - \text{C} - \\
& | & | \\
& \text{H} & \text{CH}_3
\end{array}
\qquad
\begin{array}{c}
\text{H} \\
| \\
- \text{C} - \\
| \\
\text{H}
\end{array}
$$

(A)   (B)

Polymer molecules become non-linear as the result of branching and cross-linking. For example, polyethylene if synthesized under high pressure forms a branched structure as shown below:

$$
\begin{array}{cccc}
\text{H} & \text{H} & \text{H} & \text{H} \\
| & | & | & | \\
- \text{C} - \text{C} - \text{C} - \text{C} - \\
| & | & | & | \\
\text{H} & & \text{H} & \text{H} \\
& \text{H}-\text{C}-\text{H} & & \\
& | & & \\
& \text{H}-\text{C}-\text{H} & & \\
& | & &
\end{array}
$$

If a polymer consists of more than one type of monomer unit, it is called a copolymer. For example,

$$-A-B-B-A-B-A-.$$

As can be surmized easily, biological polymers are extreme cases of copolymers. Detailed discussions of the structure of polymer molecules can be found in the book by Flory (1953).

4.3 Physical States of Polymers

Conformation in solution
The conformation of polymer molecules depends upon many factors, such as the rigidity of chains and the presence of charged groups. It also depends upon solvents, temperature and other extrinsic factors. Gaussian statistics are often used for the analysis of polymer configuration assuming that polymer chains are perfectly flexible.

Crystalline state
Like low molecular weight compounds, polymer molecules have a tendency to crystallize below certain critical temperatures. However, because of steric hindrance, polymer molecules cannot form large crystals, and the so-called crystalline state of polymers consists of small crystalline domains and amorphous regions. Nylon, polyethylene and cellulose are known for their tendency to crystallize. Many biological polymers, particularly globular proteins, have been obtained in crystalline form.

Glassy state
Amorphous polymers are flexible above their transition temperature

because of thermal motion. If the temperature is lowered, the thermal motion decreases and the structure of the polymer chains becomes frozen. Because of steric hindrance, these polymers do not crystallize, even under this condition. This is called the glassy state.

Rubber elasticity
Amorphous polymer molecules which are extensively cross-linked exhibit rubber elasticity above the critical temperature. Under these conditions, polymer chains undergo vigorous thermal motion without, however, a shift in the mass centre or orientation of the entire molecule because of cross-linkages. Micro-Brownian motion or local thermal motion is possible because the chain segments between cross-linkages are winding or twisted. It is well known that rubber has a larger entropy when it is relaxed than when it is stretched. This is because the segments between cross-linkages have more freedom of micro-Brownian motion when they are relaxed than when they are stretched.

4.4 Average Dimension (Flory 1953, Tanford 1961)

End-to-end distance
The average end-to-end distance of a flexible polymer is the root-mean-square average of the separation between the two ends. If h is the distance between two ends of a given chain, the mean end-to-end distance is defined by

$$h_{av} = \langle h^2 \rangle^{1/2}. \tag{4.1}$$

In vector notation, this equation can be written as follows:

$$h_{av} = \langle h \cdot h \rangle^{1/2}. \tag{4.2}$$

Radius of gyration
The radius of gyration r is the weight average of the distance r_i between the centre of mass and the mass element i. Namely, the radius of gyration can be defined by the following equation:

$$R_G^2 = \sum_i m_i r_i^2 \left(\sum_i m_i \right)^{-1} \tag{4.3}$$

where m_i is the mass of element i. If the masses m_i are all identical, then (4.3) can be simplified as

$$R_G^2 = \sum_i r_i^2 / N \tag{4.4}$$

where N is the total number of mass elements per chain. The average of the radius of gyration for all chain configurations is given by

$$R_G = \left(\langle \textstyle\sum r_i^2 \rangle / N \right)^{1/2}. \tag{4.5}$$

We can replace $\langle \sum r_i^2 \rangle$ with $\sum \langle r_i^2 \rangle$, namely

$$R_G = \left(\textstyle\sum \langle r_i^2 \rangle / N \right)^{1/2}. \tag{4.6}$$

Chain configuration with free bond rotation

If a hydrocarbon chain consists of $N + 1$ atoms, the chain configuration can be characterized by N bond vectors. If we assume complete freedom of the bond angle as shown in figure 4.1, the mean end-to-end length of the chain can be defined by

$$\langle h^2 \rangle = \sum_i^N \sum_j^N \langle l_i \cdot l_j \rangle. \tag{4.7}$$

Since

$$\langle l_i \cdot l_j \rangle = l_i l_j \langle \cos \theta \rangle \tag{4.8}$$

and

$$\begin{aligned}\langle \cos \theta \rangle &= 0 \qquad \text{for } i \neq j \\ &= 1 \qquad \text{for } i = j \end{aligned} \tag{4.9}$$

equation (4.7) reduces to

$$\langle h \rangle^2 = \sum_i^N l_i^2 = N l_{\text{av}}^2. \tag{4.10}$$

Therefore, the mean square end-to-end length of an ideally flexible polymer chain is given as the mean square length of individual bonds multiplied by the number of bond segments.

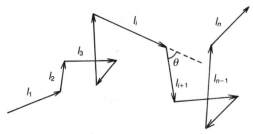

Figure 4.1 The unrestricted polymer chain. Bond lengths are arbitrary and non-uniform. The angle θ may take on all possible values arbitrarily between $0°$ and $360°$. (From Tanford 1961.) Reproduced by permission of John Wiley & Sons, Inc. © 1961.

The discussion presented above is a purely hypothetical case. Actually, bond angles are fixed, as depicted by figure 4.2 where three successive bonds of a polymethylene chain are illustrated. The first two bonds l_i and l_{i+1} lie in the plane of the paper and the last bond l_{i+2} is freely rotating, maintaining, however, the angle θ constant. Analogous to the previous case, obviously, there are N terms with $i = j$, namely $l_i \cdot l_i$. Secondly, the number of $l_i \cdot l_{i+1}$ terms is $(2N - 1)$, counting each one of them twice. Each pair contributes $l^2 \cos \theta$ to the end-to-end length of the entire chain.

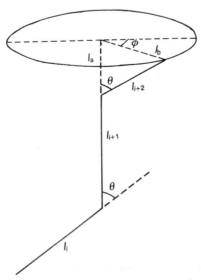

Figure 4.2 A schematic representation of a single bond carbon chain. The first two bonds are in the plane of the figure. The third bond l_{i+2} is rotating freely, i.e. the angle ϕ varies from $0°$ to $360°$. However, the angle θ is kept constant. (From Tanford 1961.) Reproduced by permission of John Wiley & Sons, Inc. © 1961.

The number of $l_i \cdot l_{i+2}$ terms is likewise $2(N - 2)$ and the mean value is

$$\langle l_i \cdot l_{i+2} \rangle = l^2 \cos^2 \theta. \tag{4.11}$$

In general, there are $2(N - k)$ terms for $l_i \cdot l_{i+k}$ and their values are evaluated to be $l \cos^k \theta$. The end-to-end length, therefore, is given as the sum of these terms:

$$\langle h^2 \rangle = l^2[N + 2(N - 1)\cos \theta + 2(N - 2)\cos^2 \theta + \ldots$$
$$+ 2(N - k)\cos^k \theta + \ldots + 2 \cos^{N-1} \theta]. \tag{4.12}$$

Since N is a large number, only those terms in which $N \gg k$ contribute significantly to the length of the entire chain. Thus, we can factor out N and obtain

$$\langle h^2 \rangle \simeq Nl^2(1 + 2\cos\theta + 2\cos^2\theta + \ldots). \tag{4.13}$$

Since $\cos\theta$ is smaller than one, we can always sum up the RHS of (4.13) using the method of Eyring (1932):

$$\langle h^2 \rangle = Nl^2\frac{1 + \cos\theta}{1 - \cos\theta}. \tag{4.14}$$

For polymethylene, for example, the angle is 70°.32′ and, therefore, $\cos\theta = 0.333$. Using this value in (4.14), we obtain the mean square end-to-end length: $\langle h^2 \rangle = 2Nl^2$.

The effect of restricted rotation

Equation (4.14) is derived using an assumption that the bond angle can be changed between 0° to 360°. In practice, the rotation of the bond is not unrestricted but certain angles are more favoured than others. With this new feature, (4.14) must be modified by introducing an additional term, i.e.

$$\langle h^2 \rangle = Nl^2\frac{1 + \cos\theta}{1 - \cos\theta}\frac{1 + \langle\cos\phi\rangle}{1 - \langle\cos\phi\rangle}. \tag{4.15}$$

This equation reduces to (4.14) for free rotation where $\langle\cos\phi\rangle = 0$ (see Benoit and Sadron 1949).

We calculated, in the foregoing, the mean end-to-end distance and the mean radius of gyration for polymers having various degrees of restriction. A detailed statistical discussion on the chain configuration of polymer chains can be found in the book by Flory or in Chapter 3 of Tanford's book.

4.5 The Mean Dipole Moment of Linear Polymers

Let us look at the structure of polyvinyl chloride (PVC) as an example. As shown in the figure below, one of the hydrogens is replaced by chlorine. This replacement creates a non-zero dipole moment in PVC.

The mean square dipole moment of a polymer having N monomer units can be written as follows:

$$\langle\mu^2\rangle = \sum_m \mu_m^2 + \sum_m \sum_{n\neq m} \boldsymbol{\mu}_m \cdot \boldsymbol{\mu}_n \tag{4.16}$$

where μ_m and μ_n are the dipole moments of individual bonds. Since the dipole moments of individual bonds are identical and equal to μ_0, we can rewrite (4.16) as follows:

$$\langle \mu^2 \rangle = \mu_0^2 \left(N + \sum_m \sum_{n \neq m} \cos \theta_{mn} \right) \tag{4.17}$$

where θ_{mn} is the angle between the moments m and n. This can be simplified as

$$\langle \mu^2 \rangle / N\mu_0^2 = (1 + S_N) \tag{4.18}$$

where

$$S_N = (1/N) \sum_m \sum_{n \neq m} \langle \cos \theta_{mn} \rangle. \tag{4.19}$$

Although the calculation of the mean value of $\cos \theta_{mn}$ seems to be similar to (4.13), the vector sum of the dipole moments is more complex, as can be seen from figure 4.2. Namely, if dipole moment vectors do not coincide with the bond vector a_m, additional angles β and γ are required for the complete characterization of dipole vectors. As shown in figure 4.3, we can derive the following equation:

$$\langle \cos \theta_{mn} \rangle = \langle \cos(a_n \mu_m) \cos \beta + \sin(a_n \mu_m) \sin \beta \cos \phi_k \rangle. \tag{4.20}$$

If free rotation of the bond a_{n+1} is allowed, the angle ϕ_k varies from $0°$ to $360°$ and the average value of $\cos \phi_k$ is zero. Thus, the second term on the RHS will disappear. Therefore, (4.20) reduces to

$$\langle \cos \theta_{mn} \rangle = \cos \beta \langle \cos(a_n \mu_m) \rangle. \tag{4.21}$$

Using a similar logic, $\langle \cos(a_n \mu_m) \rangle \cos \beta$ is equal to $\cos \beta \cos \alpha \langle \cos(a_{n-1} \mu_m) \rangle$. Repeating the same procedure, we obtain

$$\langle \cos \theta_{mn} \rangle = \cos \beta (\cos \alpha)^{n-m-1} \langle \cos(a_{m+1}, \mu_m) \rangle. \tag{4.22}$$

Substituting this equation into (4.19) and calculating the sum, we find

$$NS_N = \frac{2 \cos \alpha \cos \beta \cos \gamma}{1 - \cos^2 \alpha}. \tag{4.23}$$

Combining this and (4.18), we obtain (Debye and Bueche 1951)

$$\langle \mu^2 \rangle = \mu_0^2 N \left(1 + \frac{2 \cos \alpha \cos \beta \cos \gamma}{1 - \cos^2 \alpha} \right). \tag{4.24}$$

Assuming the angles β and γ satisfy the following relations:

$$\cos \beta = 1 \qquad \cos \gamma = -\cos \alpha \tag{4.25}$$

the mean square moment then becomes

$$\langle \mu^2 \rangle = \mu_0^2 N \left(1 - \frac{2 \cos^2 \alpha}{1 - \cos^2 \alpha} \right). \tag{4.26}$$

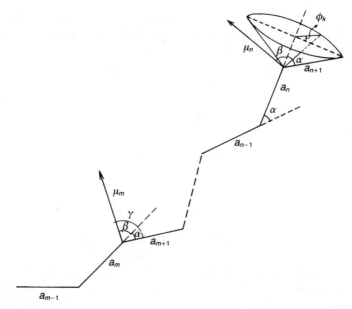

Figure 4.3 A diagram showing the spatial configuration of a polymer chain. The dipole moments are along side chains, therefore, three angles are needed to identify the direction of the dipole moments with respect to the bond vectors. α is the angle between bond vectors a_m and a_{m+1}. β is the angle between the bond vector a_m and dipole vector μ_m. γ is the angle between the dipole vector μ_m and bond vector a_{m+1}. (After Debye and Bueche 1951.)

This is the equation derived by Fuoss and Kirkwood in 1941 which preceded the calculation by Debye and Bueche. The condition stipulated by (4.25) means that the dipole moment vectors of polymer chains must be parallel to the bond vectors. This condition does not hold for some polymers such as PVC. In contrast to (4.26), (4.24) is a general theory which can be applied to any polymer chain.

4.6 The Effect of Hindered Rotation on Dipole Moment (Debye and Bueche 1951)

Equation (4.24) was derived assuming that the angle ϕ_k varies freely from 0° to 360° (see figure 4.2). With this assumption, the second term on the RHS of (4.20) disappears. However, in reality, the vector a_{n+1} does not undergo free rotation but there are a few orientations in which the polymer chains are stable. Figure 4.4 illustrates the potential profile for the rotation of the bond a_{n+1}. The broken curve depicts the smooth

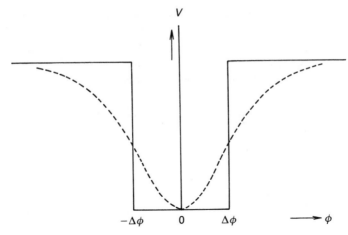

Figure 4.4 A profile of potential energy for the rotation of the bond a_{n+1}. The broken curve shows the realistic shape of the potential well and the full lines depict a simplified square potential. When $\Delta\phi$ is small, the position of the bond a_{n+1} is restricted rigidly and no rotational freedom exists. On the other hand, if the bond is freely rotating, the angle $\Delta\phi$ varies from 0° to 360° and the potential well has no minimum. (From Debye and Bueche 1951.) Reproduced by permission of the American Institute of Physics.

and realistic potential well, and the full lines illustrate a simplified square potential curve. Namely, the probability that the vector a_{n+1} is outside this square potential well is zero, and that the vector is inside the well is unity. The angle ϕ_k must be so chosen as to make the calculated dipole moment agree with the observed value. Figure 4.5 illustrates three typical curves showing the variation of dipole moment with the degree of hindrance of bond rotation. The abscissa is the degree of hindrance. In stiff chains, the angle $\Delta\phi$ is limited to a very small value, therefore, $\sin \Delta\phi = \Delta\phi$. Thus, the value of $\sin \Delta\phi/\Delta\phi$ reduces to 1. On the other hand, for free rotation, the angle $\Delta\phi$ will be 360° and, therefore, $\sin \Delta\phi/\Delta\phi$ is zero. For a polymer such as polypara-chlorostyrene, $\cos \alpha = \cos \beta = - \cos \gamma = \frac{1}{3}$ and (4.24) reduces to

$$\langle \mu^2 \rangle / N\mu_0^2 = 11/12. \tag{4.27}$$

Therefore, the value of the ordinate is 0.97 for free rotation. As the degree of hindrance increases, the value of $\langle \mu^2 \rangle / N\mu_0^2$ either increases or decreases, depending upon the orientation of dipole moments in the polymer chain. If the dipole moments are oriented in such a way as to accumulate with increased stiffness, the value of $\langle \mu^2 \rangle N\mu_0^2$ would increase with the increase in stiffness. Curve A shows the behaviour of this type of polymer. However, it was observed experimentally that $\langle \mu^2 \rangle / N\mu_0^2$ for polyparachlorostyrene is 0.56. This indicates that the dipole moments of this type of polymer have an orientation which alternates in opposite directions so that stiffening causes partial cancellation of the dipole moments. Obviously, polyparachlorostyrene belongs to the type C polymers. On the other hand, in some polymers, stiffening of the chain causes a nearly complete cancellation of dipole moment as shown by curve B.

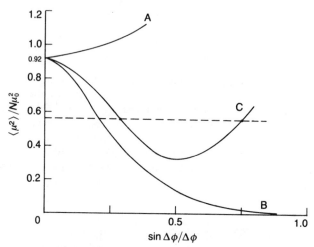

Figure 4.5 Three typical curves showing the variation of dipole moment of polymer chains with the degree of rotational freedom. Curve A shows the case where dipole moments are aligned in such a way as to accumulate with increasing rigidity. Curve B shows the polymers in which dipoles cancel with one another with an increase in rigidity. Curve C shows the case where the increase in rigidity causes partial cancellation of dipole moments. (From Debye and Bueche 1951.) Reproduced by permission of the American Institute of Physics.

4.7 The Mean Dipole Moment of Copolymers

In copolymers which consist of two different monomer units 1 and 2, the distribution of these units is statistically random. Therefore, the configuration of these copolymers is difficult to define. Reflecting the irregularity of the distribution of monomer units, the average dipole moment of copolymers requires a theoretical treatment which is different from those discussed above.

Let us assume that the dipole moments of monomer units are μ_1 and μ_2 respectively. If we ignore the interaction between bonds which are not directly connected, the mean moment is given by

$$\langle \mu^2 \rangle = \langle \sum_m \mu_m \cdot (\mu_{m-1} + \mu_m + \mu_{m+1}) \rangle. \tag{4.28}$$

If we assume that the probability of finding type 1 and type 2 units in the chain depends on the molar fractions x_1 and x_2, and that the probability of a dipole μ_j being between dipoles μ_k and μ_l is P_j^{kl}, then (4.28) can be rewritten as

$$\langle \mu^2 \rangle = \langle \sum_{j,k,l}^{2} x_j P_j^{kl} \mu_j \cdot (\mu_j + \mu_k + \mu_l) \rangle. \tag{4.29}$$

Assuming the angles between $\mu_j - \mu_k$ and $\mu_j - \mu_l$ to be θ_{jk} and θ_{jl} respectively, (4.29) can be written using scalar notation:

$$\langle \mu^2 \rangle = \sum_{j,k,l}^{2} x_j P_j^{kl} \mu_j (\mu_j + \mu_k \langle \cos \theta_{jk} \rangle + \mu_l \langle \cos \theta_{jl} \rangle). \tag{4.30}$$

Noting that

$$\mu_j + \mu_k \langle \cos \theta_{kj} \rangle + \mu_l \langle \cos \theta_{jl} \rangle =$$

$$\tfrac{1}{2}(\mu_j + 2\mu_k \langle \cos \theta_{kj} \rangle) + \tfrac{1}{2}(\mu_j + 2\mu_l \langle \cos \theta_{jl} \rangle$$

and introducing new symbols:

$$\alpha_{11} = p_1^{11} + 0.5 \, p_1^{12} = P_{11}$$
$$\alpha_{22} = p_1^{22} + 0.5 \, p_2^{12} = P_{22}$$
$$\alpha_{12} = p_1^{22} + 0.5 \, p_1^{12} = 1 - P_{11} \tag{4.31}$$
$$\alpha_{21} = p_2^{11} + 0.5 \, p_2^{12} = 1 - P_{22}$$

(4.30) becomes

$$\langle \mu^2 \rangle = x_1 \mu_1 (\mu_1 + 2\alpha_{11}\mu_1 \langle \cos \theta_{11} \rangle + 2\alpha_{12}\mu_2 \langle \cos \theta_{12} \rangle$$
$$+ x_2 \mu_2 (\mu_2 + 2\alpha_{22}\mu_2 \langle \cos \theta_{22} \rangle + 2\alpha_{21}\mu_1 \langle \cos \theta_{21} \rangle. \tag{4.32}$$

The quantity P_{11}, for example, represents the probability that a monomer unit of type 1 is added to another type 1 unit. This is defined as

$$P_{11} = \frac{k_{11}/k_{12}}{k_{11}/k_{21} + M_2/M_1} \tag{4.33}$$

where k_{11} and k_{12} are rate constants for the addition of a type 1 or type 2 unit to a type 1 monomer unit. M_1 and M_2 are the concentrations of monomer types 1 and 2. Using a relation between x and P:

$$x_1/x_2 = \frac{(1 - P_{22})}{(1 - P_{11})} \tag{4.34}$$

(4.31) reduces to

$$\langle \mu^2 \rangle = L + Tx_1 + Sx_1 P_{11} \tag{4.35}$$

where

$$L = \mu_2^2 + 2\mu_2^2 \langle \cos \theta_{22} \rangle$$
$$T = \mu_1^2 - \mu_2^2 - 4\mu_2^2 \langle \cos \theta_{22} \rangle + 4\mu_1\mu_2 \langle \cos \theta_{12} \rangle \tag{4.36}$$

and

$$S = 2\mu_1^2 \langle \cos \theta_{11} \rangle + 2\mu_2^2 \langle \cos \theta_{22} \rangle - 4\mu_1\mu_2 \langle \cos \theta_{12} \rangle.$$

Experimentally, the mean dipole moment of a copolymer must be determined as a function of the mole fraction of either component. Figure 4.6 illustrates the dipole moment of the copolymer of chlorostyrene and styrene at various mole fractions. Using this curve, we can extract the values of L, T and S, and determine the values of $\langle \cos \theta_{11} \rangle$, $\langle \cos \theta_{22} \rangle$ and $\langle \cos \theta_{12} \rangle$. If we use 0.64 D and 1.04 D for the dipole moments of chlorostyrene and styrene respectively, and using proper values for the rate constants k_{11} and k_{12}, we obtain

$$\langle \cos \theta_{11} \rangle = -0.46 \qquad \langle \cos \theta_{12} \rangle = -0.31 \qquad \text{and} \qquad \langle \cos \theta_{22} \rangle = -0.34.$$

Using these values, the angles θ_{11}, θ_{12} and θ_{22} are found to be 117°, 108° and 109° respectively.

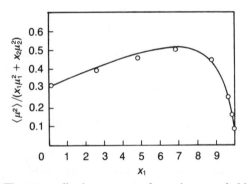

Figure 4.6 The mean dipole moment of copolymers of chlorostyrene and styrene. The abscissa is the fraction of styrene x_1. The open circles are the measured values and the full curve is calculated using (4.35). (From Work and Trehu 1956.) Reproduced by permission of the American Institute of Physics.

4.8 The Dielectric Relaxation of Polymer Molecules

So far we have discussed the mean dipole moment of polymer molecules. The orientation of polymer molecules in the presence of an electrical field has a very large relaxation time because of their large size. Figure 4.7 shows the frequency dependence of the dielectric constant of polymethyl acrylate in organic solvent (Williams and Ferry 1955). As shown, the characteristic frequency is found between 100 Hz and 1 kHz. In addition, the dispersion curve encompasses a wide frequency range between 1 Hz and almost 1 MHz. This broad dispersion curve indicates a wide distribution of relaxation time. The distribution of relaxation time arises partly because of the non-uniformity of polymer chains. However, there is another, more fundamental, reason for the distribution of time constants, namely, as has been discussed, the segments in polymer molecules have an internal freedom of rotation. This gives rise to a number of different configurations and effective lengths of these polymer chains, which entails a wide distribution of relaxation time.

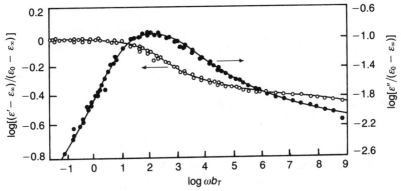

Figure 4.7 The dielectric constant and dielectric loss of polymethyl acrylate solution. b_T is an empirical correction parameter used to adjust both ε' and ε'' along the abscissa for different temperatures. b_T is unity at 25 °C. Note the broad dispersion curve indicating the wide distribution of relaxation times. (From Williams and Ferry 1955.)

The dielectric relaxation of polymer molecules, particularly the distribution of relaxation time, has been treated by Kirkwood and Fuoss (1941). In the following, only a rough outline of this theory is given, as the original theory is very complex and difficult to describe in full. The interested reader may find the detail in the original paper.

A polymer molecule has monomer units of the type

$$-\overset{\displaystyle H}{\underset{\displaystyle H}{C}} - \overset{\displaystyle X}{\underset{\displaystyle H}{C}} -$$

Let us assume that polar polymer chains possess r internal degrees of freedom. When a field $Ee^{j\omega t}$ is applied, the polymer chains will acquire a non-vanishing dipole moment $\mu(q_1 \ldots q_r)$ where q_j specifies the internal configuration of the polymer. If the direction of the applied field is represented by e, the mean component of the dipole moment in the direction of the field is defined by

$$\langle \mu \cdot e \rangle = \int e \cdot \mu(q) f(q, t) \, dq \qquad (4.37)$$

where $f(q, t)$ is the distribution function of the internal configuration. For a small field, we can expand the distribution function as follows:

$$f(q, t) = f_0(q)(1 + f_1 Ee^{j\omega t} + O(E^2)) \qquad (4.38)$$

where $f_0(q)$ is the distribution function in the absence of electric field. $O(E^2)$ collects all the terms higher than second order. Using this, $\langle \mu \cdot e \rangle$ becomes

$$\langle \mu \cdot e \rangle = pEe^{j\omega t} + O(E^2) \qquad (4.39)$$

where p is the electric moment induced by a unit electrical field as a result of dipole orientation. Using (4.37) and (4.39), we obtain

$$p = \int e \cdot \mu(q) f_0(q) f_1(q) \, dq. \qquad (4.40)$$

In order to solve (4.40), the distribution functions f_0 and f_1 must be found by solving the diffusion equation

$$\nabla \cdot \mathbf{D} \cdot (\nabla f + f \nabla V / kT) = \partial f / \partial t \qquad (4.41)$$

where $\mathbf{D}$ is the internal diffusion tensor, ∇ is the gradient operator in the r dimension configuration space, and V is the potential of internal and external forces. In the absence of a field, the distribution function $f_0(q)$ is given by

$$f_0(q) = \exp(- V_0/kT)\left(\int \exp(- V_0/kT) dq\right)^{-1} \qquad (4.42)$$

where V_0 is the potential of internal forces. If a field $Ee^{j\omega t}$ is applied, the total potential energy V changes to

$$V = V_0 - \mu \cdot Ee^{j\omega t}. \qquad (4.43)$$

Substituting (4.42) into (4.38), we obtain

$$f(q, t) = A_0 \exp(- V_0/kT)(1 + f_1(q)Ee^{j\omega t} + O(E^2)) \qquad (4.44)$$

where

$$A_0 = \left(\int \exp(- V_0/kT) dq\right)^{-1}.$$

Substituting these equations into (4.41), the following expression for f_1

is derived:

$$f_1(q) = \frac{1}{\sqrt{3A_0}} \frac{1}{kT} \sum_\lambda \frac{\mu_\lambda \psi_\lambda \exp(V_0/2kT)}{1 + j\omega\tau_\lambda} \tag{4.45}$$

where $\tau_\lambda = 1/\lambda$ and μ_λ is defined by

$$\mu_\lambda = \sqrt{3A_0} \int (\boldsymbol{\mu} \cdot \boldsymbol{e}) \exp(-V_0/2kT)\psi_\lambda^* \, dq. \tag{4.46}$$

$-\lambda$ and ψ_λ are the eigenvalue and eigenfunction respectively associated with the diffusion equation (4.41). (The calculation of eigenvalues is discussed by Kirkwood and Fuoss). Substituting these into (4.40), we obtain an expression for polarization p:

$$p = (1/3kT) \sum_\lambda \frac{|\mu_\lambda|^2}{1 + j\omega\tau_\lambda}. \tag{4.47}$$

Let us define the normalized polarization Q:

$$Q = P/P_0 \tag{4.48}$$

where P_0 is the polarization at zero frequency. Using Q, (4.47) can now be written as

$$Q(\omega) = \sum_\lambda \frac{Q(\lambda)}{1 + j\omega\tau_\lambda} \tag{4.49}$$

where

$$Q(\lambda) = |\mu_\lambda|^2 / \langle \mu_\lambda^2 \rangle. \tag{4.50}$$

Assuming that eigenvalues λ are closely spaced, we can replace (4.49) with

$$Q(\omega) = \int_0^\infty \frac{G(\tau) \, d\tau}{1 + j\omega\tau} \tag{4.51}$$

where $G(\tau)$ is the distribution function for relaxation time.

The next step is to determine the density function $G(\tau)$ using a polymer molecule of the type $(CH_2\text{–}CHX)$, in which X is a dipolar substituent. The dipole moments lie along the bond C–X. The successive C–C bonds of the polymer skeleton can be represented by unit vectors $\boldsymbol{a}$ and the directions of the C–X bonds branching from the skeleton by another set of vectors $\boldsymbol{b}$. The distribution of relaxation time arises from the internal freedom of chain segments and numerous conformations occur as the result of internal diffusion. The calculation is exceedingly complex and, therefore, only the final result is given without all the detail:

$$G(\tau) = \frac{\tau_0}{(\tau + \tau_0)^2} \qquad \text{for } 4\tau_0/n\pi^2 \leqslant \tau \leqslant n\tau_0/6 \tag{4.52}$$

$$G(\tau) = 0 \qquad \text{for } \tau > n\tau_0/6 \qquad \text{or} \qquad \tau < 4\tau_0/n\pi^2 \tag{4.53}$$

where

$$\tau_0 = N\tau \qquad (4.54)$$

$$\tau = 3\pi a^2 b\eta/kT \qquad (4.55)$$

and

$$b = (3\Delta/8\pi)^{1/3} \qquad (4.56)$$

where a is the C–C distance, and Δ is the volume of the monomers. Also, τ is the relaxation time of the monomers and τ_0 the relaxation time of the polymer.

$Q(\omega)$ can be separated into real and imaginary parts $J(x)$ and $H(x)$:

$$Q(\omega) = J(x) - jH(x) \qquad (4.57)$$

where $x = \omega\tau_0$ or $N\omega\tau$. Substituting J and H into (4.51) and integrating, we obtain

$$J(x) = \frac{1}{(1 + x^2)^2}\left((1 + x^2) + \frac{\pi x}{2}(x^2 - 1) - 2x^2 \log x\right) \qquad (4.58)$$

and

$$H(x) = \frac{x}{(1 + x^2)^2}((x^2 - 1)\log x - x^2 + \pi x - 1). \qquad (4.59)$$

Using (4.59), we can find the value of the function $H(x)$ at $x = 1$, i.e. $\omega = 1/\tau_0$. At this frequency, $H(x)$ has a maximum value and $J(x)$ is 0.5, of the total dielectric increment. We found the value of $H(x)$ to be 0.286, which is considerably lower than that predicted by the Debye theory, i.e. 0.5, for single relaxation time. Therefore, the dielectric relaxation of flexible polymers is characterized by multiple time constants or continuous spectra of relaxation times. The value of $H(x)$ for a polyvinyl chloride–diphenyl mixture was found experimentally to be 0.175; this is even smaller than the theoretical value (see figure 4.8).

The calculation discussed above was performed with the assumption that the polymer sample is monodisperse, i.e. the degree of polymerization is uniform. However, polymer samples are usually polydisperse unless they are carefully fractionated using proper techniques. The distribution of molecular weights causes further spreading of the relaxation times. Consideration of this distribution, however, can improve the agreement slightly. In other words, the calculated value of $H(x)$ is still substantially higher than the observed value.

In this calculation, the interaction between segments is ignored or assumed negligible. The interaction between segments of polymer chains tends to broaden the dispersion curve and lower the value of the $H(x)$ function. Therefore, the disagreement between the theoretical calculation and the experimental results may be attributed to the neglect of the intra- or inter-chain interactions.

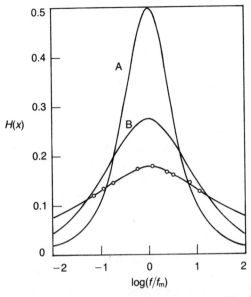

Figure 4.8 Dielectric loss $H(x)$ plotted against normalized frequency f/f_m. Curve A shows the Debye function with one relaxation time. Curve B shows the loss curve calculated for linear polymers using (4.59). The open circles are the measured dielectric loss for a polyvinyl chloride: diphenyl mixture (80 : 20) at 50 °C. (From Kirkwood and Fuoss 1941.) Reproduced by permission of the American Institute of Physics.

4.9 Rouse's Model for Flexible Polymers and Zimm's Theory of Dielectric Properties

The viscoelastic properties of a solution of linear polymer molecules are unusual in many ways because of (i) the length of the polymers, (ii) the flexibility of the polymer chains and (iii) intra- and inter-chain segment interaction. In the measurement of the viscoelasticities of polymer chains, only the resultant of the coordinated segmental motions is observed. Theories can be developed for the polymers only as the sum of the motion of chain segments.

In the theory developed by Rouse (1953), a velocity gradient continuously alters the distribution of configurations of the polymer molecules and the coordinated thermal motion of the chain segments causes configurations to drift toward the most stable distribution. Rouse's paper actually deals only with viscoelastic properties, but his model plays a very important role for the later development of the theories of polymer dielectric properties by Zimm and other investigators.

A complete description of the configuration of a polymer molecule requires the specification of the spatial orientation of each bond in the chain. However, the general theoretical treatment of viscoelastic properties of polymer chains using this approach is exceedingly complex, and the mathematical difficulty is staggering as demonstrated by Kirkwood and Fuoss in the previous section. In Rouse's model, polymer molecules are approximated as a number of flexible submolecules. The submolecules are long enough to be flexible so that we can apply Gaussian statistics to each segment. Namely, if one end of a submolecule is fixed at the origin of its own coordinate system (see figure 4.9), the probability $\psi(x, y, z)dxdydz$ that the other end will be within the volume element $dxdydz$ lying between x and $x + dx$, y and $y + dy$ and z and $z + dz$ is given by

$$\psi(x, y, z)\,dxdydz = (\beta/\pi)^{3/2}\exp[-\beta(x^2 + y^2 + z^2)]\,dxdydz \quad (4.60)$$

where $\beta = 3/(2\sigma^2)$ and σ^2 is the mean square distance between two ends of a submolecule. Since there are a number of configurations for each submolecule which satisfy the above equations, the total probability of a

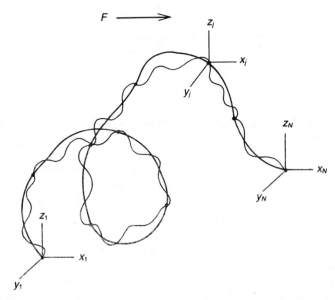

Figure 4.9 Rouse's linear polymer model used for the calculation of viscoelasticity. This is a schematic presentation of a simplified model consisting of submolecules. Each submolecule is flexible enough to follow Gaussian statistics. The subcoordinates of a few segments are shown.

polymer molecule having a requisite end-to-end separation is given by

$$\Psi_i d\phi_i = \prod_i^N \psi(x_i, y_i, z_i) \, dx_i dy_i dz_i$$

$$= (\beta/\pi)^{3N/2} \exp\left(-\beta \sum_{i=1}^N (x_i^2 + y_i^2 + z_i^2)\right) dx_1 dy_1 dz_1 \ldots dx_N dy_N dz_N.$$

$$(4.61)$$

When we apply a velocity gradient α, one of the effects is to carry each segment of the individual polymer chains along with the liquid. The second effect is the diffusional motion by which the distribution of chain configuration moves toward the new equilibrium state in the presence of a velocity gradient. This calculation requires the evaluation of the effect of perturbation of the configuration on the thermodynamic potential of polymer chains.

Zimm's theory (1956) for the dielectric relaxation of polymers is based on a model which is similar to that of Rouse. Like Rouse's model, Zimm's polymer chains consist of a number of submolecules which behave as Gaussian-type flexible chains. In the following, the outline of Zimm's theory will be given in a simplified manner.

In Zimm's model, based on hydrodynamical considerations, the average rate of the jth bead (a representative point in submolecules) moving in the x direction is given by

$$\partial x/\partial t = v_x + (1/\rho)\mathbf{H} \cdot \mathbf{F}_x \qquad (4.62)$$

where v_x is the x-component of the velocity that the fluid would have if forces from the polymer chain do not exist. ρ is a frictional coefficient. The function $\mathbf{F}_x$ consists of two parts.

(i) The force which a chain segment $j - 1$ to j exerts on bead j

$$- (3kT/b^2)(x_j - x_{j-1}) \qquad (4.63)$$

where b^2 is the mean of the chain length l^2. Likewise, the force which segment j to $j + 1$ exerts on bead j is

$$-(3kT/b^2)(x_j - x_{j+1}). \qquad (4.64)$$

Therefore the sum of these forces is

$$-(3kT/b^2)(- x_{j-1} + 2x_j - x_{j+1}). \qquad (4.65)$$

(ii) An additional force on the beads due to Brownian motion:

$$-kT \, \partial \ln \psi/\partial x_j$$

where ψ is the probability of finding each bead j between x and $x + dx$, and correspondingly for y_j and z_j. Combining these two components, total force F_{xj} is given by

$$F_{x0} = -kT\partial \ln \psi/\partial x_0 - (3kT/b^2)(x_0 - x_1) \qquad \text{for } j = 1$$

$$F_{xj} = -kT\partial \ln \psi/\partial x_j - (3kT/b^2)(-x_{j-1} + 2x_j - x_{j+1})$$
$$\text{for } j = 2 \text{ to } N - 1$$

$$F_{xN} = -kT\partial \ln \psi/\partial x_N - (3kT/b^2)(x_N - x_{N-1}) \qquad \text{for } j = N. \qquad (4.66)$$

The matrix $\mathbf{H}$ has elements $H_{jj} = 1$ and $H_{ij} = \rho T_{ij}$ if $i \neq j$ where T_{ij} is the interaction coefficient. Combining $\mathbf{H}$ and F and repeating the calculation from bead 1 to N, (4.66) becomes

$$\partial x/\partial t = v_x - D\mathbf{H}\cdot(\partial/\partial x)\ln \psi - \sigma\mathbf{H}\cdot\mathbf{A}x \qquad (4.67)$$

where $D = kT/\rho$, $\sigma = 3kT/b^2\rho$ and ρ is the frictional coefficient. $\partial/\partial x$ represents a column vector $\{\partial/\partial x_0, \partial/\partial x_1, \partial/\partial x_2, \dots\}$ and

$$\mathbf{A} = \begin{pmatrix} 1 & -1 & 0 & \dots & 0 & 0 & 0 \\ -1 & 2 & -1 & \dots & 0 & 0 & 0 \\ \vdots & \vdots & \vdots & & \vdots & \vdots & \vdots \\ 0 & 0 & 0 & \dots & -1 & 2 & -1 \\ 0 & 0 & 0 & \dots & 0 & -1 & 1 \end{pmatrix}. \qquad (4.68)$$

The eigenvalue (λ_i) of matrix $\mathbf{H}\cdot\mathbf{A}$ is given, without detailed discussion, in the following. Assuming that hydrodynamic interactions are very small (free draining)

$$\lambda_i = \pi^2 i^2/N^2. \qquad (4.69)$$

The unknown function ψ is the probability of finding a bead in a small volume $dxdydz$ and is to be determined from the equation of continuity:

$$\partial\psi/\partial t = -\text{div } \psi[(\partial x/\partial t) + (\partial y/\partial t) + (\partial z/\partial t)]. \qquad (4.70)$$

With this fomulation, we are ready to calculate dielectric relaxation.

If we apply an electrical field in the x direction, the extension x of segments produces a proportionate electric moment μx. Averaging with the Boltzmann factor, we can find the mean moment in a field E:

$$\mu\langle x \rangle = \mu^2 E b^2/3kT. \qquad (4.71)$$

Hence, the average extension in the x direction is

$$\langle x \rangle = \mu E b^2/3kT.$$

Adding this term to (4.62), the mean force exerted on j by the segment from $j - 1$ to j is given by

$$-(3kT/b^2)(x_j - x_{j-1} - \langle x \rangle). \qquad (4.72)$$

Using this, we obtain a force equation in a matrix form:

$$F_x = -kT(\partial/\partial x)\ln \psi - (3kT/b^2)(\mathbf{A} \cdot x + \langle x \rangle e) \qquad (4.73)$$

where e is a vector which is given by

$$e = \begin{pmatrix} -1 \\ 0 \\ \vdots \\ 0 \\ 1 \end{pmatrix}. \qquad (4.74)$$

Here it is assumed that the application of a field exerts forces on the charges at both ends of a segment with equal magnitude and opposite sign. The quantity of central interest is the mean extension of the whole chain $\langle x_N - x_0 \rangle$. For periodic fields, $E = E_0 \exp(j\omega t)$, the result of the calculation of mean extension is given by

$$\langle x_N - x_0 \rangle = (b^2 E / 3kT) \sum_{i=\text{odd}} \frac{8\mu/N\mu_i}{1 + j\omega/\sigma\lambda_i}. \qquad (4.75)$$

The mean electric moment p is defined as

$$p = \mu \langle x_N - x_0 \rangle / E. \qquad (4.76)$$

Therefore

$$p = \mu^2 b^2 / 3kT \sum_{i=\text{odd}} \frac{8/N\mu_i}{1 + j\omega/\sigma\lambda_i}. \qquad (4.77)$$

For free-draining polymers, the mean polarization per unit field is given by

$$p = (\mu^2 b^2 / 3kT) \sum_{i=\text{odd}} \frac{8N/\pi^2 i^2}{1 + j\omega/\sigma\lambda_i}. \qquad (4.78)$$

The relaxation time $\tau_i (= 1/\sigma\lambda_i)$ is given, after inserting values for σ and λ_i, by

$$\tau = \frac{\rho N^2 b^2}{3kT\pi^2 i^2}. \qquad (4.79)$$

It should be noted that the relaxation time is proportional to N^2 rather than to N. This means that the contribution of chain segments to the relaxation of the polymer increases rapidly as the number of monomer units increases. This means that the contribution of the largest chain segment would dominate over other smaller segments.

In equation (4.76), it was assumed that the mean electric moment is proportional to the mean extension of the entire chain $\langle x_N - x_0 \rangle$. This assumption is valid only when each bond of the chain has a moment parallel to the chain. In this case, the moment of the entire chain is the vector sum of the moments of each segment. However, if the moment is in the side chains with respect to which the chain is symmetrical, the mean moment of the chain at moderate extension must vanish. In this case, (4.76) is not applicable. Polyvinyl chloride, for example, is one of these cases.

4.10 Dielectric Properties of Flexible Polymers in Laminar Flow

The necklace model of Rouse (1953) and Zimm (1956) is made of infinitely deformable elastic links and perfectly flexible joints. If laminar flow is applied to this model chain, the flexible coil will extend in the direction of the flow. Because of its infinite flexibility, the chain configuration will be fully extended at large shear rates and its dipole moment will approach a limiting value of extended chain. However, real polymers do not have infinitely flexible joints and the similarity between real polymers and Rouse's model ceases as the shear rate of laminar flow increases. Only at very small velocity gradients is the deformation of polymer segments so slight that polymers may behave as an ideally flexible chain. The effect of laminar flow on the configuration of flexible polymers has been studied by light scattering (Peterlin and Reinhold 1964). It was observed that the effect of laminar flow on the chain configuration of random-coil polymers was only slight in contrast to the expected behaviour of the idealized Rouse's chain. The effect of laminar flow on the dielectric constant of a polymer has been investigated by several authors (Saito and Kato 1957, Jacobsen 1953, Takashima 1970, Barisas 1974). Their research dealt mostly with rigid rod polymers and their orientation behaviour in the presence of both electric fields and mechanical shear. Another investigation of this sort was performed by Wendish (1964) who observed experimentally that the increase in dielectric constant of flexible nitrocellulose was quadratic with shear rate. The results indicate that laminar flow not only affected the orientation behaviour but also changed the configuration of flexible chain segments.

The theory given by Peterlin and Reinhold (1965) discusses quantitatively the effect of laminar flow on the polarization of flexible polymer chains in the limit of zero velocity gradient. The distribution function of chain segments in the presence of both laminar flow and electrical fields is given by

$$\psi \prod_i \exp(\mu_i E / kT) \tag{4.80}$$

where ψ accounts for the distribution due to laminar flow and the exponential function is due to the interaction of dipole with applied fields. Since the Boltzmann factor is, in general, very small, we can expand this function in a power series:

$$\psi\left(1 + \sum_i \mu_j E / kT + \ldots\right). \tag{4.81}$$

If the dipole moment is parallel to the bond, the mean dipole moment of polymer segments is given by

$$\langle \mu_{j\parallel} \rangle = \langle \mu_{j\parallel} \cos \theta_j \rangle \tag{4.82}$$

where θ_j is the angle between the segment and the field vector. In the presence of laminar flow, the average contribution of dipoles in the direction of the field is

$$\langle \mu_{j\parallel} \rangle = \langle \mu_{j\parallel} \cos \theta_j \rangle_\psi + (E/kT) \langle \mu_{j\parallel}^2 \cos^2 \theta_j \rangle_\psi. \tag{4.83}$$

The angle brackets with the subscript ψ mean that average values are obtained with the distribution function ψ. Since $\langle \cos \theta_j \rangle$ is zero, the first term on the RHS of (4.83) disappears. Thus the mean moment is solely due to the second term.

If the moment is perpendicular to the segment, the direction of dipole moment cannot be fixed since the azimuthal angle perpendicular to the segment direction remains free. However, for a small shear rate, where the increase in dipole moment is linear, we can derive an approximate equation, i.e.

$$\langle \mu_{j\perp} \rangle = - \langle \mu_{j\perp} \sin \theta_j \sin \alpha_j \rangle$$

$$= - \langle \mu_{j\perp} \sin \theta_j \sin \alpha_j \rangle_\psi + (E/kT) \langle \mu_{j\perp}^2 \sin^2 \theta_j \sin^2 \alpha_j \rangle_\psi. \tag{4.84}$$

The mean values in (4.83) and (4.84) have been evaluated by Reinhold and Peterlin (1965). With these, we obtain the equation for orientational polarization. If dipole moment is parallel to the bond, (4.83) reduces to

$$P_\parallel = E\mu^2 Z/3kT. \tag{4.85}$$

In this case, laminar flow has no effect on the dipole moment of a flexible chain. On the other hand, if dipole moment is perpendicular to the segment, we obtain

$$P_\perp = (E\mu^2 Z/3kT)\left(1 + (\beta^2/Z)\sum_{j=1}^{Z} F(j)\right) \tag{4.86}$$

where Z is the number of segments in the chain and β is the normalized velocity gradient, given by

$$\beta = M[\eta]\eta_0 G/NkT \tag{4.87}$$

where M is the molecular weight, $[\eta]$ is the intrinsic viscosity of polymers, η_0 is the viscosity of the solvent and G is the shear rate. $F(j)$ is a function which has a value of approximately 0.4 for large Z, hence

$$P_\perp = (E\mu^2 Z/3kT)(1 + 0.4\beta^2/Z). \tag{4.88}$$

This equation indicates that the effect of laminar flow on rotational polarization becomes less for larger Z. However, M is proportional to Z and $[\eta]$ is proportional to $Z^{1/2}$ for random-coil polymers. Therefore, β is actually proportional to Z^2. Therefore, if we use G instead of β, the orientational polarization P will show a rapid increase with Z. The ratio $\Delta P/P$ can be translated to an observable quantity, $\Delta\varepsilon_{or}/\varepsilon_{or}$, and we can compare the theoretical value with the experimental result using a relative dielectric constant. Figure 4.10 shows theoretical curves (diagram (a)) and experimental results by Wendish (1964) (diagram (b)). These two curves are in agreement in that both the observed and calculated $\Delta\varepsilon/\varepsilon$ have a similar dependence on velocity gradient. However, there is a serious discrepancy between the shear rate used for the theoretical calculation and that used for the experiment by Wendish. In spite of this, the theory is, at least qualitatively, in agreement with observation.

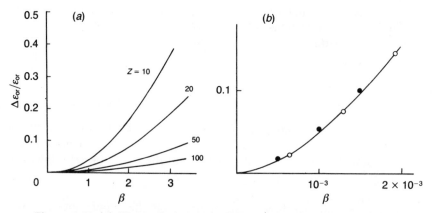

Figure 4.10 (a) Theoretical curves of $\Delta\varepsilon_{or}/\varepsilon_{or}$ as a function of the flow parameter β. The segment dipole is perpendicular to the chain direction. Z indicates the number of segments. (b) Extrapolated values of the relative increase in dielectric constant $(\Delta\varepsilon_{or}/\varepsilon_{or})_{c=0}$ at 20 °C (open circles) and 40 °C (full circles). (From Peterlin and Reinhold 1965.) Reproduced by permission of Steinkopff-Verlag Darmstadt.

4.11 The Effect of Charges on Polymer Dipoles

It has been mentioned briefly that the configuration of polymer chains is affected by the presence of charged groups in the molecule. In this section, the dielectric relaxation of charged polymers will be discussed based on the theory by Van Beek and Hermans (1957). The model used by these authors is similar to Rouse's polymer chain, but they also assumed that the dipole vectors are parallel to the bond vectors along the chain.

Let us consider a string of N beads 1, 2, ..., N, where N is a large even number. We assign to the ith bead a charge of $(-1)^i e$. With this arrangement, the first bead carries a negative charge $-e$ and the Nth bead carries a positive charge $+e$. The mean square moment per submolecule is $\mu^2 = e^2 l^2$ and the total moment of the polymer molecule is given by

$$P = e \sum_{k=1}^{N} (-1)^k r_k \tag{4.89}$$

where r_k is the position of the kth bead. The mean polarization $\langle p \rangle$ of this polymer is given by

$$\langle P \rangle = (\mu^2 E / 3kT) \sum_{n=\text{odd}} \frac{a_n}{1 + j\omega \tau_n} \tag{4.90}$$

where

$$\tau_n = \tau_0 \sin^{-2}(n\pi/2N) \tag{4.91}$$

and

$$\tau_0 = \zeta L^2 / 12kT. \tag{4.92}$$

Also

$$a_n = (2/N) \sin^{-2}(n\pi/N). \tag{4.93}$$

In these equations, L is the end-to-end length of the submolecules and ζ is the frictional coefficient.

For small n, e.g. $n = 1$, (4.91) reduces to

$$\tau_n = \tau_0 (2N)^2 / \pi. \tag{4.94}$$

This relaxation time represents the rotation of the entire polymer chain. On the other hand, if n is comparable to N, then

$$\tau_n \simeq \tau_0. \tag{4.95}$$

This relaxation time represents the rotational relaxation of monomer units. Therefore, the relaxation time of charged polymers ranges between these two limiting values.

4.12 Dielectric Properties of Polyelectrolytes

The theory given by Van Beek and Hermans is based on an assumption that the dielectric relaxation of charged polymers is due to the orientation of chain segments or entire polymer molecules. However, as will be discussed later, the dielectric relaxation of charged synthetic or biological polymers is due mostly to the polarization of mobile charges rather than the orientation of chains. As an example of the dielectric relaxation of synthetic polyions, the results obtained by Minakata *et al* (1972) with tetra-n-butyl-ammonium polyacrylic acid is shown in figure 4.11. Even a quick glance at this diagram is sufficient to note the exceedingly large dielectric increments. It is difficult to conceive any mechanism which might explain this unusually large dipole moment using polar orientation theories. The dielectric relaxation of polyions is generally considered to be due to the polarization of loosely bound counterions rather than the orientation of permanent dipoles. The theories of counterion polarization will be discussed in detail in Chapter 6. Several references on the dielectric properties of synthetic polyelectrolytes are given in a separate reference at the end of this chapter.

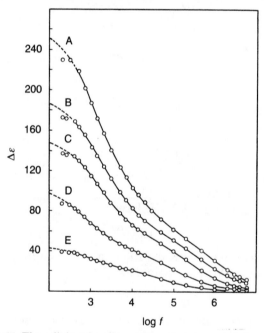

Figure 4.11 The dielectric dispersion of tetra-n-butyl ammonium polyacrylic acid as a function of concentration. Curves: A, $C_p = 5$ mN; B, 2.5 mN; C, 1.25 mN; D, 0.5 mN; E, 0.2 mN. (From Minakata *et al* 1972.) Reproduced by permission of John Wiley & Sons, Inc. © 1972.

The purpose of this chapter was to review the early development of the theories of dipole moment and the relaxation of synthetic polymer molecules; the author did not intend to review the recent literature in this field. For the interested reader, the *Digest of Literature on Dielectrics* series by the National Research Council, Washington DC provides easy access to recent papers on this topic. One of the review articles on polymer dielectrics is given in the references (Hedvig 1977).

Suggested Reading

Flory P 1953 *The Principles of Polymer Chemistry* (Ithaca: Cornell University Press)

Oka S and Nakada O 1959 *Discussion of the Dielectric Properties of Solids* (Tokyo: Iwanami)

Tanford C 1961 *Physical Chemistry of Macromolecules* (New York: Wiley) ch 3

References

Barisas B G 1974 Effect of shearing laminar flow on dielectric polarization of suspension of rigid particles *Macromolecules* **7** 930–3

Debye P and Bueche F 1951 Electric moments of polar polymers in relation to their structure *J. Chem. Phys.* **19** 589–94

Eyring H 1932 Resultant electric moment of complex molecules *Phys. Rev.* **39** 746–8

Fuoss R M and Kirkwood J G 1941 Electrical properties of solids. VIII. Dipole moments in polyvinyl chloride-diphenyl systems *J. Am. Chem. Soc.* **63** 385–94

Hedvig P 1977 Electrical properties of polymers *Dig. Lit. Dielectr.* **41** 310

Jacobsen B 1953 Method for obtaining streaming orientation and simultaneous determination of dielectric properties in macromolecular solution *Rev. Sci. Instrum.* **24** 949–54

Kirkwood J G and Fuoss R M 1941 Anomalous dispersion and dielectric loss in polar polymers *J. Chem. Phys.* **9** 329–40

Peterlin A and Reinhold C 1964 Light scattering by linear macromolecules oriented in laminar flow. III. Perfectly flexible necklace model with no hydrodynamic interaction *J. Chem. Phys.* **40** 1029–32

—— 1965 The influence of laminar flow on the dielectric constant of dilute polymer solution *Kolloid Z., Z. Polym.* **204** 23–8

Reinhold C and Peterlin A 1965 Light scattering by linear macromolecules oriented in laminar flow. V. Intensity and depolarization by a perfectly flexible coil with high solvent immobilization *J. Chem. Phys.* **42** 2172–6

Rouse P 1953 Theory of the linear viscoelastic properties of dilute solutions of coiling polymers *J. Chem. Phys.* **21** 1270–80

Saito N and Kato T 1957 Viscoelasticity and complex dielectric constant in the presence of an electric field and a shearing laminar flow in solutions *J. Phys. Soc. Japan* **12** 1393–402

Takashima S 1970 Dielectric behaviour of helical polyamino acids in shear gradients *J. Phys. Chem.* **74** 4446–52

Tanford C 1961 *Physical Chemistry of Macromolecules* (New York: Wiley)

Van Beek L K H and Hermans J J 1957 Dielectric relaxation in dilute solutions of polar chain molecules *J. Polym. Sci.* **23** 211–21

Wendish P 1964 The change produced in the dielectric constant of high polymer solutions by shearing strain *Kolloid Z., Z. Polym.* **199** 27–31

Williams M L and Ferry J D 1955 Dynamic mechanical properties of polymethyl acrylate *J. Colloid Sci.* **10** 474–81

Work R N and Trehu Y M 1956 Detailed structure of copolymers from dielectric measurements *J. Appl. Phys.* **27** 1003–11

Zimm B 1956 Dynamics of polymer molecules in dilute solution: viscoelasticity, flow birefringence and dielectric loss *J. Chem. Phys.* **24** 269–80

References for dielectric properties of polyions

Dintzis H M, Oncley J L and Fuoss R M 1954 Dielectric increments in aqueous solutions of synthetic polyelectrolytes *Proc. Natl. Acad. Sci. USA* **40** 62–9

Kuhn W and Kuhn H 1961 Dielektrisches Relaxationszeitspektrum von Methyl-zellulose in Wasser *Z. Elektrochem.* **65** 649–51

Mandel M and Jenard A 1963a Dielectric behavior of aqueous polyelectrolyte solutions *Trans. Faraday Soc.* **59** 1–8

—— 1963b Dielectric behavior of aqueous polyelectrolyte solutions *Trans. Faraday Soc* **59** 2158–69

Mandel M and Van der Touw F 1974 in *Polyelectrolytes* ed. E Selegny (Dordrecht: Reidel) pp285–300

Minakata A 1972 Dielectric properties of polyelectrolytes. III. Effect of divalent cations on dielectric increment of polyacids *Biopolymers* **11** 1567–82

Minakata A and Imai N 1972 Dielectric properties of polyelectrolytes. I. A study of tetra-n-butyl ammonium polyacrylate *Biopolymers* **11** 329–46

Minakata A, Imai N and Oosawa F 1972 Dielectric properties of polyelectrolytes. II. A theory of dielectric increment due to ion fluctuations by a matrix method *Biopolymers* **11** 347–59

Van der Touw F and Mandel M 1974a Dielectric increment and dielectric dispersion of solutions containing simple charged linear macromolecules. I. Theory *Biophys. Chem.* **2** 218–30

—— 1974b Dielectric increment and dielectric dispersion of solutions containing simple charged linear macromolecules. II. Experimental results with synthetic polyelectrolytes *Biophys. Chem.* **2** 231–41

5 Dielectric Properties of Proteins, Peptides and Amino Acids

5.1 Introduction

The dielectric behaviour of proteins, peptides and polyamino acids will be discussed in this chapter. Unlike simple organic molecules, the origin of dipole moment of charged macromolecules is not clearly understood. The polarization mechanism of proteins is believed to be primarily due to dipole orientation. However, other mechanisms have been proposed and debated vigorously. This chapter will discuss these alternate interpretations in addition to classical theories.

5.2 Dielectric Increment and Dipole Moment

The equations which relate dipole moment to dielectric constant have already been discussed in Chapter 2. However, these theories were developed for pure gases and pure liquids. In this section, the theory of binary mixtures such as amino acids in aquous media will be presented. When polar molecules are dissolved in a solvent, the dielectric constant of the solution increases linearly with the concentration of the solute. Therefore, the dielectric constant of binary mixtures is given by the following equation:

$$\varepsilon = \varepsilon_0 + \delta c \tag{5.1}$$

where ε and ε_0 are the dielectric constant of solution and solvent respectively. c is the concentration of the solute in moles per litre and δ is the specific dielectric increment, i.e. $\Delta\varepsilon$ divided by moles per 1000 g of water for polar molecules. As will be mentioned later, δ is defined as $\Delta\varepsilon$ divided by a gram per litre for protein molecules because of their exceedingly large molar specific increments. The dielectric increment will be used for the calculation of the dipole moment of amino acids, peptides and proteins. In the following, Kirkwood's theory (1943) for binary mixtures, which relates the specific increment δ to the dipole moment of amino acids and simple peptides, will be discussed.

In general, the following linear equation holds for binary mixtures, assuming there is no interaction between the components:

$$\frac{\varepsilon - 1}{3} = \frac{3\varepsilon}{2\varepsilon + 1}(c_1 P_1 + c_2 P_2) \tag{5.2}$$

where P_1 and P_2 are the polarizations of two components and are given by the following equations:

$$P_1 = \frac{4\pi N}{3}\alpha_1 \tag{5.3}$$

for the solvent and

$$P_2 = \frac{4\pi N}{3}\left(\alpha_2 + \frac{\mu_2 \bar{\mu}_2}{3kT}\right) \tag{5.4}$$

for the polar solute. If ε is very large as is the case for aqueous solutions, (5.2) can be simplified as

$$\frac{2\varepsilon}{9} = c_1 P_1 + c_2 P_2. \tag{5.5}$$

It should be noted that the concentrations c_1 and c_2 are not independent of each other but are bound by the following equation:

$$c_1 V_1 + c_2 V_2 = 1 \tag{5.6}$$

where V_1 and V_2 are the molar volumes of each component. Differentiating (5.5) with respect to c_2 and using (5.6), we obtain

$$\frac{2}{9}\frac{d\varepsilon}{dc_2} = P_2 + P_1 \frac{dc_1}{dc_2}. \tag{5.7}$$

Note that $d\varepsilon/dc_2$ is the general form of the dielectric increment $\Delta\varepsilon/c$. Therefore, we can replace the derivative on the LHS by δ. Also, converting to litres, we obtain

$$\frac{2000}{9}\delta = P_2 - P_1 V_2/V_1. \tag{5.8}$$

In this expression, P_1/V_1 is the volume polarizability of the solvent and is given by

$$P_1/V_1 = (2/9)\varepsilon_1 \tag{5.9}$$

where ε_1 is the dielectric constant of the solvent. Inserting (5.9) into (5.8) and rearranging, we obtain

$$P_2 = (2/9)(1000\delta + \varepsilon_1 V_2). \tag{5.10a}$$

It should be noted that P_2 is also given by (5.4).

Let us define the optical polarizability P_{20} of the solute as

$$P_{20} = \left(\frac{4\pi N}{3}\right)\alpha_2.$$ (5.10b)

Therefore,

$$P_2 - P_{20} = \frac{4\pi N}{9}\left(\frac{\mu_2\bar{\mu}_2}{kT}\right).$$ (5.11)

On rearranging this equation and substituting in the numerical values for the Boltzmann constant and Avogadro's number, we finally find

$$(\mu_2\bar{\mu}_2)^{1/2} = 0.0127\sqrt{(P_2 - P_{20})T} \text{ D}.$$ (5.12)

In order to calculate P_2 we have to determine the value of δ experimentally. For glycine, as mentioned later, δ is 22.58 mol 1^{-1} at 25 °C, thus 1000δ becomes 22 580. $\varepsilon_1 V_2$ is known to be about 3500, and so it may be ignored without introducing a serious error. Also the value of P_{20} is ~ 25.8 and can be ignored. With these values inserted in (5.12), we obtain

$$\mu_2 \simeq (\mu_2\bar{\mu}_2)^{1/2} = 16.7 \text{ D}.$$ (5.13)

Equation (5.12) has been used for the calculation of the dipole moments of amino acids, peptides and even higher molecular weight polyamino acids. For the calculation of the dipole moments of protein molecules in aqueous media, a different equation is used. This will be discussed later in this chapter.

5.3 Amino Acids

The structure of an amino acid is shown in part (a) of figure 5.1. The unique feature of amino acids compared with simple organic acids is the presence of an amino group as well as a carboxyl group. When these groups are ionized, positive and negative charges are created. In other words, amino acids behave as zwitter- (or dipolar) ions. The presence of positive and negative charges in the same molecule is the origin of the large dipole moment for amino acids. Typically, α-amino acids in which the amino and carboxyl groups are adjacent to each other, have a dipole moment of about 15–17 D (Edsall 1943). This value is almost independent of the type of side chain R (figure 5.1 (a)). Let us compare two amino acids having different side chains. One of these is glycine and the other valine. As shown in figure 5.1, the sizes of the side chains are different; however, none of the side chains have an ionizable group. Therefore, the dipole moments of these amino acids are dominated by the charge pairs of the amino group and the carboxyl group. Thus, the dipole moments of glycine and valine are approximately the same. Obviously, if side chains carry either a negative or positive charge, this simple argument will not apply. On the other hand, β-amino acids have an intervening CH_2 group between the amino and carboxyl groups, thus increasing the distance between charges. As discussed previously, the magnitude of a dipole moment is proportional to the distance between charges (see equation (1.6)). Therefore, the dipole moment of β-amino acids is greater than that of α-amino acids, i.e. of the order of 20–21 D.

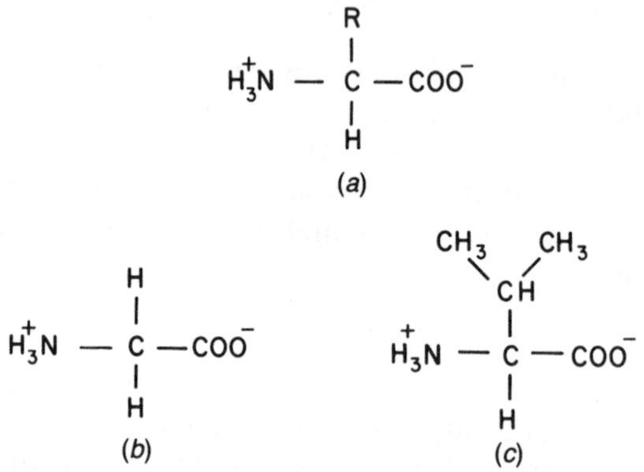

Figure 5.1 The chemical structures of (a) an amino acid, (b) gylcine and (c) valine.

Using (1.6) and measured dipole moments, we can calculate the distance between amino and carboxyl groups in α- and β-amino acids. For example, the dipole moment of α-amino acids is about 16 D or 16×10^{-18} esu cm. Substituting this value into (1.6), and using $e = 4.8 \times 10^{-10}$ esu for elementary charge, we can easily find the distance between amino and carboxyl groups to be 3.33 Å. We can use the same equation for β-amino acids as well and obtain a value of 4.16 Å as the distance between positive and negative charges. The condensation of amino acids produces peptides as shown in figure 5.2. This figure shows the structure of di- and tri-peptides. Obviously, the distance between the amino and carboxyl groups increases as the degree of polymerization increases. If peptide molecules have a linear configuration, the distance would increase linearly. If this is the case, the dipole moments of peptides also increase linearly with the degree of polymerization. Figure 5.3 shows the dipole moments of peptides plotted as a function of the number of amino acid residues. Obviously, the increase in dipole moment is not linear. The implication of this

result is that peptide chains become bent as the degree of polymerization increases.

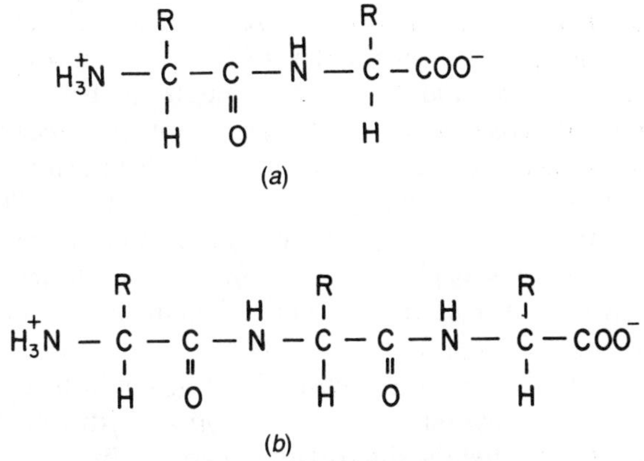

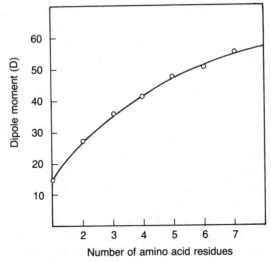

Figure 5.2 The chemical structures of di-peptide (a) and tri-peptide (b).

Figure 5.3 The dependence of the dipole moment of peptides on the number of monomer units.

The conformation of peptides at the initial stage of polymerization is not well known. However, as the degree of polymerization reaches a range of several hundred or more, a few distinct configurations begin to emerge. In the following, the configurations and the dielectric behaviour of polyamino acids will be discussed in some detail.

5.4 The Structure of Polyamino Acids

After disregarding, eventually, other structures, there remain two distinctly different conformations for polyamino acids: the α-helix and a random coil. The α-helix has been of central interest for protein chemists for many years and extensive research has been performed on this conformation using a variety of experimental techniques. The dielectric properties of polyamino acids in the helical conformation were investigated extensively by Wada and his colleages (1958, 1962, 1967).

The model which was proposed by Pauling and Corey (1951) for helical polyamino acids is illustrated in figure 5.4 for left- as well as right-handed α-helices (see Edsall and Wyman 1958). In these configurations, each NH group is hydrogen bonded to a C=O group three residues away from it. Likewise, each C=O group is linked to an NH group three residues beyond it. The direction of hydrogen bonds is, as shown in this figure, slightly tilted with regard to the helical axis. The dimension of the N–H $\cdots$ O group is believed to be about 2.7–2.9 Å and the distance along the axis of the helix is 5.4 Å. Thus, five complete turns will correspond to 18 residues. The whole α-helix will repeat itself at this interval. The hydrogen bonds in the α-helix may be regarded as closing the ring in which three amino acid residues intervene between the C=O and the NH groups that are involved in the hydrogen bond.

There are two ways of forming the helical conformation: one is the left-handed helix and the other is the right-handed helix. If all the R groups of the side chains were hydrogen, as in polyglycine, then it is readily seen from the figure that the left- and right-handed helices would be a mirror image of each other. However, for other polyamino acids, this mirror image situation does not exist. In general, the right-handed helix is more stable than the left-handed helix.

Strong evidence that the α-helix exists as a major configuration of natural or synthetic peptides was obtained from x-ray analyses. In β-keratin and synthetic polyamino acids such as poly-γ-methyl-L-glutamate, a strong x-ray reflection is observed which corresponds to a diffraction from the planes normal to the fibre axis with a spacing of 1.5 Å. This corresponds to the distance along the axis per amino acid residue in the helix. This spacing is compatible with the Pauling–Corey (1951) model and provides evidence for the presence of an α-helical configuration in polyamino acids and proteins.

Although many polyamino acids can now be obtained easily poly-γ-benzyl-L-glutamate (PBLG) is most probably the most extensively investigated one. Optical rotatory dispersion studies (Doty *et al* 1956, 1957), infrared spectroscopy (Miyazawa 1962) and dielectric measurements have been conducted using PBLG (see Stahman 1962, Fasman 1967). Because of these studies, its configuration is well known in solution

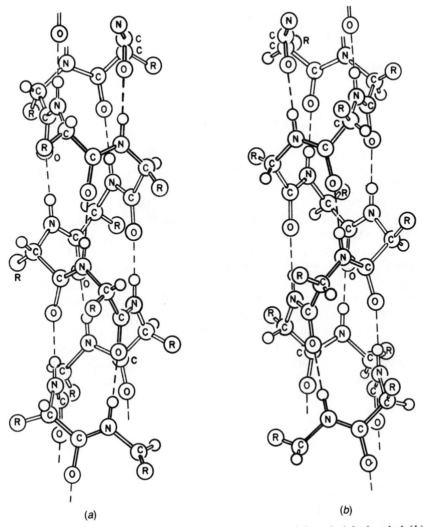

Figure 5.4 The Pauling–Corey models for left-handed (*a*) and right-handed (*b*) α-helices. (From Pauling and Corey 1951.)

under various conditions. PBLG is soluble in non-polar solvents such as dioxane and ethylene dichloride because of the presence of large non-polar benzyl groups in the side chains. It is known that PBLG maintains the helical configuration in these non-polar solvents and is stabilized by intra-chain hydrogen bonds. A helix–coil transition will take place if PBLG is transferred to mixed solvents containing polar solvents such as dichloroacetic acid. If benzyl groups are removed from the side chains of PBLG, we obtain poly-L-glutamic acid or PGA which is soluble in aqueous media. Polyglutamic acid has a helical configuration at acidic pH and has a random-coil conformation at basic pH. The helix–coil transition occurs at about pH 6.0 as detected by optical rotation and viscosity measurements (Doty *et al* 1957).

5.5 The Dipole Moment of the Peptide Group

The structural unit of polypeptides is illustrated in figure 5.5, with the notation for bond lengths and angles. Azimuthal angles through which the bond 1–3′ rotates are left as adjustable parameters. After fixing the internal coordinates, the dipole moment of the peptide group can be calculated as the vector sum of bond moments. The dipole moments of various chemical bonds are listed in table 1.2, and those which are necessary for this calculation can be found in this table. However, the dipole moments of isolated bonds are not identical to those found in large molecules. In order to calculate the dipole moment of a molecule as a vector sum of bond moments, various corrections must be made. For example, the dipole moment of N-methylacetoamide was found to be 4.39 D while the simple vector summation of bond moments gives a value of only 3.47 D. One of the reasons for this discrepancy is the neglect of the resonance effect of the C–N bond. Thus the difference between the observed and calculated dipole moments, i.e. 0.9 D can be attributed, at least partially, to a resonance structure.

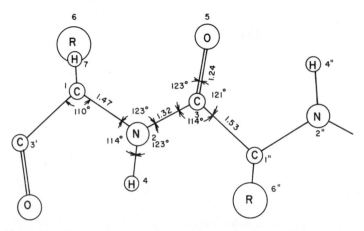

Figure 5.5 The repeating unit in peptides. Bond lengths and bond angles are shown. Reproduced from Wada 1967, p 369, by courtesy of Marcel Dekker, Inc.

In addition, C=O and NH groups are connected by intra-chain hydrogen bonds. The three structures of the N–H ··· O=C hydrogen bond are shown below:

$$N–H ··· O=C$$
$$N^+–H^- ··· O=C$$
$$N^-–H \qquad O^+=C.$$

The electronegative oxygen tends to pull hydrogen away from nitrogen. Thus, the formation of hydrogen bonds deforms slightly the charge distribution in peptide groups and this entails the change in the dipole moments of the N–H as well as the C=O bonds.

The direction of the overall dipole moment of the peptide group is illustrated in figure 5.6. This result was obtained experimentally by microwave spectroscopy using gaseous formamide as a model substance (Kurland and Wilson 1957). The net group moment of the peptide was found to be 3.71 D and the angle between the dipole vector and the C–N bond is 39° 6'. Hydrogen bonds play a crucial role for the orientation of peptide bonds in the α-helix. As shown in figure 5.4, the helical configuration is stabilized by intra-chain hydrogen bonds. N–H and C=O groups which are bridged by a hydrogen bond are three amino acid residues away from one another. Since 3.6 residues are necessary to complete an α-helical unit, the direction of the hydrogen bonds is only slightly tilted relative to the helical axis. This means that dipole moments are aligned with a high degree of regularity along the helical axis. Thus, the vector sum of these moments will produce a large longitudinal net dipole moment in helical polyamino acids.

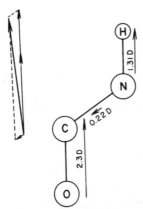

Figure 5.6 Bond moments of the peptide group. The inset shows the vectorial relationship of bond moments and the net dipole moment as their sum. (From Wada 1962.) Reproduced by permission of The University of Wisconsin Press © 1962.

5.6 Dielectric Properties of Polyamino Acids

Figure 5.7 illustrates the frequency profile of the dielectric constant of PBLG in ethylene dichloride (EDC) (Wada 1959). In this solvent, PBLG is in the helical configuration. In general, the effective length of PBLG ranges between 100 and 5000 Å. The rotary diffusion constant θ of PBLG can be calculated using these numbers in the following equation:

$$\theta = \frac{3kT}{16\pi\eta a^3}[2\ln(2a/b) - 1] \qquad (5.14)$$

where a and b are the major and minor axes of the cylinder and η is the viscosity of the solvent. This calculation gives a rotary diffusion constant for helical PBLG of between 10^{-4} and 10^{-1} s^{-1}. Assuming that rotary diffusion and electric polarization are similar processes, we can calculate the dielectric relaxation frequency. The result of this calculation ranges between 1 Hz and 1 kHz. Inspecting figure 5.7, we find the relaxation to be at about 5×10^3 Hz; this is comparable to the calculated value. Dipole moment was calculated using the following equation:

$$\mu^2 = (9kT/4\pi N)P^{00}M_w \qquad (5.15)$$

where P^{00} is the specific polarization extrapolated to zero concentration. M_w is the weight average molecular weight and T is the temperature. It

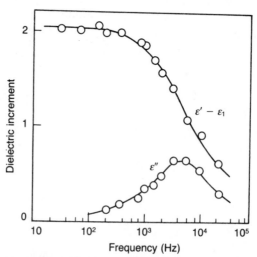

Figure 5.7 The dielectric dispersion of poly-γ-benzyl-L-glutamate. ε' is the dielectric constant, ε'' is a dielectric loss factor and ε_1 is the dielectric constant of the solvent. (From Wada 1959.) Reproduced by permission of the American Institute of Physics.

should be noted that this equation is essentially the same as (5.12). In this experiment, the molecular weight of PBLG is 154 000. Using this and the value of specific polarization ($P^{00} = 805$ cc), the dipole moment is calculated to be 2460 D. This large value reaffirms the high degree of regularity of the alignment of peptide moments in the α-helix. Figure 5.8 illustrates the molecular weight dependence of the dipole moment of PBLG. As shown, a linear relationship is clearly observed between dipole moment and molecular weight, or the degree of polymerization. Figure 5.9 shows the double logarithmic plot of relaxation frequencies and molecular weight. As shown, the linear plot has a slope ~ -2.5. Note that the rotary diffusion of rigid rod-like molecules is inversely proportional to the cube of length (see equation (5.14)). All these results are in accord with the postulate that the helical PBLG is reasonably rigid and that the dielectric relaxation of PBLG is due to a longitudinal moment. Since dipole vectors are slightly tilted, there may be a small dipole moment in the transverse direction. However, the magnitude of this transverse moment may be very small and can be ignored.

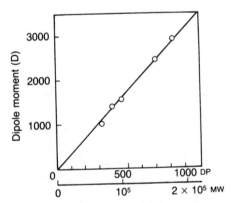

Figure 5.8 The dependence of the dipole moment of PBLG on the degree of polymerization (DP). Molecular weights are also shown on the abscissa. The solvent is ethylene dichloride at 25 °C. (From Wada 1962.) Reproduced by permission of the University of Wisconsin Press © 1962.

In order to calculate the overall dipole moment of helical PBLG, we still have to consider the dipole moments of the side chains. The configuration of the side chains of PBLG is illustrated in figure 5.10. As shown, the side chain contains a polar group C=O. The overall dipole moment depends upon the relative directions of the main chain and side

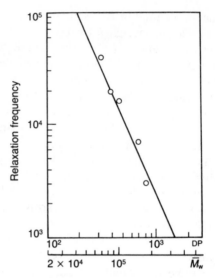

Figure 5.9 The dependence of the relaxation time of PBLG on the degree of polymerization and/or molecular weight. Both ordinate and abscissa are on a logarithmic scale. (From Wada 1962.) Reproduced by permission of the University of Wisconsin Press © 1962.

chain moments. However, no definitive information is available with regard to the orientation of side chains and so, under these circumstances, this orientation can best be treated as a statistical problem (Wada 1962).

If we designate the main chain and side chain dipole moments by μ_m and μ_s, and assume that side chain dipole moments have a certain degree of freedom of rotation, then the mean square dipole moment of a structural unit is given by the following equation:

$$\langle \mu^2 \rangle = \left\langle \sum_{i,j}(\mu_{mi} + \mu_{si})\cdot(\mu_{mj} + \mu_{sj})\right\rangle \qquad (5.16)$$

where the orientation of the main chain moments μ_{mi} and μ_{mj} is fixed by hydrogen bonds along the major helical axis and only side chain moments have either complete or restricted freedom of rotation. Multiplying the two terms in the brackets, we generate the following terms:

$$\sum_{i,j}\langle \mu_{mi}\cdot\mu_{mj}\rangle = n^2\mu_m^2(\|) \qquad (5.17)$$

$$\sum_{ij}\langle \mu_{mi}\cdot\mu_{sj}\rangle = n\mu_m(\|)\sum_{i}\langle \mu_{si}\cdot\mu_n\rangle \qquad (5.18)$$

where μ_n is a unit normal vector along the helical axis. The term $\langle \mu_{si}\cdot\mu_{sj}\rangle$ is ignored assuming the magnitude of these terms is small.

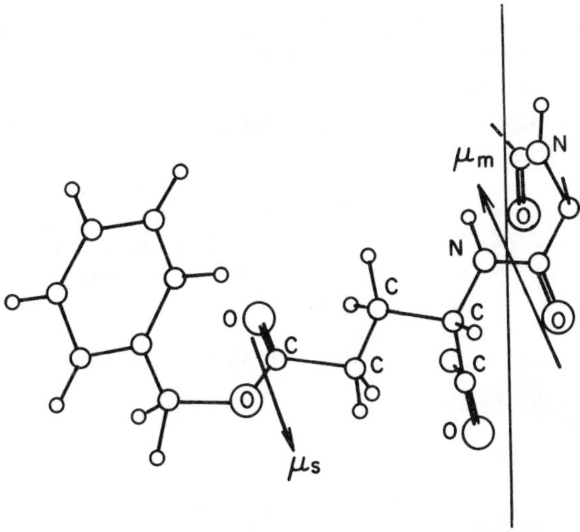

Figure 5.10 The configuration of the side chain of PBLG. μ_m and μ_s are the dipole moments of the main chain and the side chain respectively. Reproduced from Wada 1967, p 369, by courtesy of Marcel Dekker, Inc.

Since we cannot find the exact spatial orientation of side chains, we assume only two extreme cases, complete freedom and zero freedom.

Perfect flexibility of the side chains with no interaction between them leads to random orientation. The net dipole moment of the structural unit, in this case, is given by

$$\langle \mu^2 \rangle = n^2 \mu_m^2 (\|) + 2n\mu_m(\|)\langle \mu_s \cdot \mu_n \rangle \qquad (5.19)$$

where μ_s is the vector sum of μ_{si}. Because of random distribution, $\langle \mu_s \cdot \mu_n \rangle$ becomes very small and the mean square moment is dominated by the main chain moment μ_m^2.

If side chains are rigidly fixed in a certain direction, then the contribution of side chain moments is either positive or negative depending on the sense of the helices. It was found that the side chain moments make a negative contribution in a right-handed helix of L isomer while the contributon of the side chain moments is positive if the helix is left-handed.

As discussed above Wada and co-workers treated the configuration of side chains as a statistical problem. However, we can find the configuration of side chains more explicitly with the aid of substitution techniques. Applequist and Mahr (1966) investigated the dipole moment of poly-L-tyrosine. The effect of bromination of the tyrosine group was studied in attempts to find the configuration of side chains and also the sense of the helix. As stated before, in general, right-handed helices are more common for poly-L-amino acids than the left-handed helix. Figure 5.11 illustrates the geometry of group moments in the helical poly-L-tyrosine. The dipole moment of each residue, including the side chain, is given by the following equation:

$$\mu_r = \mu/N = \mu_b + \langle \mu_z \rangle \qquad (5.20)$$

where μ is the dipole moment of the entire helix and N is the number of residues. $\langle \mu_z \rangle$ is the z component of the side chain moment μ_s. The instantaneous value of μ_z is given by

$$\mu_z = \mu_s \cos\alpha \cos\beta + \mu_s \sin\alpha \sin\beta \cos\phi. \qquad (5.21)$$

Averaging over ϕ, the second term in (5.21) vanishes, hence

$$\langle \mu_z \rangle = \mu_s \cos\alpha \cos\beta. \qquad (5.22)$$

The angle β is taken to be 52.5° for a right-handed helix and 135.2° for a left-handed helix. The angle α is known to be 90° for a phenol group (Antony et al 1964), whence $\cos\alpha = 0$ and $\langle \mu_z \rangle$ vanishes. Therefore, for polytyrosine, the dipole moment of each residue is represented solely by the backbone moment, i.e. $\mu_r = \mu_b$. Under these conditions, the dipole moment of each residue can be determined by dividing the total moment of the entire helix by the number of residues, and a value of 4.94 D was found. This value is in agreement with the one obtained by Wada, i.e. 4–5 D for polybenzyl-L-glutamate. Bromination of the tyrosine group changes the dipole moment slightly from 1.55 to 1.38 D. This decrease would not have any effect on the overall dipole moment unless there is a change in the angles α and/or β. Actually the angle α

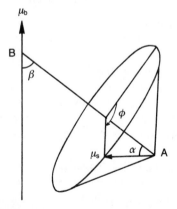

Figure 5.11 The configuration of the side chain of polytyrosine. μ_b and μ_s are the dipole moments of the backbone and the side chain respectively. Reproduced with permission from Applequist and Mahr (1966). Copyright (1966) American Chemical Society.

was found, by Antony *et al* (1964), to change from 90° to 54.7°. Hence, bromination of the phenol group in the side chains will cause a decrease in the residue dipole moment μ_r if the helical sense is left-handed and an increase if it is right-handed. Figure 5.12 illustrates the change in residue moment μ_r with the extent of bromination. The broken lines A and B in this figure are predicted for the right- and left-handed helical senses. As shown, bromination of the side chain causes a decrease in the residue dipole moment, a clear indication that the helix of polytyrosine is left-handed. As stated earlier, it is generally believed that a right-handed helix is more stable than a left-handed helix for most of the polyamino acids. Therefore this result is somewhat unexpected. Actually, Fasman *et al* (1964) concluded from their optical rotatory dispersion data in aqueous solution that the helix of polytyrosine is right-handed. The cause of this discrepancy remains to be found. However, the results obtained by Applequist and Mahr are so unambiguous that there could be no other interpretation.

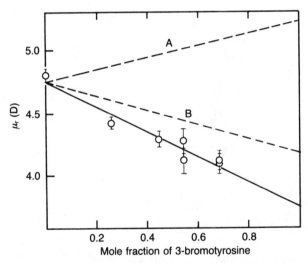

Figure 5.12 The effect of bromination of the tyrosine groups on the dipole moment per residue. Lines A and B are the prediction for right-handed and left-handed helices. The full line with open circles shows the experimental results. Reproduced with permission from Applequist and Mahr (1966). Copyright (1966) American Chemical Society.

5.7 The Helix–Coil Transition

As discussed previously, PBLG undergoes a transition from a helical conformation to a random-coil configuration as the polarity of solvents changes. As an example, we can use the mixture of ethylene dichloride and dichloroacetic acid in various ratios. As the fraction of dichloroacetic acid increases, the polarity of the mixture increases. Under this condition, intra-chain hydrogen bonds will be disrupted because of the increased affinity of CO and NH groups to solvent molecules. The disruption of intra-chain hydrogen bonds causes collapse of the helices. The effect of a helix–coil transition on the dipole moment of polyamino acids has been investigated by Wada *et al* (1972) and Marchal *et al* (1967, 1974); only the results obtained by Marchal *et al* will be discussed here.

In the papers by Wada *et al* and by Applequist and Mahr, the problem of aggregate formation of polyamino acids in non-polar solvents is not addressed explicitly. As illustrated in figure 5.13, the transition of helical polybenzyl aspartate in mixed solvents is preceded by a sharp decrease in the dielectric constant. This result indicates the presence of

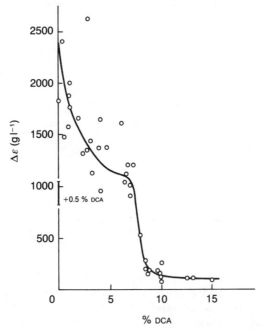

Figure 5.13 The effect of a helix–coil transition on the dielectric increment of PBLA. The initial decrease in the increment at low DCA concentration is due to deaggregation of the PBLA helices. The vertical bar indicates the range of dielectric increment for completely deaggregated PBLA. The second decrease in the dielectric increment is due to a helix–coil transition. (From Marchal 1974.) Reproduced by permission of John Wiley & Sons, Inc. © 1974.

conformation changes which are not related to the helix–coil transition. The formation of aggregates of PBLG and other polyamino acids in helix forming solvents has been known since their discovery by Doty *et al* (1956) and Tinoco (1957). Marchal and Marchal (1967) and Gupta *et al* (1974) found, using dielectric techniques, that aggregated PBLG helices still behave as a rigid rod in spite of the increase in length. These investigators further observed that aggregation causes an increase in the dielectric increment and a decrease in the relaxation frequency (an increase in the relaxation time). These observations indicate head-to-tail association of PBLG helices rather than side-by-side aggregation. Dissociation of these aggregates can be implemented by the addition of dichloroacetic acid to solution or, though incomplete, by elevated temperatures.

The helix–coil transition of poly-DL-phenylalanine (PDLPA) and poly-γ-benzyl-L-aspartate (PBLA) was investigated by Marchal (1974) using a mixed solvent of chloroform and dichloroacetic acid (DCA). The results obtained with PBLA are shown in figure 5.13. In spite of the scatter of measured points, the dielectric increment obviously decreases with increasing DCA concentration between 0 and 8%. A curve is drawn through these points to indicate a plateau owing to the completion of deaggregation by DCA. The bold vertical bar on the ordinate is the dielectric increment of an unaggregated PBLA helix showing that the deaggregation by DCA is nearly complete at this concentration. Further addition of DCA causes another sharp decrease in the dielectric increment which is due to the helix–coil transition. As shown in figure 5.13, the large dipole moment of helical PBLA is replaced by a much smaller dipole moment after the helix–coil transition.

5.8 Dielectric Properties of Polyglutamic Acid

PBLG is soluble only in organic solvents because of the presence of large non-polar benzyl groups in the side chains. We can convert PBLG into polyglutamic acid by debenzylating these side chains. The terminal group of the side chains of PGA is the carboxyl group which becomes a COO^- ion at basic pH. In ionized form, PGA is unable to form helices because of the electrostatic repulsion between charges and so it will take on a random-coil configuration.

When we decrease the pH of a medium, the charges of the carboxyl groups are neutralized and this in turn causes the transition of PGA from a random-coil configuration to a helical structure. This transition was investigated by optical rotation and viscosity measurement (Doty *et al* 1957). The transition was also investigated by dielectric measurement by Takashima (1963) and reinvestigated by Muller *et al* (1974). Only the results obtained by Muller *et al* are shown in figure 5.14. As stated, PGA is in a random-coil conformation at high pH. At high pH, the side chains are all charged. As shown in this figure, the dielectric constant of random-coil PGA is quite high. However, as the pH of the solution decreases, the dielectric constant decreases gradually, starting around pH 7 and reaching a lower limiting value at about pH 5.5.

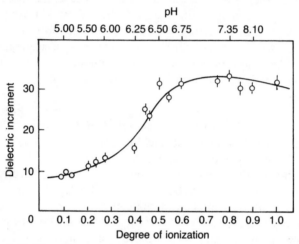

Figure 5.14 The dielectric increment of polyglutamic acid at various degrees of ionization. Note that the large dielectric increment of the random coil of PGA is replaced by a small increment after a helix–coil transition. (From Muller *et al* 1974.) Reproduced by permission of Elsevier Science Publishers BV.

This result demonstrates that random-coil PGA has a large dipole moment, contrary to random-coil PBLG, owing, perhaps, to the high charge density of ionized side chains. Because of the lack of chain regularity in the random-coil configuration, the magnitude of the permanent dipole moment may be small. Therefore, the origin of the large dipole moment of random-coil PGA is likely to be an ionic process rather than the orientation of permanent dipoles. A summary of the dipole moments and relaxation times of PGA in the helical and random-coil forms is shown in table 5.1. There has been a more recent investigation by Nakamura and Wada (1981) on the helix–coil transition of PGA. Their results revealed that the pH induced change in dielectric constant precedes a change in conformation. This indicates that a decrease in charge density of a PGA molecule causes the initial decrease in dielectric constant, even without conformation changes. The dielectric behaviour and origin of the dipole moment of polyelectrolytes will be discussed in detail in Chapter 6.

Table 5.1 Dielectric dispersion data for PGA and PBLG.†

	Form	MW	δ (g l^{-1})	Dipole moment (D)	τ (10^{-6} s)
PGA	Coil	90 000	460 ± 70	2540 ± 200	9.95–12.7
PGA	Helix	90 000	120 ± 70	1320 ± 100	—
PBLG‡	Helix	150 000		2460	99.52

†Taken from Takashima (1963).
‡Taken from Wada (1959).

5.9 Dielectric Properties of Protein Molecules

There are two very important biological macromolecules: protein and nucleic acid. Without these, biological systems would stop functioning or even stop existing. Proteins are extremely interesting polymers, not only because of their biochemical functions, but also because of their intricate structures. In order to investigate the structure and/or conformation of proteins, every possible chemical and physical methodology has been employed, including dielectric measurement. Although dielectric techniques could not be used, at an early stage, for the elucidation of the internal structure of protein molecules, a combination of this method with other techniques, such as x-ray crystallography, now enables us to study the spatial arrangement of individual amino acids in protein molecules.

As discussed previously, the dipole moment of a complex molecule is the vector sum of a number of bond moments to a first approximation. When an electrical field is applied, only this summed dipole vector will respond to the field unless the molecule is very flexible and there is a high degree of internal freedom. Actually, there are some molecules in which internal groups rotate almost independently. These molecules exhibit clearly distinguishable relaxation processes, one due to the orientation of an entire molecule and the other due to the rotation of flexible side groups. For most of the proteins, however, dielectric measurement does not reveal the internal structure of the protein molecule due to the lack of flexibility. Protein molecules consist of peptide chains as mentioned before. Long peptide chains may be flexible individually; however, when they become a part of protein molecules, usually the freedom of rotation of the individual groups is reduced drastically. When we apply an electrical field to a protein solution, proteins will behave only as a large dipole instead of several hundred small dipoles. We must remember that we detect the presence of dipoles only if they respond to applied fields individually. The small groups in protein molecules may undergo a variety of vibro-rotational motions. However, the characteristic frequencies of these motions are much higher than the frequencies of dielectric measurements, i.e. the millimetre wavelength region or even the far-infrared region. Therefore, dielectric techniques are not able to detect these motions unless the measurement is performed at extremely high frequencies. Ultra-high frequency dielectric measurement with protein samples will be discussed in Chapter 8.

Let us now turn to a discussion of experimental results obtained so far and the principles of analysis of these data. The purpose of this section is not to review all the data existing at present but rather to discuss a few fundamental observations and the analytical methods which were developed to interpret the results.

Figure 5.15 illustrates the frequency profile of the real and imaginary parts of the dielectric constant of protein solution schematically. In the low RF region, at around 100 kHz to 1 MHz, the dielectric constant of the solution is considerably higher than that of water or a dilute salt solution (broken line). As the frequency increases, the dielectric constant of the solution decreases gradually. The dispersion of dielectric constant in this region is due to the relaxation of protein dipoles. At the same time, the imaginary part (dielectric loss) exhibits a maximum value at the centre of the dispersion curve. As the frequency of the applied field further increases and reaches the 1 GHz region, there appears another dispersion which is due to the relaxation of bulk water in the solution. Between these two regions, we observe a small dielectric

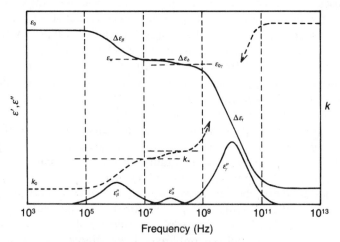

Figure 5.15 A schematic representation of the dielectric constant, conductivity and dielectric loss of protein solution. The three relaxation processes are due to protein dipoles, bound water and bulk water.

dispersion which is presumably due to the water of hydration or bound water. The discussion in this section focuses only on the dispersion in low RF regions.

One of the recent data which was obtained with human haemoglobin is shown in figure 5.16 (Takashima 1986). From this figure, we obtain three pieces of information which are needed for a quantitative description of protein molecules: (i) the dielectric increment; (ii) the relaxation frequency and (iii) the distribution of relaxation times.

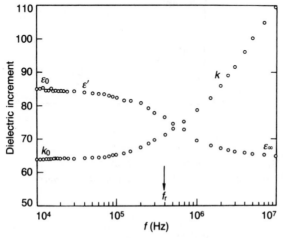

Figure 5.16 A frequency profile of the dielectric constant and conductivity of human haemoglobin at 25 °C. (The pH is 6.8).

Dielectric increment

The dielectric increment is defined as the difference between the low-frequency and high-frequency dielectric constants ε_0 and ε_∞. This quantity is called the total increment while the quantity $\varepsilon_0 - \varepsilon_{water}$ is called a dielectric increment and $\varepsilon_{water} - \varepsilon_\infty$ a dielectric decrement. The total increment is the quantity which is related to dipole moment. However, in order to calculate dipole moment, it must be ensured that the dielectric increment is linearly proportional to the concentration of protein. In other words, the specific increment (the increment divided by concentration) must be independent of protein concentration. Figure 5.17 illustrates the dependence of dielectric increment and decrement on protein concentration. As shown, there is a linear relationship between these quantities and the concentration of the protein. Thus, the specific dielectric increment can be calculated at any finite concentration by dividing $\Delta\varepsilon$ by C. If the specific increment depends on concentration,

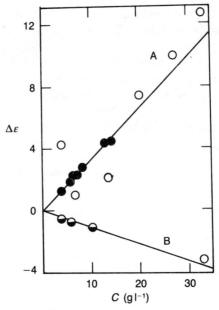

Figure 5.17 The dielectric increment (curve A) and decrement (curve B) of protein solution at various concentrations. Different symbols indicate the results of separate measurements. (From Oncley 1938.)

then the limiting value of δ must be determined by interpolation to zero concentration. The dipole moment of proteins can be calculated using the following equation (Oncley 1943):

$$\mu^2 = [9000kT/4\pi Nh][M\delta] \qquad (5.23)$$

where the quantities k, N and T have their usual meeting. M is molecular weight and h is an empirical parameter. Since the specific increment is expressed as grams per litre, we have to multiply δ by the molecular weight to be consistent with molar units. The value of the empirical parameter h has been discussed by several investigators. Of these, a value of 5.8, which was determined empirically using glycine (the dipole moment is 15.8 D), is most frequently used. The dielectric increments and calculated dipole moments of various globular proteins, along with other parameters, are tabulated in table 5.2. As shown, the dipole moments of globular proteins range between 170 and 700 D. Although the values are numerically large, these extraordinary dipole moments are solely due to the large size of protein molecules, rather than to a unique arrangement of charged groups. Paradoxically, these dipole moments are small for the large size of protein molecules, as demonstrated by a simple discussion.

Table 5.2 The dielectric increment and dipole moments for protein molecules (at 25 °C).†

Protein	Solvent	MW	$\Delta\varepsilon_0$ (g l^{-1})	$-\Delta\varepsilon_\infty$ (g l^{-1})	δ	D
Horse CO haemoglobin	Water	64 000	0.33	0.09	0.42	480
Myoglobin	Water	17 000	0.15	0.06	0.21	170
Insulin	a	40 000			0.38	360
β-lacto-globulin	b	40 000	1.51	0.07	1.58	730
Egg albumin	Water	44 000	0.10	0.07	0.17	250
Horse serum albumin	Water	70 000	0.17	0.07	0.24	380
Horse-γ-pseudo-globulin	Water	142 000	1.08	0.06	1.14	1100
Edestin	c	310 000	0.7	0.1	0.8	1400

† Taken from Oncley (1943).
a 80% propyrene glycol.
b 0.5 mol and 0.25 mol glycine.
c 2 mol glycine.

For example, the dipole moment of horse haemoglobin is about 480 D. Let us use this value in (1.6) and find how many charge pairs are needed to generate this dipole moment in haemoglobin. The diameter of haemoglobin is about 50 Å and the magnitude of elementary charge e is 4.8×10^{-10} esu. Therefore, the product of these two is 240×10^{-18} esu cm. Thus, we need only two charge pairs across the molecule to produce a dipole moment of 480 D. At isoelectric pH, there are more than 100 charge pairs distributed on the surface of a haemoglobin molecule. If we align them diametrically across the molecule, there will be an enormous dipole moment. The fact that the dipole moment of haemoglobin is only 480 D leads us to a conclusion that the charge distribution in protein molecules has a high degree of spherical symmetry.

This observation raises a question as to whether this symmetrical charge distribution is accidental or whether proteins 'choose' a peptide folding in such a way as to minimize net dipole moment. Barlow and Thornton calculated the net dipole moments of a variety of proteins whose 3D structures are known (1987). The results of these calculations show that most proteins have a highly symmetrical distribution of charged groups. There are only a few exceptions in which charge distributions are asymmetric. This clearly indicates that the 'small' dipole moments of protein molecules are not accidental and that attractive electrostatic interactions between charges and dipoles play an important role for the folding and stability of protein molecules. The detail of these calculations will be discussed again later in this chapter.

Experimentally, the relationship between protein folding and net dipole moment was studied by Takashima some years ago (1964). The dielectric increments of native, urea denatured and renatured bovine serum albumin are shown in table 5.3. As shown, the dielectric increment of bovine serum albumin (BSA) increases markedly with urea denaturation, indicating the loss of charge symmetry upon unfolding of the peptide chains. However, the removal of urea and subsequent renaturation of BSA results in a small dielectric increment which is nearly the same as that of native protein. This result may offer evidence that protein molecules gain stability by minimizing the net dipole moment.

Table 5.3 The dielectric increment, relaxation time and intrinsic viscosities of native, denatured and renatured bovine serum albumin.[†]

	Dielectric increment	Dipole moment (D)	Relaxation time (s)	Viscosity (100 ml g^{-1})
Native	0.09	229	2×10^{-7}	0.043
Urea denatured	0.79	679	4×10^{-7}	0.125
Renatured	0.10	241	2×10^{-7}	0.045

†Taken from Takashima (1964).

Relaxation frequency
The relaxation frequencies and/or relaxation times of various protein molecules are shown in table 5.4. This table shows that the relaxation

Table 5.4 Relaxation frequency, relaxation time and geometry asymmetry of various proteins.[†] The solvents for these proteins are listed in table 5.2. The relaxation frequencies and relaxation times are corrected for viscosity.

Protein	η/η_{H_2O} (cps)	f_{obs} (MHz)	$\tau \times 10^{-8}$ (s)	$\tau_0 \times 10^{-8}$ (s)	a/b	$\Delta\varepsilon_a/\Delta\varepsilon_b$	θ
Horse CO haemoglobin	1.0	1.9	8.4	6.6	1.6		
Insulin	1.7	0.59	1.6				
β-lacto-globulin	1.0	1.0;3.0	15;5.1	4.3	4.3	0.25	63
Egg albumin	1.0	0.86;3.4	18;3.4	3.7	5.0	1.5	40
Horse serum albumin	1.0	0.44;2.1	36;7.5	6.0	6.0	1.0	45
Horse-γ-pseudo-globulin	1.0	0.064;0.57	250;28	22.0	9.0	1.0	45
Edestin	1.33	0.05;0.44	240;27	21.0	9.0	1.0	45

†Taken from Oncley (1943).

times of globular proteins lie between 1.6×10^{-8} and 2.5×10^{-6} s. Figure 5.18 illustrates the relationship between relaxation time and the molecular weight of these proteins. Some proteins show two different relaxation times. For these cases, the relaxation time of an equivalent sphere (a sphere having the same volume as that of protein) was calculated and used in this figure. Clearly there is linear proportionality between relaxation time and the cube of molecular weights. This tendency is particularly noticeable with small spherical proteins.

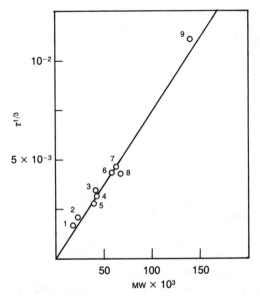

Figure 5.18 The size dependence of the relaxation time of protein molecules: 1, myoglobin; 2, ribonuclease; 3, β-lacto-globulin; 4, egg albumin; 5, insulin; 6, human haemoglobin; 7, horse haemoglobin; 8, horse serum albumin; 9, horse-γ-pseudo-globulin.

Let us analyze the dielectric dispersion data of myoglobin as an example. This protein is a small spherical molecule having a molecular weight of 17 000. Figure 5.19 shows the frequency profile of the dielectric constant of myoglobin solution. The relaxation frequency is found at about 5×10^6 Hz and this gives rise to a relaxation time of 3.18×10^{-8} s. Using this value in Debye's equation:

$$\tau = 4\pi a^3 \eta / kT \qquad (5.24)$$

where a is the radius of a spherical molecule and η is the viscosity of solution, we can calculate the radius 'a' assuming that the viscosity of myoglobin solution is identical to that of water, i.e. 1 centipoise. The calculation shows the radius of myoglobin to be about 30 Å (Lumry and

Yue 1965) which is very close to that observed using x-ray crystallography.

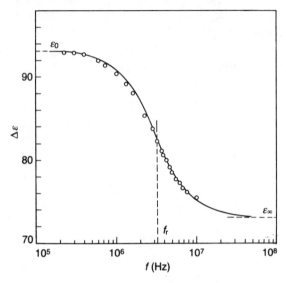

Figure 5.19 The dielectric dispersion of myoglobin. The full curve is calculated using the Debye equation. f_r is the relaxation frequency.

An analysis of the slope of the dispersion curve is even more informative. The full curve in this figure was calculated using Debye's equation (equation (3.19)). As demonstrated, measured values follow the Debye theory very closely. This observation clearly indicates that the relaxation of myoglobin is characterized by only one time constant and that myoglobin can be considered as a spherical molecule as a first approximation. This result shows that the orientation of a myoglobin molecule in aqueous media is amazingly a simple process. Intermolecular and solvent–solute interactions do not play any significant role in the orientation of a myoglobin molecule. As discussed above, in general, the analysis of dielectric data for spherical protein molecules is relatively easy. However, this simplistic interpretation does not hold for non-spherical proteins.

Distribution of relaxation times (Oncley 1943)
Figure 5.20 shows the dielectric dispersion curve of a flagellin, the monomer form of bacterial flagella. This protein is known to have a prolate ellipsoidal shape with an axial ratio of about 10:1. The broken curve was calculated using Debye's equation, assuming one relaxation

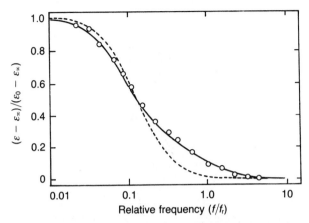

Figure 5.20 The dielectric dispersion of flagellin. The broken curve is calculated using the Debye equation and the full curve is calculated assuming two relaxation times with an axial ratio of $1 : 10$. (From Gerber *et al* 1972.) Reproduced by permission of Academic Press Inc.

time. Obviously measured points do not follow the Debye theory this time. In general, proteins which are known to be non-spherical show a similar deviation from Debye theory. We can readily postulate that the deviation stems from the presence of more than one relaxation process. As a next step, let us speculate as to what is the origin of the distribution of relaxation times. The orientation of spherical molecules is isotropic, i.e. the rotation of the molecule is always the same regardless of the direction of the dipole, hence, there is only one relaxation time. However, for non-spherical molecules, the relaxation time depends on the direction of the dipole moment. If the direction of a dipole moment coincides with the direction of the major axis of an elongated ellipsoidal molecule, then orientation would occur along the major axis with a long relaxation time. On the other hand, if the dipole is directed along the minor axis, the relaxation time will be shorter. If the dipole vector is tilted, as shown in figure 5.21, the molecule will rotate along the dipole vector rather than along the molecular axes. This will result in complex tumbling motions of the entire molecule. However, we can simplify the analysis by assuming that the dipole is a composite of three vectors, i.e. the vectors along the axes a, b and c. Therefore, the orientation of ellipsoidal molecules can be described sufficiently accurately by three relaxation processes. However, we can further simplify the problem, without the loss of generality, by assuming that the two minor axes b and c are equal, thus, we have two relaxation processes to deal with. So long as the lengths of axis a and axis b (or c) are discretely different, we can assume that the relaxation times are also distinctly different.

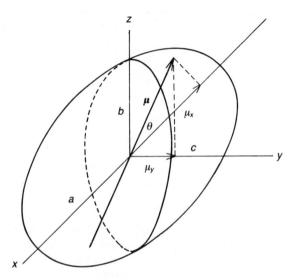

Figure 5.21 The ellipsoid model used for the analysis of a two relaxation time problem. θ is the angle between the major axis and the dipole vector μ. The two minor axes b and c are assumed to be equal.

Under these circumstances, we can generalize the Debye equation as follows:

$$\varepsilon = \varepsilon_\infty + \frac{\Delta\varepsilon_a}{1 + j\omega\tau_a} + \frac{\Delta\varepsilon_b}{1 + j\omega\tau_b} \tag{5.25}$$

$$= \varepsilon_\infty + \sum_{i=1}^{2} \frac{\Delta\varepsilon_i}{1 + j\omega\tau_i} \tag{5.26}$$

where τ_i are the relaxation times for rotation along the molecular axes. $\Delta\varepsilon_i$ is the dielectric increment associated with these rotations. Note that this equation is a special case of (3.53). When the distribution of relaxation time becomes continuous, (5.26) reduces to (3.53). Equation (5.25) contains four unknowns, i.e. $\Delta\varepsilon_a$, $\Delta\varepsilon_b$, τ_a and τ_b, and two known quantities, ε and ε_∞. Therefore, it is impossible to determine these unknowns uniquely using this equation alone. In the following we will discuss how we can construct theoretical curves using (5.25).

First of all, we normalize the sum $\Delta\varepsilon_a$ and $\Delta\varepsilon_b$. Therefore, the total amplitude of dielectric dispersion at sufficiently low frequency is assumed to be unity regardless of its absolute magnitude, i.e.

$$\Delta\varepsilon_a + \Delta\varepsilon_b = 1. \tag{5.27}$$

Secondly, we define an angle θ between the dipole vector and the major axis of the molecule by the following equation:

$$\theta = \tan^{-1}\left(\frac{\Delta\varepsilon_a}{\Delta\varepsilon_b}\right)^{1/2} = \tan^{-1}\frac{\mu_a}{\mu_b}. \tag{5.28}$$

For example, if the dipole vector makes an angle of 45° with the molecular axes, then the relative magnitudes of $\Delta\varepsilon_a$ and $\Delta\varepsilon_b$ must be equal and are 0.5. The next step is to reduce τ_a and τ_b to some parameters which are common to both of them. In general, we can assume that dipolar orientation is similar to rotary diffusion. Under these circumstances, the use of Perrin's equation (1934) for rotary diffusion enables one to find the relationship between relaxation time and molecular geometry. Perrin's equation of rotary diffusion for prolate ellipsoids of revolution (an ellipsoid with three axes a, $b=c$) is given by

$$\frac{\tau_a}{\tau_0} = \frac{2(1 - q^4)}{[3q^2(2 - q^2)/(1 - q^2)^{1/2}]\{\ln[1 + (1 - q^2)^{1/2}]/q\} - 3q^2} \tag{5.29}$$

where q is the axial ratio b/a and τ_0 is the relaxation time of an equivalent sphere, i.e. a sphere which has the same volume as that of an ellipsoid. If we assume an axial ratio of 1:10, the RHS of (5.25) gives a numerical value of 8.73. This means that the relaxation time τ_a of an ellipsoid having an axial ratio of 1:10 is 8.73 times larger than that of an equivalent sphere.

Likewise, the relaxation time τ_b can be calculated using the following equation:

$$\frac{\tau_b}{\tau_0} = \frac{4(1 - q^4)}{[3q^2(1 - 2q^2)/(1 - q^2)^{1/2}]\{\ln[1 + (1 - q^2)^{1/2}]/q\} + 3}. \tag{5.30}$$

Using this equation and a value of b/a of 1:10, we find that the relaxation time τ_b is larger than that of an equivalent sphere by a factor of 1.276.

With these transformations of four parameters, (5.25) can be reduced to the following equation:

$$\varepsilon - \varepsilon_\infty = \frac{0.5}{1 + j8.7\omega\tau_0} + \frac{0.5}{1 + j1.2\omega\tau_0}. \tag{5.31}$$

The remaining task is simply to assume some value for τ_0 arbitrarily. For globular proteins, we can assume a relaxation frequency of 1 MHz and this can be converted to 1.6×10^{-7} s. Substituting this value into (5.31), we can readily calculate the value of $\varepsilon - \varepsilon_\infty$. Repeating the calculation at various frequencies between $\omega\tau_0 = 1$ and $\omega\tau_0 = 100$, we can construct a dispersion curve as shown by the full curve in figure 5.20. We can repeat the calculation for different axial ratios and dipole angles θ. Having done so, we can construct a family of theoretical curves as shown by figure 5.22. Note that only the axial ratio is varied in this figure. Figure 5.23 shows another set of theoretical curves in which the dipole angle θ is changed while the axial ratio is kept constant.

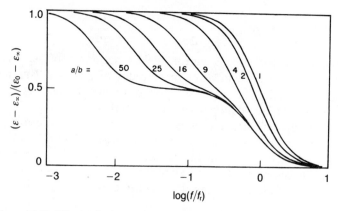

Figure 5.22 Theoretical curves constructed using a two relaxation time model for various axial ratios. The numbers for each curve indicate the axial ratio (a/b). The ordinate is the normalized dielectric increment. (From Oncley 1943.)

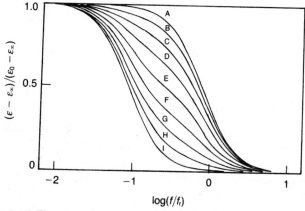

Figure 5.23 Theoretical curves constructed using a two relaxation time model for various dipolar angles. Curves A–I are for 0°, 5.7°, 14°, 26.5°, 45°, 63°, 76°, 84°, 90°. (From Oncley 1943.)

Fitting the measured dispersion curves into one of the calculated curves, we can determine the axial ratio of protein molecules and the angle between the dipolar and molecular axes. The discussion so far has been focused on prolate ellipsoids of revolution. However, the calculation can be extended to an oblate ellipsoid of revolution using another set of Perrin's equations:

$$\frac{\tau_a}{\tau_0} = \frac{2(1 - q^4)}{[3q^2(2 - q^2)/(q^2 - 1)^{1/2}]\tan^{-1}\sqrt{q^2 - 1} - 3q^2} \quad (5.32)$$

for axis a and

$$\frac{\tau_b}{\tau_0} = \frac{4(1 - q^4)}{[3q^2(2q^2 - 1)/(q^2 - 1)^{1/2}]\tan^{-1}\sqrt{q^2 - 1} + 3} \tag{5.33}$$

for axis b. The results of these calculations are summarized in table 5.4.

As already discussed, the method is straightforward. However, the actual procedure used to obtain definitive information with regard to molecular dimension using this technique is tedious and by no means simple. In particular, there is no simple way to differentiate prolate ellipsoids and oblate ellipsoids using dielectric data alone. This method is very effective if the axial ratio is far from unity. However, if axial ratios approach unity, calculated curves, as shown in figure 5.22, become closer together and curve fitting is extremely difficult. For example, if the axial ratio of a protein is 1:2, the theoretical curve is not distinguishable from that for a sphere. Since many globular proteins have a shape which is not markedly different from a spherical shape, this technique may have a limited usefulness. This method was, however, effectively used by Rosseneau-Motreff *et al* (1971, 1973) for several proteins.

5.10 Proton Fluctuation Theory

The theory discussed above is based on an assumption that the mechanism of dielectric relaxation of protein molecules is due to orientational polarization. Under these circumstances, the polarization process is similar to that of rotary diffusion and the solution of (5.26) is facilitated by the use of Perrin's equations. However, there are some experimental observations which may be considered as an indication that the polarization of protein molecules may not be due to orientation of the permanent dipoles alone. Based on these observations, alternate theories were proposed. One of these theories postulates that the polarization of protein is solely due to the orientation of bound water molecules (Jacobsen 1955). Another theory was proposed by Kirkwood and Shumaker (1952) and is commonly called proton fluctuation theory. Whereas Jacobsen's theory was dismissed on the grounds that the amount of bound water of protein molecules may not be enough to produce a dipole moment of several hundred D, the proton fluctuation theory could not be refuted readily and vigorous debates continued for several years. In what follows, an outline of the theory and subsequent controversies will be discussed.

Charged polyelectrolytes contain a large number of loosely bound protons. For example, in protein molecules, there are a number of neutral and negatively charged groups such as NH_2 and COO^- to which protons are bound. Except at very acidic pH, the number of basic sites exceeds the average number of bound protons. Therefore, many possible configurations of proton distribution in protein can exist with little free energy differences. If an electrical field is applied, the fluctuation of proton distribution will produce a non-vanishing mean square electric moment even if the mean dipole moment is zero.

The mean square dipole moment averaged over all mobile proton configurations can be given by the following equation:

$$\langle \mu^2 \rangle = \langle \mu \rangle^2 + \Delta\mu^2 \tag{5.34}$$

where $\langle \mu \rangle$ is the mean permanent dipole moment and $\Delta\mu$ is the dipole moment fluctuation associated with charges and mobile protons and is given by

$$\Delta\mu^2 = \langle (\mu - \langle \mu \rangle)^2 \rangle. \tag{5.35}$$

We assume that n numbers of basic groups with a charge e_i are situated at R_i from the centre of molecular mass. Furthermore, we define a proton occupation variable x_i for each basic group to be unity when it is occupied by a proton and zero otherwise. The electric moment μ of a molecule, its mean value $\langle \mu \rangle$ and the fluctuation $\Delta\mu^2$ are given by the following equations:

$$\mu = \sum_i (e_i + ex_i)R_i \tag{5.36}$$

$$\langle \mu \rangle = \sum_i (e_i + e\langle x_i \rangle)R_i \tag{5.37}$$

$$\Delta\mu^2 = e^2 \sum_{i,k} (\langle x_i x_k \rangle - \langle x_i \rangle \langle x_k \rangle)R_i R_k \tag{5.38}$$

where averages are taken over all distributions $x_1, \ldots, x_n$ of the protons among the basic sites of the molecules. The values of $\langle x_i \rangle$ and $\langle x_i x_k \rangle$ can be calculated using the theory of acid–base equilibrium of zwitterions (Kirkwood 1943). If electrostatic interaction between protons is neglected, we obtain

$$\langle x_i \rangle = \frac{1}{1 + K_i/[H^+]} \tag{5.39}$$

$$\langle x_i x_k \rangle - \langle x_i \rangle \langle x_k \rangle = 0 \qquad i \neq k \tag{5.40}$$

$$\langle x_i^2 \rangle - \langle x_i \rangle^2 = \frac{1}{2 + K_i/[H^+] + [H^+]/K_i} \qquad i = k. \tag{5.41}$$

Substituting these equations into (5.38), we obtain an expression for the dipole moment fluctuation for n_α equivalent basic groups with R_α^2 as the mean square distance from the mass centre of a molecule:

$$\Delta\mu^2 = e^2 \sum_\alpha \frac{n_\alpha R_\alpha^2}{2 + K_\alpha/[H] + [H]/K_\alpha}. \tag{5.42}$$

A further simplification is possible if we assume that all basic groups are uniformly distributed on the surface of the molecule, which is considered a prolate ellipsoid of revolution:

$$\Delta\mu^2 = e^2 f^2 b_0^2 \sum_\alpha \frac{n_\alpha}{2 + K_\alpha/[H^+] + [H^+]/K_\alpha} \tag{5.43}$$

where b_0 is the radius of an equivalent sphere and f is a shape factor. The fluctuation moments of representative proteins are calculated using (5.43) and the results are shown in table 5.5, along with experimental data. Inspection of this table reveals an important implication of this theory. Namely, except for β-lacto-globulin, the calculated proton fluctuation moment is comparable in magnitude to its observed value. In other words, the electric moment due to proton fluctuation is sufficiently large to account for the observed dipole moment of these proteins. Under these circumstances, the contribution of permanent electric moment can be assumed to be negligible. As emphasized before, Oncley's analyses are based on the assumption that the electric polarization of protein is due to the orientation of permanent dipoles. If proton fluctuation theory is proven correct, Oncley's theory would collapse completely. Thus, the proposal of proton fluctuation theory became

Table 5.5 Dipole moment fluctuation in proteins.†

Proteins	β-lacto-globulin	Egg albumin	Horse haemoglobin	Human serum albumin
MW	37 000	46 000	64 000	69 000
Dissociation groups per molecule				
Arginine (pK=12.5)	6	15	14	25
Histidine (pK=6.0)	4	7	36	16
Lysine (pK=10.5)	29	20	38	58
Tyrosine (pK=10.1)	8	9	11	18
Free COOH (pK=4.0)	59	51	53	93
pH	5.6	4.8	6.4	5.0
b^0 (Å)	24	26	30	30
a/b	5	4	5	5
μ(obs) (D)	700	260	480	380
$\Delta\mu$(cal) (D)				
(sphere)	170	320	390	430
(ellipsoid)	270	440	620	680

†Taken from Kirkwood and Shumaker (1952).

quite controversial and stirred vigorous debates that lasted for some years.

Figure 5.24 illustrates the pH profile of proton fluctuation moment for serum albumin. Clearly, ionizable groups contribute a large moment when the pH of the solution approaches the pK of these groups. The large peaks at pH 4 and near pH 10 are clearly due to carboxyl and amino groups. Also the small hump at around pH 6 is due to the imidazole group of histidine residues. This curve is obtained with an assumption that electrostatic interaction between protons is negligible. The effect of the interaction between protons was investigated by the same authors. In short, the effect of electrostatic interactions is not negligible and the pH profile of the calculated fluctuation moment is significantly different from the values obtained without inter-proton interactions. The crosses in figure 5.24 are the observed dipole moments of serum albumin (Takashima 1965). The discrepancy between these results and the theoretical curve is substantial. This result suggests either the importance of electrostatic interactions or that the theory by Kirkwood *et al* overestimates the contribution of proton fluctuation to the total dipole moment.

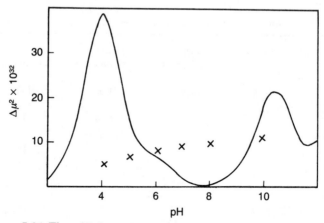

Figure 5.24 The pH dependence of the proton fluctuation moment calculated by Kirkwood and Shumaker (1952) assuming no electrostatic interaction between protons. The crosses are experimental results. (From Takashima 1965.)

As already discussed, Kirkwood *et al* have calculated the mean square fluctuation moment due to mobile protons; however, it must be noted that their calculation is performed only for static fields. It remains to be seen under what conditions the fluctuation moment manifests itself and can be distinguished from the relaxation of permanent dipoles. In order to analyze the moment due to proton fluctuation, we have to consider at least two relaxation times, i.e. the orientation time τ_i and the fluctuation time τ_δ. The equation derived by Scheider (1965) elegantly expresses the dielectric increment of protein solution as shown below:

$$\varepsilon(\omega) = \frac{4\pi N}{3kT}\sum_{i=1}^{3}\left[\bar{\mu}_i^2\left(\frac{1}{1 + j\omega\tau_i}\right) + \left(\frac{\overline{\delta\mu_i^2}}{1 + \tau_i/\tau_\delta}\right)\left(\frac{1}{1 + j\omega\tau_i'}\right)\right]. \quad (5.44)$$

The first term on the RHS is due to the classical relaxation of the permanent dipole and the second term is due to the relaxation of the mean square fluctuation moment. Here the time constant τ_i' is defined by the following equation:

$$\frac{1}{\tau_i'} = \frac{1}{\tau_\delta} + \frac{1}{\tau_i}. \quad (5.45)$$

First let us analyze the equation for extreme cases. If $\tau_\delta \ll \tau_i$, i.e. if the fluctuation time is much smaller than the orientation time, then the time constant τ_i' can be represented by τ_δ alone. This is the case where the rate of proton fluctuation is much faster than the rate of orientation and the relaxation due to proton fluctuation can be found at very high frequencies. The magnitude of the dielectric increment due to the proton fluctuation moment is, in this case, very small because the increment is given by the following approximate expression:

$$\frac{\overline{\delta\mu_i^2}}{\tau_i/\tau_\delta} \quad (5.46)$$

where the denominator τ_i/τ_δ is a large number. Under this condition, we conclude that the increment due to the fluctuation moment $\overline{\delta\mu_i^2}$ will not be detected at any frequencies.

If $\tau_\delta \gg \tau_i$, the rate of proton fluctuation is much slower than that of orientation. Obviously, the time constant τ_i' reduces to τ_i and (5.42) simplifies to

$$\varepsilon(\omega) = \frac{4\pi N}{3kT}\sum_{i=1}^{3}(\bar{\mu}_i^2 + \overline{\delta\mu_i^2})\left(\frac{1}{1 + j\omega\tau_i}\right). \quad (5.47)$$

In this case, the relaxation due to mobile protons cannot be differentiated from that of dipole orientation even though the magnitude of $\overline{\delta\mu_i^2}$ may be very large. Only if two relaxation times, τ_i and τ_δ, are not far apart, or even equal, can the two relaxations be found as independent processes. In the following, evidence that the fluctuation time is much larger than the orientation time is presented.

The groups on the surface of proteins having pK values near the middle pH range are free carboxyl groups and imidazole groups of histidine. The dissociation of protons from these groups can be shown by the following schemes:

$$CH_3COO^- + H^+ \underset{k_{21}}{\overset{k_{12}}{\rightleftharpoons}} CH_3COOH$$

$$Im + H^+ \underset{k_{21}}{\overset{k_{12}}{\rightleftharpoons}} Im\,H^+$$

where k_{12} and k_{21} are rate constants for the dissociation and association of protons respectively. It can be shown that the fluctuation time τ_δ and the rate constants k_{12} and k_{21} are related by the following equation:

$$\tau_\delta = \frac{1}{k_{12} + [H^+]k_{21}}. \tag{5.48}$$

The rate constants k_{12} and k_{21} for the imidazole group, for example, were determined by Eigen *et al* (1960):

$$k_{12} = 2.3 \times 10^{10}\ M^{-1}\ s^{-1} \qquad \text{and} \qquad k_{21} = 2.3 \times 10^3\ s^{-1}.$$

Substituting these values into (5.48) we obtain the following values for fluctuation time:

$$\tau_\delta(COO^-) = 0.6 \times 10^{-5}\ s$$

$$\tau_\delta(Im) = 1.2 \times 10^{-5}\ s.$$

Since the orientation time of globular proteins is of the order of 10^{-7} s, one can readily observe that the fluctuation time is much longer than the orientation time. We can, therefore, conclude that (5.47) is a valid expression for most globular proteins.

The above discussion demonstrates the inseparability of the two relaxation processes for most globular proteins where the proton fluctuation rate is much slower than the orientation relaxation rate. The conclusion drawn from Scheider's analysis entails the inability to separate the two relaxation processes by experimentation under normal conditions. However, there are means to change the orientational relaxation time of protein molecules so that the relation $\tau_\delta \gg \tau_i$ would no longer be valid. One of them is to increase the orientation time constant by increasing the viscosity of the solvent. As shown by Debye, the orientation time of spherical molecules is proportional to the viscosity of the solvent (equation (5.24)). Figure 5.25 shows the frequency profile of the dielectric constant of egg albumin dissolved in solvents having widely different viscosities (Takashima 1962). Let us take a look at curves A and D which were obtained with solvent viscosities of 1 and 6.5 centipoises. Curve A gives a relaxation time of 2.1×10^{-7} s. This is indeed much smaller than the proton fluctuation time, i.e. 10^{-5} s. However, as the viscosity of the solvent increases, a down shift of the dispersion curve to lower frequencies is detected. For example, the

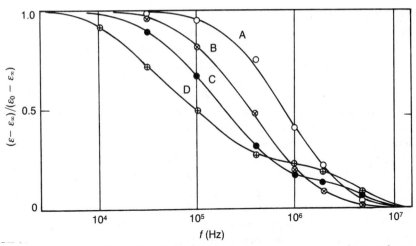

Figure 5.25 The dielectric dispersions of egg albumin at various solvent viscosities. Curves A–D are for 1 cps, 1.7 cps, 3.7 cps and 6.5 cps. (From Takashima 1962.) Reproduced by permission of John Wiley & Sons, Inc. © 1962.

centre frequency of curve D is located at 75 kHz if one ignores the small hump at high frequencies. This gives rise to a relaxation time of 2×10^{-6} s, which is not very different from the calculated proton fluctuation time. Clearly, the change in the orientation characteristics of protein molecules by a viscosity increase produced a separation of the dispersion curve into two components. The one at low frequencies is clearly a viscosity sensitive dipole orientation and the other is a viscosity insensitive component due to some unknown relaxation process. Although this result proves that the relaxation curve of egg albumin consists of two components, there is yet no proof that the second component is indeed due to proton fluctuation. The relaxation time we obtain from the second dispersion is 4.8×10^{-8} s. This is a very small value compared with the calculated proton fluctuation time. An interesting observation is that small proteins such as myoglobin show only viscosity sensitive or orientational relaxation, indicating that the ionic process which produces the viscosity insensitive relaxation is either negligibly small or entirely lacking. There is some indication, however, that the relaxation due to an ionic process increases markedly as the size of the protein molecule increases.

5.11 A Theoretical Calculation of the Dipole Moment of Proteins

Calculation of the dipole moment of complex molecules such as proteins cannot be performed without detailed information on the coordinates of charged sites. Fortunately, owing to x-ray crystallographic studies, the 3D structure of small protein molecules is now well known. Because of this, it became possible to compute the dipole moments of small proteins such as myoglobin and haemoglobin. The calculation of the dipole moment of myoglobin was first carried out by Schlecht (1969). First of all, the number of charged sites and their signs must be known at an isoelectric point at which the measurement of dipole moment is usually carried out. Table 5.6 shows the ionizable amino acid residues, their numbers, and their signs for myoglobin. As shown, there are 23 pairs of positive and negative charges on the surface. Actually, there are altogether 12 histidines in myoglobin but only six of them are titratable. Only those which are exposed to the surface are needed for the calculation. Secondly, the coordinates of all of these amino acid residues must be known. The dipole moment value of multiple charge pairs can be calculated using the following equation:

$$\mu = nRe \qquad (5.49)$$

where n is the number of charge pairs and R is the distance between the centres of positive and negative charges. The separation between charge centres was found to be as small as 1.55 Å for myoglobin. Since n is already known, we can readily calculate the dipole moment and a value of 170 D is obtained. The dipole moment determined experimentally is 165–170 D and the agreement between the calculated and observed values is excellent.

Table 5.6 The number, pK and sign of charge of the ionizable groups in sperm whale myoglobin.†

Group	Number	pK	Sign
Glutamic acid	14	4.4	−
Aspartic acid	6	4.4	−
Carboxyl group	1	4.4	−
Propionic acid	2	4.4	−
Hemic acid	1	8.9	−
Lysine	19	10.6	+
Arginine	4	7.8	+
α-amino end	1	7.8	+
Histidine	6	6.6	+

†Taken from Schlecht (1969).

The agreement between theoretical and experimental values indicates that the dipole moment of myoglobin is dominated by the moment due to fixed surface charges. The dipole moment due to bond moments and

the one due to proton fluctuation can, therefore, be considered negligibly small. However, Schlecht's calculation was carried out only at
isoelectric pH where the numbers of positive and negative charges are
equal. The calculation of dipole moment including the components due
to core moments (the sum of bond moments) and proton fluctuation was
carried out by South and Grant (1972) at pH values other than that at
the isoelectric point.

The first part of their calculation includes the moment due to surface
charges and bond moments. The equation used for this calculation is
given below:

$$\mu = e\sum\langle x_i\rangle r_i + \int_{v'} \rho(r)r\,dv \qquad (5.50)$$

where the first term on the RHS is the electric moment due to surface
charges and the second term is due to the partial charge distribution.
$\rho(r)$ is the average volume charge density at any point r in the
molecule. x_i has a value of ± 1, 0 or -1, depending on the charge on
site i. The probability $p_i(\mathrm{pH})$ of a given group, i, having a bound
proton depends on the pH of the solvent and the pK_i of the group as
shown by

$$p_i(\mathrm{pH}) = \frac{1}{1 + 10^{\mathrm{pH}-\mathrm{pK}_i}}. \qquad (5.51)$$

The average charge on group i is, therefore

$$e\langle x_i\rangle = e[p_i(\mathrm{pH}) - \delta_i] \qquad (5.52)$$

where δ_i is 0 for basic groups and 1 for acidic groups. Thus, the first
term of (5.50) becomes

$$e\sum_{i=1}^{n}\langle x_i\rangle r_i = e\sum_{i=1}^{n}[p_i(\mathrm{pH}) - \delta_i]r_i. \qquad (5.53)$$

The origin of r_i is taken as the centre of molecular mass which was
computed using the coordinates of all atoms in the molecule. The result
of this calculation is shown in figure 5.26. Secondly, the net dipole
moment due to partial charge (core moments) was computed using a
value of 3.5 D for each amino acid residue. The net dipole moment due
to core moments was found to oppose the one due to surface charges.
Curve B in figure 5.26 shows the sum of these two components. It
should be noted that the contribution of the core dipole moment is
small, if not negligible, compared with the one due to surface charges.

The third component is the mean square fluctuation of protons as
discussed by Kirkwood *et al* (see equation (5.42)). In the original
calculation, Kirkwood *et al* assumed a uniform distribution of ionizable
groups on the surface of ellipsoids. However, the coordinates of
ionizable groups are now accurately known and the mean square

moment can be estimated with higher accuracy. The result of the calculation of $\Delta\mu^2$ for myoglobin is also shown in figure 5.26. The result of these calculations demonstrates that (i) inclusion of the core moments causes a wider discrepancy between the measured and calculated values; (ii) without core moments, the sum of the surface charge and fluctuation moments agrees well with the measured values at a few chosen pH values—for whale myoglobin, the agreement is particularly good at a pH between 7 and 8; (iii) the contribution of fluctuation moment increases below pH 6 but the discrepancy between theoretical and measured values increases. As has been discussed, attempts to calculate the dipole moment of myoglobin have limited success only at and near isoelectric pH. While Schlecht's calculation indicates that a permanent dipole moment due to fixed charges is the dominant component, the calculation by South *et al* demonstrates that the mean square fluctuation moment $\Delta\mu^2$, as well as core moments due to partial charges, are by no means negligible, particularly in the low pH region.

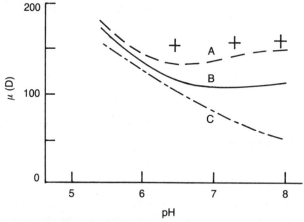

Figure 5.26 The calculated dipole moment of myoglobin at various pH. The crosses are measured moments. Curves: A, the moment due to fixed surface charge and fluctuation dipole; B, the sum of the moments due to surface charges, core moment, and fluctuation dipole; C, the mean square moment due to proton fluctuation. (From South and Grant 1972.) Reproduced by permission of The Royal Society.

The calculation of dipole moment discussed in the foregoing is limited to myoglobin which is one of the smallest proteins. The internal structure of myoglobin is well documented and the coordinates of individual amino acid residues are known in detail. As the size of the protein molecule increases, the calculation of dipole moment becomes more difficult and time consuming. However, in so far as the coordinates of amino acid residues are known accurately, the computation can be performed even with protein molecules which are larger than myoglobin. As a matter of fact, a calculation of the dipole moment of haemoglobin had been carried out by Orttung (1969a, b, 1970), earlier than the works of Schlecht (1969) and South and Grant (1972). The outline of Orttung's work will be given in the following.

Using Onsager's formalism (see equation (2.53)) the dielectric constant of polar molecules is given by

$$\varepsilon - 1 = \frac{4\pi N}{3}\left(\frac{3\varepsilon}{2\varepsilon + 1}\right)(\alpha_{\alpha\alpha}^0 + \langle\mu_\alpha\mu_\alpha\rangle_0/kT) \tag{5.54}$$

where ε is the dielectric constant of solution and $\alpha_{\alpha\alpha}$ and μ_α are the polarizability tensor and dipole moment of the molecule respectively. N is the number of molecules. Onsager's theory as applied to globular proteins is

$$\mu_\alpha = \frac{2\varepsilon + 1}{2\varepsilon + \varepsilon_p}\left(\frac{\varepsilon_p + 2}{3}\mu_{v\alpha} + \mu_{e\alpha}\right) \tag{5.55}$$

where ε_p is the effective dielectric constant of the protein and $\mu_{v\alpha}$ is the vacuum dipole moment of the molecule without proton binding sites. The second term in brackets, $\mu_{e\alpha}$, is the embedded dipole moment arising from the proton binding sites. $\mu_{e\alpha}$ is defined as follows:

$$\mu_{e\alpha} = \sum_{j=1}^{n}(e_j + ex_j)R_{j\alpha} \tag{5.56}$$

where e_j is the charge at the jth site and x_j is the proton occupation variable, having values of either 0 or 1. R_j is the position vector of the jth site relative to the centre of mass.

The mean square moment $\langle\mu_\alpha\mu_\alpha\rangle_0$ is defined as the sum of the square mean moment and the fluctuation component:

$$\langle\mu_\alpha\mu_\alpha\rangle_0 = \langle\mu_\alpha\rangle_0\langle\mu_\alpha\rangle_0 + \Delta(\mu_\alpha\mu_\alpha). \tag{5.57}$$

If the fluctuation of $\mu_{v\alpha}$ is negligible, then we have

$$\Delta(\mu_\alpha\mu_\beta) = \left(\frac{2\varepsilon + 1}{2\varepsilon + \varepsilon_p}\right)^2 \Delta(\mu_{e\alpha}\mu_{e\beta}). \tag{5.58}$$

This equation indicates that the fluctuation of dipole moment arises solely from the fluctuation of proton occupation variables. The theoretical calculation requires, therefore, the calculation of $\langle\mu_{v\alpha}\rangle$, $\langle\mu_{e\alpha}\rangle$ and $\Delta(\mu_{e\alpha}\mu_{e\alpha})$. The method of calculation is discussed in an earlier paper by Orttung in detail (Orttung 1968), so only the results will be summarized below.

The centre of mass and the intrinsic dipole moment were calculated using computer generated coordinates. The centre of mass was found at

$z = -1.52$ Å and the intrinsic dipole moment, which was calculated with an assumption that only the planar peptide groups make a net contribution, was found to be 87 D and 80 D for the α- and β-units of haemoglobin (haemoglobin consists of two α-units and two β-units). The vector components are

$$(\mu_x, \mu_y, \mu_z) = (-62, 59, -12)$$

for the α-chain and

$$(\mu_x, \mu_y, \mu_z) = (77, 23, -2)$$

for the β-chain. The net moment of the tetramer is directed along the z axis with $\mu_z = -28$ D. Interestingly, deoxygenation changes the centre of mass to $z = 1.51$ Å and the net dipole moment to $\mu_z = 115$ D.

The calculated mean dipole moment $\langle \mu_{e\alpha} \rangle$ of oxy-haemoglobin is shown in figure 5.27 at various pH. Curve A was calculated using the independent site approximation, i.e. the interaction between sites is considered negligible. Curve B was computed with the group average approximation, i.e. the difference between individual sites of the same type is ignored. Curve C was obtained by computing the occupation

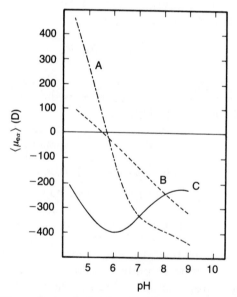

Figure 5.27 The calculated dipole moment of horse haemoglobin. Curves: A, the independent site approximation; B, the group average approximation; C, the individual site average approximation. Reproduced with permission from Orttung (1969b). Copyright (1969) American Chemical Society.

variable individually, considering interaction with other sites of the same and different type. It should be noted that the outcome of the calculation of $\langle \mu_{e\alpha} \rangle$ critically depends upon the method of computing the site occupation variables. The large difference emphasizes the importance of correlation and variations among the sites of a given type.

Figure 5.28 illustrates the root-mean-square fluctuation dipole moment at various pH. As shown, the group average approximation gives rise to a result which is similar to that obtained by individual site average calculations. It should also be noted that curve A, which was calculated using the independent site approximation, resembles the curve obtained by Kirkwood *et al* (see figure 5.24) which was calculated assuming no electrostatic interaction between protons.

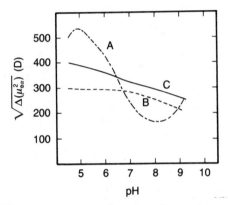

Figure 5.28 The root-mean-square proton fluctuation moment calculated using the same method as for figure 5.27. Reproduced with permission from Orttung (1969b). Copyright (1969) American Chemical Society.

Although the haemoglobin molecule is four times larger than myoglobin and its structure more complex, the 3D configuration of this protein is known. Using available data for the site coordinates of charged groups, Orttung demonstrated that the mean square dipole moment and proton fluctuation moment can still be calculated despite its complex structure.

5.12 More Discussion on the Charge Distribution in Proteins

The calculation by Schlecht demonstrated that the separation between positive and negative charge centres is only 1.55 Å in myoglobin. This finding clearly shows that the distribution of surface charges in proteins is remarkably uniform and this symmetry results in a small net dipole moment. The calculation of dipole moment has been limited to myoglobin and haemoglobin. It is, therefore, not possible to generalize this finding and draw a definitive conclusion with regard to the charge distribution in proteins.

As mentioned briefly before, the distribution of charges on the surface of protein molecules has been analyzed in detail by Barlow and Thornton for 32 proteins, the 3D structures of which are known (1987). Barlow and Thornton calculated three parameters for these proteins: (i) the total charge density; (ii) the charge polarity and (iii) protein dipole moment using the model shown in figure 5.29.

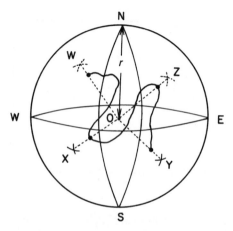

Figure 5.29 The positions of charged groups, labelled w–z are projected to the surface of a sphere of radius r. r is calculated as the mean distance between all charged groups and the centroid of the protein (0). The protein is illustrated as a thread. (From Barlow and Thornton 1987.) Reproduced by permission of John Wiley & Sons, Inc. © 1987.

The total charge density
The total charge density D_t is defined as the number of charges (disregarding the sign) divided by the total accessible contact area in Å^2. The range of D_t was found to be 1–2 charges per 100 Å^2 and the mean value turned out to be 1.42 ± 0.31 . This indicates that the total charge density does not vary significantly from one protein to another. In addition, the local charge density D_l was analyzed. D_l is defined as the number of charges divided by the area of a circle with a radius of 8 Å centred at each charged site. Although D_l was calculated for small

circles with a radius of 8 Å, the value obtained was normalized to an area of 100 Å². The value of the local charge density varied widely between 0.5 and 25 charges per 100 Å² with a mean value of 2.08 per 100 Å². This clearly indicates that charges often form clusters although they appear to avoid pair formation between charges with equal sign.

Charge polarity
To determine the extent of asymmetry of the charge distribution, a vector is used which is defined as

$$R = \left| \sum_{i}^{N} r_i \right| \qquad (5.59)$$

where r_i is a unit vector drawn in the direction of the ith charge from the centroid of the protein, as shown by figure 5.29. As a reference, random distributions of the charged groups are generated using a computer and the predicted values of charge polarity are computed. Namely, 1000 sets of randomly selected atomic sites are generated and the value of R is calculated. The value of R varies for each random distribution generated by the computer. Some of the randomly selected charges are statistically symmetric and some others can be asymmetric. Therefore, the R-values of computer generated charge distribution fluctuate statistically depending on the symmetry or asymmetry of the randomly selected distribution. The R-value for real charges, $R[\text{obs}]$, is compared with the R-value for random distributions, $R[\text{cal}]$. The percentage of $R[\text{cal}]$ larger than $R[\text{obs}]$ is recorded. Table 5.7 shows the results of this calculation for some selected proteins. For example, ferredoxin has only 5% of $R[\text{cal}]$ exceeding $R[\text{obs}]$. This result can be interpreted as an indication that the charge distribution in ferredoxin is very asymmetric. On the other hand, $R[\text{obs}]$ of egg white lysozyme is smaller than $R[\text{cal}]$ for almost all of the time, indicating that the R-vector for lysozyme is very small. This indicates that the charge distribution in lysozyme is extremely symmetric. Table 5.7 lists protein molecules with descending asymmetry (the percentage of $R[\text{cal}] > R[\text{obs}]$). The value of $R[\text{obs}]$, however, does not necessarily follow the same order as that of the second column because of the difference in the size of these proteins.

Protein dipole moment
The dipole moment of protein is defined by the following equation:

$$\mu = \sum_{i}^{N} q_i r_i \qquad (5.60)$$

where q is the charge and r is the distance. To determine the extent of polarization of the positive and negative charges, electric dipole moments are calculated using this equation. Three dipole moments were

Table 5.7 Charge polarity and asymmetry in the charge distribution in proteins.†

Protein	R[obs]	R[cal] > R[obs] (%)
Ferredoxin	5.04	5
Actinidin	8.33	10
Concanavalin A	8.44	17
Ribonuclease A	5.97	22
Trypsin	5.96	23
Liver alcohol dehydrogenase	11.34	26
High potential iron protein	4.54	30
Pancreatic trypsin inhibitor	3.59	42
Thermolysin	7.29	44
Carboxypeptidase	6.53	50
L-arabinose binding protein	7.51	50
Lactase dehydrogenase	7.43	66
Phospholipase A2	3.80	66
Phage T4 lysozyme	4.80	69
Subtilisin	3.33	76
Haemoglobin α-chain	2.68	85
Bence-Jones protein	1.94	87
Carbonic anhydrase C	3.43	92
Flavodoxin	2.41	94
Lysozyme	1.45	98

†Taken from Barlow and Thornton (1987). Reproduced by permission of John Wiley & Sons, Inc. © 1987.

calculated: (a) the moment due to formal charges of Asp, Glu and C-termini (e^-) and N-termini (e^+); (b) the moment due to partial charges of all atoms; (c) the moment due to partial charges excluding C- and N-terminal charges. Analogous to charge polarity, the expected moments due to random charge distribution are calculated and compared with the moment due to real charges. Table 5.8 shows the calculated real moments of various proteins and the percentage of μ[cal] exceeding μ[obs]. As before, ferredoxin ranks the first in the asymmetry list, namely, the dipole moment of this protein is disproportionately large for its size. The rank order of dipole moment is not identical to the similar list for charge polarity.

The values of dipole moments for proteins range from 100 to 700 D. However, these large values are due to the size of the proteins and are not due to the unique arrangements of charged sites in these molecules.

Table 5.8 The calculated dipole moment of various proteins.†

Protein	Dipole moment (D)	μ[cal] > μ[obs] (%)
Ferredoxin (FDX)	238	12
Actinidin (ACT)	675	23
Ribonuclease A (RNS)	481	24
Pancreatic trypsin (PTI) inhibitor	269	29
Flavodoxin (FXN)	523	35
Subtilisin (SBT)	508	45
Thermolysin (TLN)	711	47
Liver alcohol (ADH) dehydrogenase	823	58
Trypsin (PTN)	356	61
Phage T4 lysozyme (LZM)	407	68
Carbonic anhydrase C (CAC)	525	71
Carboxypeptidase (CPA)	492	73
Bence-Jones protein (REI)	177	74
Haemoglobin α-chain (MHB)	235	80
High potential iron (HIP) protein	144	83
Lactase dehydrogenase (LDH)	420	91
Concanavalin A (CNA)	187	97
Lysozyme (LYZ)	111	97
L-arabinose binding (ABP) protein	239	98
Phospholipase A2 (BP2)	103	98

†Taken from Barlow and Thornton (1987). Reproduced by permission of John Wiley & Sons, Inc. © 1987.

It is a simple task to demonstrate that these moments can be generated by placing one or two charge pairs diametrically across protein molecules. Figure 5.30 shows the calculated dipole moments of proteins plotted against molecular diameters. In addition, the dipole moments which are generated by hypothetical charge pairs located diametrically are shown. Line A is due to one charge pair, line B due to two charge pairs and line C is due to three charge pairs. As shown, all of the dipole moments for protein fit within the area bounded by lines A–C. There are a few exceptions such as lysozyme whose dipole moments are even below the hypothetical moment due to one charge pair. This means that the centres of positive and negative charges in this protein are almost identical.

In spite of Barlow and Thornton's detailed calculation of the dipole moments of a variety of proteins due to formal and partial charges, no attempt was made by them to compute the mean square moment due to proton fluctuation. Since this component can not be completely ignored, particularly for large proteins, the addition of this information will enhance the significance of their calculations.

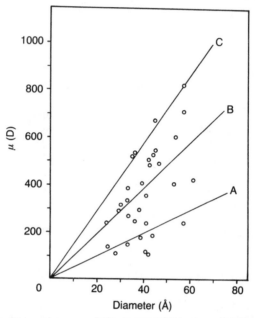

Figure 5.30 The calculated dipole moments of proteins plotted against their mean molecular diameters. Lines on the plot indicate the variation of the dipole moment for proteins of increasing size. Line A is calculated with one charge pair, line B with two charge pairs and line C with three charge pairs. (From Barlow and Thornton 1987.) Reproduced by permission of John Wiley & Sons, Inc. © 1987.

5.13 Charge Pair Formation and Protein Structure

The charge distribution in protein molecules has been analyzed using charge polarity and dipole moment as measures of charge symmetry or dissymmetry. The result of this analysis clearly indicates that charged sites are distributed, except for a few cases, nearly uniformly over the entire surface. Because of this, the centres of positive and negative charges are almost identical, resulting in a relatively small net dipole moment. This result raises a question as to whether or not there are certain rules for the distribution of charged side chains. The analyses by Barlow and Thornton indicate that charges with different signs tend to form clusters. This is a clear indication that electrostatic attraction between unlike charges plays a significant role in the folding of peptide chains in protein molecules.

Attempts to find certain rules in charge distribution in proteins have been made by at least two groups. Of these, Wada and Nakamura (1981) are perhaps the first ones to carry out a systematic investigation of the distribution of surface charges in a variety of proteins. These authors used a parameter which is defined by the following formula:

$$\alpha_{ij} = e_i e_j / r_{ij} \tag{5.61}$$

where e_i and e_j are the charges of the ith and jth residues and r_{ij} is the distance between them. The calculation of this parameter was carried out for attractive charge pairs (ion pairs with opposite charge signs) and repulsive pairs (ion pairs with the same charge signs). The sum of attractive $(\Sigma_{+-}\alpha_{ij})$ and repulsive $((\Sigma_{++}\alpha_{ij})$ and $(\Sigma_{--}\alpha_{ij}))$ pairs is plotted against the distance r_{ij}. Figure 5.31 shows the results of this calculation.

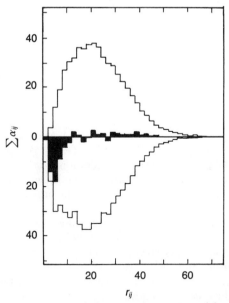

Figure 5.31 The number of charge pairs separated by a distance of r_{ij}. The upper histogram shows the distribution of repulsive charge pairs and the lower histogram the attractive charge pairs. The shaded histogram shows the difference between them. (From Wada and Nakamura 1981.) Reprinted by permission from *Nature*, vol 293, p757. Copyright © 1981 Macmillan Journals Limited.

The repulsive and attractive interactions are plotted individually as histograms in a positive as well as in a negative direction. The shaded histogram indicates the difference between attractive and repulsive pairs. The difference histogram clearly indicates that an attractive charge interaction dominates at distances below 10 Å. Above 10 Å, however, a repulsive interaction seems to dominate, indicating that charge pairs are surrounded by charges of the same sign as that of the central ion. Another analysis by Barlow and Thornton (1983) using 34 proteins led to a conclusion which is similar to that reached by Wada *et al*.

Suggested Reading

Fasman G D (ed.) 1967 *Poly-α-amino Acids. Protein Models for Conformational Studies* (New York: Dekker)

Oncley J L 1943 in *Proteins, Amino Acids and Peptides* ed. E J Cohn and J T Edsall (New York: Reinhold) p543

Stahman M A (ed.) 1962 *Polyamino Acids, Polypeptides and Proteins* (Madison, Wis: University of Wisconsin Press)

Takashima S and Minakata A 1973 Dielectric behaviour of biological macromolecules *Dig. Lit. Dielectr.* **37** 602–53

Wyman J and Edsall J T (ed.) 1958 *Biophysical Chemistry* vol I (New York: Academic) ch 6

References

Antony A A, Fong F K and Smyth C P 1964 Microwave absorption and molecular structure in liquids. LVIII. *J. Phys. Chem.* **68** 2035

Applequist J and Mahr T G 1966 The conformation of poly-L-tyrosine in quinolin from dielectric dispersion studies *J. Am. Chem. Soc.* **88** 5419–29

Barlow D J and Thornton J M 1983 Ion pairs in proteins *J. Mol. Biol.* **168** 867–85

—— 1987 Charge distribution in proteins *Biopolymers* **25** 1717–33

Doty P, Bradbury J H and Holtzer A M 1956 Polypeptides. IV. The molecular weight, configuration and association of poly-γ-L-benzyl glutamate in various solvents *J. Am. Chem. Soc.* **78** 947–54

Doty P, Wada A and Yang J T 1957 Polypeptides. VIII. Molecular configurations of poly-L-glutamic acid in water-dioxane solution *J. Polym. Sci.* **23** 851–61

Edsall J T 1943 in *Proteins, Amino Acids and Peptides* ed. E J Cohn and J T Edsall (New York: Reinhold) p140

Edsall J T and Wyman J 1958 in *Biophysical Chemistry* vol I (New York: Academic) p323

Eigen M, Hammes G C and Kustifin K 1960 Fast reactions of imidazole studies with relaxation spectrometry *J. Am. Chem. Soc.* **82** 3482–3

Fasman G D, Bodenheimer E and Lindblow C 1964 Optical rotatory dispersion of poly-L-tyrosine and copolymers of L-glutamic acid and L-tyrosine *Biochemistry* **3** 1665

Gerber B R, Routledge L M and Takashima S 1972 Self-assembly of bacterial flagellar protein: dielectric behavior of monomers and polymers *J. Mol. Biol.* **71** 317–37

Gupta A K, Dufour C and Marchal E 1974 Structure of aggregates and helix-coil transition of polypeptides by dielectric measurements. I. Poly-γ-benzyl-L-glutamate in dioxane and dioxane-dichloroacetic acid *Biopolymers* **13** 1293–308

Jacobson B 1955 On the interpretation of dielectric constants of aqueous macromolecular solution. Hydration of macromolecules *J. Am. Chem. Soc.* **77** 2919–26

Kirkwood J G 1943 in *Proteins, Amino Acids and Peptides* ed. E J Cohn and J T Edsall (New York: Reinhold) pp294–7

Kirkwood J G and Shumaker J B The influence of dipole moment fluctuations on the dielectric increment of proteins in solution *Proc. Natl. Acad. Sci. USA* **38** 855–62

Kurland R J and Wilson E B 1957 Microwave spectrum, structure, dipole moment and quadrupole coupling constants of formamide *J. Chem. Phys.* **27** 585–90

Lumry R and Yue R H S 1965 Dielectric dispersion of protein solutions containing small zwitterions *J. Phys. Chem.* **69** 1162–74

Marchal E 1974 Structure of aggregates and helix-coil transition of polypeptides by dielectric measurements. II. Poly-D,L-phenylalanine and poly-γ-benzyl-L-aspartate in chloroform-dichloroacetic acid mixtures *Biopolymers* **13** 1309–16

Marchal E and Marchal J 1967 Etude du comportment hydrodynamique du poly-γ-glutamate de benzyl et de la poly-DL-phenylalanine en helice par absorption dielectrique en solution diluée *J. Chim. Phys.* **64** 1607–15

Miyazawa T 1962 in *Polyamino Acids, Polypeptides and Proteins* ed. M A Stahman (Madison, Wis: University of Wisconsin Press) p201

Muller G, Van der Touw F, Zwolle S and Mandel M 1974 Dielectric properties of poly-L-glutamic acid in salt free aqueous solution *Biophys. Chem.* **2** 242–54

Nakamura H and Wada A 1981 Dielectric properties of protein solutions of poly-L-glutamic acids *Biopolymers* **20** 2567–82

Oncley J L 1938 Studies of the dielectric properties of protein solution I. carboxyhaemoglobin *J. Am. Chem. Soc.* **60** 1115–23

Orttung W H 1968 Anisotropy of proton fluctuations and the Kerr effect of protein solution. Theoretical considerations *J. Phys. Chem.* **72** 4058–67

—— 1969a Calculation of the mean-square dipole moment and proton fluctuation anisotropy of hemoglobin at low ionic strength *J. Phys. Chem.* **73** 418–23

—— 1969b The Kerr effect optical dispersion of hemoglobin. A molecular interpretation *J. Phys. Chem.* **73** 2908–15

Pauling L and Corey R B 1951 Configurations of polypeptide chains with favored orientations around single bonds: two new pleated sheets *Proc. Natl. Acad. Sci. USA* **37** 205–85, 729–40

Perrin F 1934 Mouvement Brownien d'un ellipsoide. Dispersion dielectrique pour des molecules ellipsoidales *J. Phys. Radium* **5** 497–511

Rosseneau-Motreff M Y, Soetewey F, Lamote R and Peeters H 1971 Size and shape determination of apotransferrin and transferrin monomers *Biopolymers* **10** 1039–48

—— 1973 Dielectric study of the urea denaturation of delipidated and relipidated bovine serum albumin *Biopolymers* **12** 1259–67

Scheider W 1965 Dielectric relaxation of molecules with fluctuating dipole moment *Biophys. J.* **5** 617–28

Schlecht P 1969 Dielectric properties of hemoglobin and myoglobin. II. Dipole moment of Sperm whale myoglobin *Biopolymers* **8** 757–65

South G P and Grant E H 1972 Dielectric dispersion and dipole moment of myoglobin in water *Proc. R. Soc.* A **328** 371–87

Takashima S 1962 Dielectric dispersion of protein solutions in viscous solvents *J. Polym. Sci.* **56** 257–65

—— 1963 Dielectric dispersion of poly-glutamic acid solution *Biopolymers* **1** 171–82

—— 1964 Dielectric dispersion of albumins. Studies of denaturation by dielectric measurement *Biochem. Biophys. Acta* **79** 531–8

—— 1965 A study of proton fluctuation in protein. Experimental study of the Kirkwood-Shumaker theory *J. Phys. Chem.* **69** 2281–6

—— 1986 unpublished data

Tinoco I Jr 1957 Dynamic electrical birefringence studies of poly-γ-benzyl-L-glutamate *J. Am. Chem. Soc.* **79** 4336–8

Wada A 1958 Dielectric properties of polypeptide solutions. I. The electric dipole moment of α-helix in dioxane *J. Chem. Phys.* **29** 674–5

—— 1959 Dielectric properties of polypeptide solutions. II. Relation between the electric dipole moment and the molecular weight of α-helix *J. Chem. Phys.* **30** 328–30

—— 1962 in *Polyamino Acids, Polypeptides and Proteins* ed. M A Stahman (Madison, Wis: University of Wisconsin Press) p131

—— 1967 in *Poly-α-amino Acids. Protein Models for Conformational Studies* ed. G D Fasman (New York: Dekker) pp369–90

Wada A and Nakamura H 1981 Nature of the charge distribution in proteins *Nature* **293** 757–8

Wada A, Tanaka T and Kihara H 1972 Dielectric dispersion of the α-helix at the transition region to random coil *Biopolymers* **11** 587–605

6   The Theory of Interfacial Polarization

6.1 Introduction

The theories for polar molecules and their application to synthetic polymers and proteins have been discussed in some detail in the previous chapters. The electric polarization of these polymers is known to be due to the orientation of permanent dipoles. However, some biological polymers such as nucleic acids do not have a permanent dipole moment and their polarization is believed to be due to some other mechanisms. A theory which is based on the concept of interfacial polarization of inhomogeneous materials has been developed in order to explain the unusual dielectric properties of these substances. The history of inhomogeneous dielectric theories is actually longer than that of polar molecules. Although dielectric research has been dominated by the theory of polar molecules for a long time, the development of interfacial polarization theories and their application to biological systems have progressed steadily over many years. This development was accelerated by the cognizance of the fact that the dielectric properties of many charged biological polyions cannot be fully explained by polar theories. The dielectric behaviour of DNA is one of these examples. The discussions which follow begin with the classical Maxwell–Wagner polarization mechanism and conclude with recent counterion polarization theories.

6.2 The Maxwell–Wagner Theories

Many biological materials do not have a permanent moment, yet when dissolved or suspended in aqueous media, they exhibit a dielectric constant which is far larger than that of the solvent. Not only biological systems, but also simple systems such as the suspension of latex particles which obviously do not have a permanent dipole moment exhibit an exceedingly large dielectric constant and an anomalous dispersion. These observations cannot be explained by the theory of polar molecules which was discussed in the previous chapters.

Let us define what are inhomogeneous dielectric materials. First of all, these systems must have discrete domains. In the case of the suspension of particles, for example, there is a sharp boundary at their surface. Although these suspended particles are of microscopic size, still they are much larger than solvent molecules so that the solvent phase can be considered a dielectric continuum. Secondly, the matters on both sides of the boundary must have different electrical properties, i.e. different dielectric constants and/or conductivities.

The simplest example of an inhomogeneous system is two slabs having different dielectric properties in contact with each other, as shown by figure 6.1. In this figure, ε_1, ε_2, k_1 and k_2 are the dielectric constants and conductivities of these slabs respectively. The equivalent circuit of this system is shown in figure 6.1(b). Note that each slab is represented by a condenser in which a capacitor and resistor are connected in parallel. (This type of condenser is called a leaky condenser.) The admittance of these slabs between A and B is given by

$$1/y = 1/y_1 + 1/y_2 \tag{6.1}$$

where y is the overall admittance and y_1 and y_2 are the admittances of each slab. Admittance, in general can be defined by the following formula:

$$y = G + j\omega C \tag{6.2}$$

where ω is the angular frequency and is equal to $2\pi f$. If we substitute this expression into (6.1) and separate real and imaginary parts, we obtain the following equation:

$$G + j\omega C = \frac{a(G_1 + G_2) + \omega^2 b(C_1 + C_2)}{(G_1 + G_2)^2 + \omega^2(C_1 + C_2)^2}$$

$$+ j\omega \frac{b(G_1 + G_2) - a(C_1 + C_2)}{(G_1 + G_2)^2 + \omega^2(C_1 + C_2)^2} \tag{6.3}$$

where a and b are defined by the following formulae:

$$a = G_1 G_2 - \omega^2 C_1 C_2 \tag{6.4}$$

$$b = (C_1 G_2 + C_2 G_1). \tag{6.5}$$

Equating the real and imaginary parts of both sides of (6.3), we find the conductivity and capacity of two slabs in series:

$$G = \frac{a(G_1 + G_2) + \omega^2 b(C_1 + C_2)}{(G_1 + G_2)^2 + \omega^2(C_1 + C_2)^2} \tag{6.6}$$

$$C = \frac{b(G_1 + G_2) - a(C_1 + C_2)}{(G_1 + G_2)^2 + \omega^2 (C_1 + C_2)^2}. \tag{6.7}$$

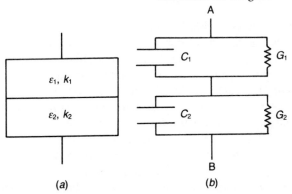

Figure 6.1 (*a*) Two dielectric slabs in series. (*b*) The equivalent circuit, ε and k are the relative dielectric constant and conductivity respectively.

Equations (6.3) to (6.7) can be put into a simple and more elegant form by converting capacitance C and conductance G into permittivity ε and conductivity k using $C = (\varepsilon/4\pi)A/d$ and $G = kA/d$ where A and d are the electrode area and distance:

$$\varepsilon = \varepsilon_\infty + \frac{\Delta\varepsilon}{1 + (\omega\tau)^2} \tag{6.8}$$

where ω is angular frequency, ε_∞, $\Delta\varepsilon$ and relaxation time τ are defined by the following equations:

$$\varepsilon_\infty = \frac{\varepsilon_1\varepsilon_2}{(\varepsilon_1 + \varepsilon_2)} \tag{6.9}$$

$$\Delta\varepsilon = \frac{(\varepsilon_2 k_1 - \varepsilon_1 k_2)^2}{(k_1 + k_2)^2(\varepsilon_1 + \varepsilon_2)} \tag{6.10}$$

$$\tau = \frac{\varepsilon_1 + \varepsilon_2}{4\pi(k_1 + k_2)}. \tag{6.11}$$

Although (6.8) is identical to Debye's equation, the parameters ε_∞, $\Delta\varepsilon$ and τ have a different origin. Likewise, the conductivity can be expressed by similar equations:

$$k = k_0 + \frac{\Delta k(\omega\tau)^2}{1 + (\omega\tau)^2} \tag{6.12}$$

where

$$k_0 = \frac{k_1 k_2}{k_1 + k_2} \tag{6.13}$$

$$\Delta k = \frac{(\varepsilon_2 k_1 - \varepsilon_1 k_2)^2}{(\varepsilon_1 + \varepsilon_2)^2(k_1 + k_2)}. \tag{6.14}$$

This simple analysis clearly demonstrates that an inhomogeneous dielectric material can give rise to a frequency dependent dielectric constant and conductivity which are quite similar to those due to the orientation of polar molecules. The analysis presented above can be extended to multilayered dielectric slabs. However, the mathematical treatment becomes increasingly cumbersome as the number of slabs increases. The mathematical treatment is much more complex even for three layers. Unlike a two-layer system which is characterized by only one time constant, three-layer systems require two relaxation times for a complete description of the dielectric properties.

Although the discussion presented above explains the basic principle of inhomogeneous dielectric theory, the model used is not generally useful for the analysis of the dielectric behaviour of biological polymers and cells unless stratified macromolecular structures are dealt with. Other models such as spherical and/or ellipsoidal particles suspended in a solvent are much more realistic for the analysis of the dielectric behaviour of charged macromolecules and biological cells. In what follows, the theory which was developed by Maxwell will be discussed. This theory is the foundation of inhomogeneous dielectric theories and can be extended or generalized to other more complex and biologically relevant models.

6.3 Maxwell's Mixture Equation for Spherical Particles

This theory was first developed by Maxwell as early as 1892 and was later generalized by Wagner (1914). Hence the name of Maxwell–Wagner theory. Although Maxwell's derivation has been discussed by a number of authors in a variety of reviews and books already, it is still necessary to repeat the essence of his mathematical formulation as a starting point. Originally, the theory was developed for dilute suspensions. However, it was later confirmed experimentally that Maxwell's equation is valid even for highly concentrated particle suspensions.

The derivation consists of two steps. The first step is to calculate the electrical potential inside and outside a spherical particle suspended in a solvent. Figure 6.2 illustrates the model used for this calculation. In this figure, ε_1, ε_2, k_1 and k_2 are the dielectric constants and conductivities of the suspending medium and particle respectively. E is the intensity of the applied field. The potentials are calculated by solving the Laplace equation for each region, i.e. inside and outside the particle. Since a spherical particle is dealt with in this case, spherical polar coordinates are the most appropriate coordinate system for the potential calculation. The Laplace operator in spherical coordinates is given by

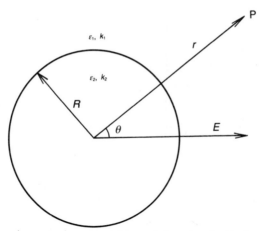

Figure 6.2 The spherical particle used for the derivation of Maxwell's mixture equation. The subscripts 1 and 2 indicate the external medium and the interior of the particle respectively.

$$\nabla^2 = \frac{1}{r^2} \frac{\partial}{\partial r}\left(r^2 \frac{\partial}{\partial r}\right) + \frac{1}{r^2 \sin\theta} \frac{\partial}{\partial\theta}\left(\sin\theta \frac{\partial}{\partial\theta}\right) + \frac{1}{r^2 \sin^2\theta} \frac{\partial^2}{\partial\varphi^2} \quad (6.15)$$

where r, θ and φ are coordinate variables which are defined by figure 6.3. The potential functions inside and outside the particle are obtained as solutions of the following Laplace equations:

$$\nabla^2\psi_1 = 0 \qquad \nabla^2\psi_2 = 0. \quad (6.16)$$

The solution of these equations in spherical coordinates is given by

$$\psi = \left(Ar + \frac{B}{r^{n+1}}\right) P_n^0(\cos\theta) \quad (6.17)$$

where A and B are constants which are determined using boundary conditions. $P_n^0(\cos\theta)$ is the Legendre polynomial of the nth degree. The Legendre polynomials for $n = 0, 1, 2$ and 3 are given below:

$$P_0^0 = 1$$
$$P_1^0 = \cos\theta$$
$$P_2^0 = \tfrac{3}{2}\cos^2\theta - \tfrac{1}{2}$$
$$P_3^0 = \tfrac{1}{2}(5\cos^3\theta - 3\cos\theta).$$

In equation (6.17), all the solutions except for $n = 1$ disappear. Thus the potentials ψ_1 and ψ_2 are given by

$$\psi_1 = \frac{A}{r^2}\cos\theta - Er\cos\theta \qquad \text{for } r \geqslant R \quad (6.18)$$

$$\psi_2 = Br\cos\theta - Er\cos\theta \qquad \text{for } r \leqslant R. \quad (6.19)$$

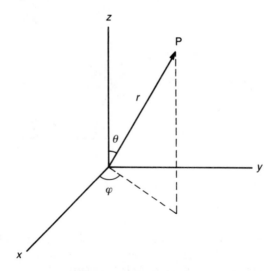

Figure 6.3 A graphic representation of spherical coordinates.

$Er \cos \theta$ is the primary potential due to applied field. The constants A and B are determined by the boundary conditions at $r = R$, i.e.

$$\psi_1 = \psi_2 \tag{6.20}$$

and

$$\varepsilon_1^* \frac{\partial \psi_1}{\partial r} = \varepsilon_2^* \frac{\partial \psi_2}{\partial r}. \tag{6.21}$$

The potential functions thus calculated are

$$\psi_1 = - \left(\frac{\varepsilon_1^* - \varepsilon_2^*}{2\varepsilon_1^* + \varepsilon_2^*} \frac{R^3}{r^3} + 1 \right) Er \cos \theta \tag{6.22}$$

$$\psi_2 = - \left(\frac{\varepsilon_1^* - \varepsilon_2^*}{2\varepsilon_1^* + \varepsilon_2^*} + 1 \right) Er \cos \theta. \tag{6.23}$$

These are the potential functions outside and inside of a particle suspended in a dielectric continuum. The second step of the derivation will now be discussed. We will use a model which is illustrated in figure 6.4 where N of the spherical particles discussed above are surrounded by a large spherical boundary of radius R'. We select a point P at a distance r from the centre of this large sphere and calculate the potential once more:

$$\psi_P = - \left(\frac{\varepsilon_1^* - \varepsilon^*}{2\varepsilon_1^* + \varepsilon^*} \frac{R'^3}{r^3} + 1 \right) Er \cos \theta \tag{6.24}$$

where ε^* is the effective homogeneous permittivity of the large sphere

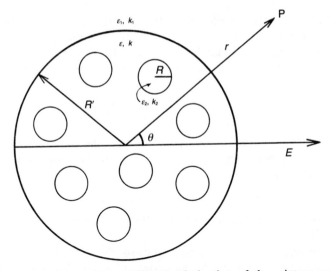

Figure 6.4 The model used for the derivation of the mixture equation. The small spheres are the same particles as shown in figure 6.2. The large sphere contains N small particles.

containing N small particles. The potential ψ_P can also be calculated as the sum of the potential due to N small particles. For the superposition theorem to hold, the density of the small particles must not be excessively high:

$$\psi_P = -\left(N\frac{\varepsilon_1^* - \varepsilon_2^*}{2\varepsilon_1^* + \varepsilon_2^*}\frac{R^3}{r^3} + 1\right)Er\cos\theta. \tag{6.25}$$

These two potentials which are calculated using different methods are, nevertheless, equal. Therefore, the RHS of (6.24) and (6.25) can be equated to obtain

$$\frac{\varepsilon_1^* - \varepsilon^*}{2\varepsilon_1^* + \varepsilon^*}\frac{R'^3}{r^3} = N\frac{\varepsilon_1^* - \varepsilon_2^*}{2\varepsilon_1^* + \varepsilon_2^*}\frac{R^3}{r^3}. \tag{6.26}$$

Noting that $N(R/R')^3$ is the volume fraction of small particles suspended in the large spherical boundary, we finally obtain Maxwell's mixture equation:

$$\frac{\varepsilon_1^* - \varepsilon^*}{2\varepsilon_1^* + \varepsilon^*} = p\frac{\varepsilon_1^* - \varepsilon_2^*}{2\varepsilon_1^* + \varepsilon_2^*}. \tag{6.27}$$

If we solve this equation for ε^*, we obtain

$$\varepsilon^* = \varepsilon_1^*\frac{(2\varepsilon_1^* + \varepsilon_2^*) - 2p(\varepsilon_1^* - \varepsilon_2^*)}{(2\varepsilon_1^* + \varepsilon_2^*) + p(\varepsilon_1^* - \varepsilon_2^*)}. \tag{6.28}$$

This equation expresses the complex dielectric constant of the mixture in terms of the dielectric constants of suspended particles and suspending media. Numerical calculation shows that the increase in ε^* is not linear with the increase in volume fraction p. Only at very low p does the permittivity of the mixture increase linearly with it. For small p (6.28) reduces to

$$\varepsilon^* = \varepsilon_1^*\left(1 + 3p\,\frac{\varepsilon_2^* - \varepsilon_1^*}{2\varepsilon_1^* + \varepsilon_2^*}\right). \qquad (6.29)$$

According to this equation, ε^* reduces to ε_1, the dielectric constant of the solvent, at zero volume fraction but it increases linearly with increasing volume fraction p.

6.4 Wagner's Generalization of the Mixture Equation

In the original derivation of the mixture theory, Maxwell used real dielectric constants for particles and suspending media. Wagner's generalization of Maxwell's theory is to replace real dielectric constants with a complex dielectric constant $\varepsilon^* = \varepsilon - j4\pi k/\omega$. Substituting this expression into (6.29) and separating real and imaginary parts, we obtain equations for the dielectric constant and conductivity of suspension. The expression for dielectric constant is

$$\varepsilon = \varepsilon_\infty + \frac{\Delta\varepsilon}{1 + (\omega\tau)^2} \qquad (6.30)$$

where

$$\varepsilon_\infty = \varepsilon_1\left(1 + 3p\,\frac{\varepsilon_2 - \varepsilon_1}{\varepsilon_2 + 2\varepsilon_1}\right) \qquad (6.31)$$

$$\Delta\varepsilon = 9p\,\frac{(\varepsilon_1 k_2 - \varepsilon_2 k_1)^2}{(2k_1 + k_2)^2(2\varepsilon_1 + \varepsilon_2)} \qquad (6.32)$$

$$\tau = \frac{2\varepsilon_1 + \varepsilon_2}{4\pi(2k_1 + k_2)}. \qquad (6.33)$$

Likewise, we obtain similar equations for the conductivity of suspension. Numerical values of the dielectric increment, conductivity Δk and relaxation time τ are shown in figure 6.5 for various values of ε_1 and k_1. Clearly, $\Delta\varepsilon$ and Δk are strongly dependent on these parameters. However, interestingly, these quantities do not depend on the size of the suspended particle, as seen from (6.30)–(6.33).

6.5 More Discussion on the Maxwell–Wagner Theory

Maxwell's equations, as noted, were derived with an assumption that the

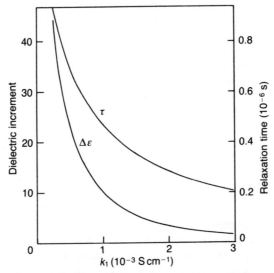

Figure 6.5 The dielectric increments calculated using Maxwell–Wagner theory. The following parameters were used for these calculations: $\varepsilon_1 = 78.5$, $\varepsilon_2 = 6.0$ and $k_2 = 10^{-3}\,\mathrm{S\,cm^{-1}}$. k_1 is treated as an adjustable parameter. $p = 0.3$.

conductivities of each phase are uniform and constant. However, due to the condensation of ions near the interface, this is not a valid approximation. The presence of space charges in the suspending medium alters the potential gradient in this region. The change in electrical potential in the presence of charges cannot be calculated accurately by Laplace's equation. A correct potential function can be obtained by solving Poisson's equation as shown below:

$$\nabla^2 \psi_{\mathrm{m}}(r,\ t) = -\frac{4\pi e}{\varepsilon_{\mathrm{m}}}[n^+(r,\ t) - n^-(r,\ t)] \tag{6.34}$$

where ε_{m} is the dielectric constant of the suspending medium and $n^+(r, t)$ and $n^-(r, t)$ are the positive and negative ion densities.

The use of Poisson's equation instead of Laplace's equation for the analysis of inhomogeneous dielectrics was first discussed by Trukhan (1967) for conductive spheres in non-conductive media. Bonincontro *et al* (1980) also presented a discussion for charged particles. Following a recent analysis by Garcia *et al* (1985), the use of Poisson's equation leads to the following modification of Maxwell's equation:

$$\Delta\varepsilon = \varepsilon_{\mathrm{m}}\left(1 + 3p\frac{\varepsilon_{\mathrm{p}} - (\varepsilon_{\mathrm{m}} + \varepsilon_{\mathrm{p}}R)}{\varepsilon_{\mathrm{p}} + 2(\varepsilon_{\mathrm{m}} + \varepsilon_{\mathrm{p}}R)}\right) \tag{6.35}$$

where ε_{m} and ε_{p} are the dielectric constants of the medium and the particle respectively. r is the radius of the particle and p is the volume fraction.

Equation (6.35) clearly demonstrates that the use of Poisson's equation instead of Laplace's equation adds the term $\varepsilon_p R$ in the numerator as well as in the denominator. This means that the dielectric increment calculated by Maxwell's equation tends to be overestimated.

6.6 Hanai's Mixture Theory

As mentioned earlier, Maxwell's equation was derived for a dilute suspension of spherical particles. Although it was later found that (6.27) holds even for highly concentrated suspensions, nevertheless, Maxwell's equation is rigorous only for low volume fractions. An extension of this theory for high volume fractions was performed by Hanai (1960) for dissipative dielectric particles. The derivation begins with (6.28) by incrementing the volume fraction p by a minute amount at each step. If p is increased by Δp, the dielectric constant of suspension ε^* must be incremented by $\Delta \varepsilon^*$. Thus, $\varepsilon^* \rightarrow \varepsilon^* + \Delta \varepsilon^*$. At the same time, the dielectric constant of the suspending medium ε_1^* is no longer that of a pure solvent but is a dielectric constant of the mixture at the time of volume increment, i.e. $\varepsilon_1^* \rightarrow \varepsilon^*$. In addition, p is replaced by $\Delta p'/(1 - p')$ where p' is the volume fraction of the disperse phase in the additional process. (The proof is discussed by Hanai.) Inserting these definitions into (6.28) and after rearrangement, we obtain

$$\frac{2\varepsilon^* + \varepsilon_2^*}{3\varepsilon^*(\varepsilon^* - \varepsilon_2^*)}\Delta \varepsilon^* = -\frac{\Delta p'}{1 - p'} \tag{6.36}$$

where ε_2^* is the dielectric constant of the particle. In this derivation, the term involving $\Delta p \Delta \varepsilon^*$ is eliminated because of its insignificant numerical value. The augmentation of volume fraction will continue until p reaches a final value. This process can be summarized by integrating both sides of (6.36):

$$\int_{\varepsilon_1^*}^{\varepsilon^*} \frac{2\varepsilon^* + \varepsilon_2^*}{3\varepsilon^*(\varepsilon^* - \varepsilon_2^*)}d\varepsilon^* = \int_0^p \frac{-dp'}{1 - p'} = \log(1 - p). \tag{6.37}$$

The function on the LHS of this equation has two poles at $\varepsilon^* = 0$ and $\varepsilon^* = \varepsilon_2^*$. However, they are not in the region of integration. Thus the integral can be evaluated and the following mixture equation is obtained:

$$\frac{\varepsilon^* - \varepsilon_2^*}{\varepsilon_1^* - \varepsilon_2^*}\left(\frac{\varepsilon_1^*}{\varepsilon^*}\right)^{1/3} = 1 - p. \tag{6.38}$$

This is actually Bruggeman's equation (1935) which is generalized to complex dielectric constants. If both sides of this equation are cubed, we find

$$(\varepsilon^*)^3 - 3\varepsilon_2^*(\varepsilon^*)^2 + \left(3(\varepsilon_2^*)^2 + \frac{[(p-1)(\varepsilon_1^* - \varepsilon_2^*)]^3}{\varepsilon_1^*}\right)\varepsilon^* - (\varepsilon_2^*)^3 = 0. \quad (6.39)$$

Using the full expression for complex dielectric constant:

$$\varepsilon^* = \varepsilon - j4\pi k/\omega$$

the values of ε and k are calculated using a computer program which enables us to find the complex roots of the polynomial equation. Some of the results of these calculations are shown in figure 6.6. The values of the phase parameters used for these calculations are $\varepsilon_2 = 2.10$ and $k_2 = 6.55 \times 10^{-11}$ S cm^{-1} for the oil phase and $\varepsilon_1 = 77.5$ and $k_1 = 24.5 \times 10^{-6}$ S cm^{-1} for the aqueous phase. The broken curves are calculated using (6.30) with one relaxation time. As shown, Hanai's equation gives rise to dispersion curves which are broader than the Maxwell–Wagner equation. Figure 6.7 shows the low-frequency limiting dielectric constant at various volume fractions. Curve A was calculated using Hanai's formula and curve B was obtained with the Maxwell–Wagner equation. Clearly, the agreement between Hanai's prediction and the experimental results is much better than that predicted by the Maxwell–Wagner equation. Hanai's formula has not been reduced to the form of (6.30) and the theoretical expression for relaxation time has not been derived. Hanai applied his mixture equation to many colloidal suspensions and biological cells.

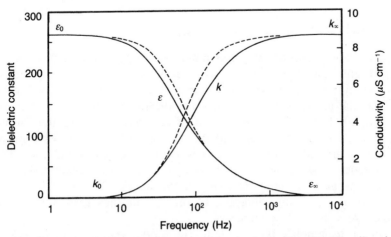

Figure 6.6 The dielectric dispersion curves calculated using Hanai's mixture equation (1960). The broken curves were calculated using Maxwell's theory. Note that Hanai's theory yields broader curves than Maxwell's equation. (From Hanai and Koizumi 1975.) Reproduced by permission of the *Bulletin of the Institute for Chemical Research*.

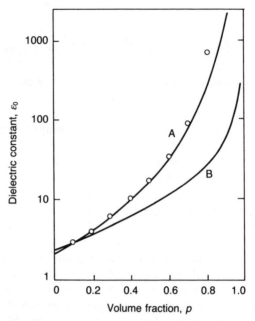

Figure 6.7 The dielectric constant calculated by Hanai's equation at various volume fractions (curve A) and by Maxwell's formula (curve B). Open circles are the low-frequency dielectric constants of a water–oil emulsion at 25 °C. (From Hanai and Koizumi 1975.) Reproduced by permission of the *Bulletin of the Institute for Chemical Research*.

6.7 Particles Surrounded by a Thin Layer

The discussion presented above assumes that suspended particles are not charged and that there are no counterion layers surrounding them. However, this situation hardly exists in real suspensions. Most biological and colloidal particles are charged and ionizable groups attract counterions around them. The condensation of counterions causes the formation of a conductive layer around the core particle. The distribution of counterions around charged particles is a very complex problem and has been discussed in numerous reviews and books (see Kruyt 1949 and Rice and Nagasawa 1961). In general, the density of counterions is a maximum at the Debye length and then diminishes gradually as the distance from the surface increases.

In order to simplify the problem, the model shown in figure 6.8 is used for the calculation of electrical potential (Miles and Robertson 1932). Namely, a core spherical particle is surrounded by a discrete layer having a uniform dielectric constant and conductivity. With this simplification, the electrical potential can be calculated for an external medium, for a thin shell of counterions and for the interior of a particle by solving the following three Laplace equations:

$$\nabla^2\psi_1 = 0 \qquad \nabla^2\psi_2 = 0 \qquad \nabla^2\psi_3 = 0 \qquad (6.40)$$

with the solutions

$$\psi_1 = A_{11}r^{-2}\cos\theta - Er\cos\theta \qquad (6.41)$$

$$\psi_2 = (A_{21}r + B_{21}r^{-2})\cos\theta - Er\cos\theta \qquad (6.42)$$

$$\psi_3 = B_{31}r\cos\theta - Er\cos\theta \qquad (6.43)$$

where A_{11}, A_{21}, B_{21} and B_{31} are constants which are determined using the following boundary conditions:

$$\psi_1 = \psi_2 \qquad \varepsilon_1\frac{\partial\psi_1}{\partial r} = \varepsilon_2\frac{\partial\psi_2}{\partial r} \qquad \text{at } r = R \qquad (6.44)$$

$$\psi_2 = \psi_3 \qquad \varepsilon_2\frac{\partial\psi_2}{\partial r} = \varepsilon_3\frac{\partial\psi_3}{\partial r} \qquad \text{at } r = R_1. \qquad (6.45)$$

A complete solution for this model is extremely tedious. However, Miles and Robertson obtained a partial solution which, nevertheless, enables one to compute the dielectric increment and relaxation time for the shell model.

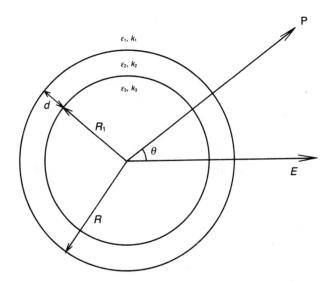

Figure 6.8 A spherical particle with a thin shell. The subscripts 1, 2 and 3 indicate the external medium, the shell and the internal fluid respectively. R and d are the radius of the sphere and the thickness of the shell respectively.

Using the boundary conditions shown above, the coefficient A_{11} can be found. Inserting this into (6.41) we obtain an equation for ψ_1, i.e.

$$\psi_1 = -\left(\frac{(\varepsilon_1^* - \varepsilon_2^*)(2\varepsilon_2^* + \varepsilon_3^*)R^3 + (\varepsilon_2^* - \varepsilon_3^*)(\varepsilon_1^* + 2\varepsilon_2^*)R_1^3}{(2\varepsilon_2^* + \varepsilon_3^*)(2\varepsilon_1^* + \varepsilon_2^*)R^3 + 2(\varepsilon_2^* - \varepsilon_3^*)(\varepsilon_1^* - \varepsilon_2^*)R_1^3}\frac{R^3}{r^2} + r\right)E\cos\theta.$$

(6.46)

On the other hand, we can obtain the potential ψ_1 in a different way. Namely, we define an effective complex dielectric constant ε^* for a particle including the core and shell. Once this is done, the potential problem reduces to Maxwell's treatment except for the use of an effective dielectric constant ε^* instead of ε_2^*. Under this condition, the potential ψ_1 is given by the following equation:

$$\psi_1 = - E\cos\theta\left(r + \frac{R^3}{r^2}\frac{\varepsilon_1^* - \varepsilon^*}{2\varepsilon_1^* + \varepsilon^*}\right).$$

(6.47)

Equating (6.46) and (6.47) we obtain

$$\frac{\varepsilon_1^* - \varepsilon^*}{2\varepsilon_1^* + \varepsilon^*} = \frac{(\varepsilon_1^* - \varepsilon_2^*)(2\varepsilon_2^* + \varepsilon_3^*)R^3 + (\varepsilon_2^* - \varepsilon_3^*)(\varepsilon_1^* + 2\varepsilon_2^*)R_1^3}{(2\varepsilon_1^* + \varepsilon_3^*)(2\varepsilon_1^* + \varepsilon_2^*)R^3 + 2(\varepsilon_2^* - \varepsilon_3^*)(\varepsilon_1^* - \varepsilon_2^*)R_1^3}.$$

(6.48)

Solving this for ε^*:

$$\varepsilon^* = \varepsilon_2^* \frac{(2\varepsilon_2^* + \varepsilon_3^*)R^3 - 2(\varepsilon_2^* - \varepsilon_3^*)R_1^3}{(2\varepsilon_2^* + \varepsilon_1^*)R^3 + (\varepsilon_2^* - \varepsilon_3^*)R_1^3}.$$

(6.49)

Introducing the relation $R = R_1 + d$ (see figure 6.8) and assuming that $R_1 \gg d$, we can simplify (6.49) as

$$\varepsilon^* = (\tfrac{3}{2}\varepsilon_3^* + (3d/R)\varepsilon_2^*)/(\tfrac{3}{2} + 3d/R)$$

(6.50)

where we assume that $R \simeq R_1 \simeq R_2$. This expression can further be simplified by neglecting terms which are numerically small:

$$\varepsilon^* = \varepsilon_3^* + \varepsilon_2^*(2d/R).$$

(6.51)

Substituting the expression $\varepsilon^* = \varepsilon - j4\pi k/\omega$ for ε_3^* and ε_2^* in (6.51), we obtain

$$\varepsilon^* = (\varepsilon_3 + \varepsilon_2 2d/R) - j(4\pi/\omega)(k_3 + k_2 2d/R).$$

(6.52)

Therefore, the effective conductivity and dielectric constant of the particle, including the shell, are

$$k = k_3 + k_2\,2d/R$$

(6.53)

$$\varepsilon = \varepsilon_3 + \varepsilon_2 2d/R.$$

(6.54)

Using (6.53) and (6.54) in the Maxwell–Wagner equations, we can calculate ε_∞, $\Delta\varepsilon$ and τ. For example, the relaxation time is given by the following equation, instead of (6.33):

$$\tau = \frac{2\varepsilon_3 + \varepsilon}{4\pi(2k_3 + k)}. \tag{6.55}$$

Substituting (6.53) and (6.54) into (6.55):

$$\tau = \frac{2\varepsilon_3 + \varepsilon_1 + \varepsilon_2 2d/R}{4\pi(2k_3 + k_1 + k_2 2d/R)}. \tag{6.56}$$

Obviously, ε_2 and k_2 would disappear if the thickness of the shell reduces to zero and (6.56) returns to the Maxwell–Wagner equation. The expressions for ε and k can be obtained using the same method. Figure 6.9 shows the numerical values of $\Delta\varepsilon$ and τ for the shell model.

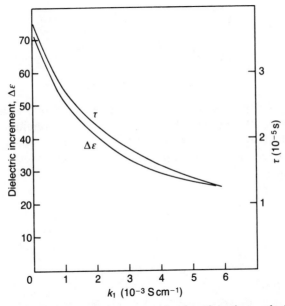

Figure 6.9 The dielectric increment and relaxation time calculated using Miles and Robertson's equation. The following phase parameters are used: $\varepsilon_1 = 78.5$, $\varepsilon_2 = 6.0$, $\varepsilon_3 = 60$, $k_2 = 10^{-3} \, \mathrm{S \, cm^{-1}}$, $R = 2 \times 10^{-4}$ and $d = 10^{-6}$ cm. $p = 0.3$.

6.8 The Calculation of the Mixture Equation for the Shell Model

As has been discussed, the treatment of Miles and Robertson led them to (6.53), (6.54) and (6.56), with which the dielectric parameters for the shell model can be calculated. However, it must be emphasized that Miles and Robertson's solution is not a complete description of the dielectric properties of particles with a shell. A complete theoretical treatment of the shell model was developed partially by Daenzer (1938) and completed subsequently by Pauly and Schwan (1959). Firstly, the mixture equation developed by Daenzer will be discussed.

The model used by Daenzer is only slightly different from the one used by Miles and Robertson, as shown by figure 6.8. The calculation of the potential for this model is similar to the previous one and therefore only the essential steps of the solution are given below.

If we apply an electrical field, a small dipole is created at the centre of a sphere. Therefore, the potential due to applied field is equivalent to the potential due to the dipole moment m or αE, as shown by figure 6.10 (where α is the polarizability). With this, the potential functions in three regions are calculated as shown below:

$$\psi_1 = E_z - \frac{m_z}{r^3} = E\left(r - \frac{\alpha}{r^2}\right)\cos\theta \tag{6.57}$$

$$\psi_2 = A_z - \frac{B_z}{r^3} = \left(Ar - \frac{B}{r^2}\right)\cos\theta \tag{6.58}$$

$$\psi_3 = C_z = Cr\cos\theta. \tag{6.59}$$

As before, E_z or $Er\cos\theta$ is the primary potential due to applied field. C in (6.59) is the internal field. The boundary conditions needed to calculate the constants A, B and C are identical to (6.44) and (6.45). Using these conditions we can eliminate the constants to obtain an equation for α:

$$\alpha = R_1^3\frac{x - y_1}{x + 2y_1} \tag{6.60}$$

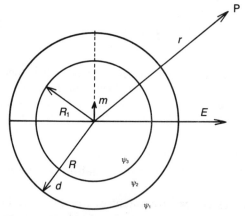

Figure 6.10 The shell model used by Daenzer. The arrow indicates the dipole moment induced by applied field.

where y_1 is the complex conductivity of the medium and is given by $y_1 = k_1 - j\omega\varepsilon_1/4\pi$. The quantity x is given by

$$x = y_2 \frac{1 - 2(R/R_1)^3(y_2 - y_3)/(2y_2 + y_3)}{1 - (R/R_1)^3(y_2 - y_3)/(2y_2 + y_3)}. \tag{6.61}$$

The polarizability can also be calculated by assuming a homogeneous sphere having an effective admittance $\bar{y}$ and a radius $R_1 + d$:

$$\alpha = (R_1 + d)^3 \frac{\bar{y} - y_1}{\bar{y} + 2y_1}. \tag{6.62}$$

The polarizabilities which are calculated using these two methods are, nevertheless, equivalent and therefore we can equate (6.60) and (6.62):

$$\frac{\bar{y} - y_1}{\bar{y} + 2y_1} = \left(\frac{R_1}{R_1 + d}\right)^3 \frac{x - y_1}{x + 2y_1}. \tag{6.63}$$

Using Maxwell's mixture theory we can also write an equation

$$\frac{y - y_1}{y + 2y_1} = p\frac{\bar{y} - y_1}{\bar{y} + 2y_1} \tag{6.64}$$

where y is the overall complex conductance of the suspension. At very low or high frequencies at which the complex conductance can be replaced by either the real dielectric constant or the conductivity, these equations can be used to determine the electrical properties of the thin layer surrounding the core particle, provided the dielectric constants of the suspension are accurately measured.

At intermediate frequencies, the full expression for complex conductance must be substituted into these equations. This extremely tedious calculation was performed by Pauly and Schwan some years ago (1959). Because of its complexity, the author will try to present the derivation as simply as possible. Also, a portion of Pauly and Schwan's formulation has been modified slightly.

To start the generalization, Pauly and Schwan used (6.64) and the following equation:

$$\frac{\bar{y} - y_2}{\bar{y} + 2y_2} = \left(\frac{R_1}{R_1 + d}\right)^3 \frac{y_3 - y_2}{y_3 + 2y_2}. \tag{6.65}$$

Equation (6.65), although it looks different, is equivalent to (6.63). Eliminating the admittance or complex conductance of the homogeneous sphere, $\bar{y}$, using (6.64) and (6.65), and substituting the expressions for admittance, $y = k + j\omega\varepsilon/4\pi$, we obtain the following equation:

$$\frac{y}{y_1} = \frac{(A - \omega^2 B) + j\omega E}{(C - \omega^2 D) + j\omega F} \tag{6.66}$$

where A, B, C, D, E and F are defined as follows:

$$A = (1 + 2p)k_2a + 2(1 - p)k_1b$$

$$B = (1/4\pi)^2[(1 + 2p)\varepsilon_2c + 2(1 - p)\varepsilon_1d]$$

$$C = (1 - p)k_2a + (2 + p)k_1b \qquad\qquad (6.67)$$

$$D = (1/4\pi)^2[(1 - p)\varepsilon_2c + (2 + p)\varepsilon_1d]$$

$$E = (1/4\pi)[(1 + 2p)(\varepsilon_2a + k_2c) + 2(1 - p)(\varepsilon_1b + k_1d)]$$

$$F = (1/4\pi)[(1 - p)(\varepsilon_2a + k_2c) + (2 + p)(\varepsilon_1b + k_1d)].$$

Furthermore, v, a, b, c and d are defined as follows:

$$v = R_1^3/(R_1 + d)^3$$

$$a = (1 + 2v)k_3 + (2 + v)k_2$$

$$b = (1 - v)k_3 + (2 + v)k_2 \qquad\qquad (6.68)$$

$$c = (1 + 2v)\varepsilon_3 + 2(1 - v)\varepsilon_2$$

$$d = (1 - v)\varepsilon_3 + (2 + v)\varepsilon_2.$$

Equation (6.66) can be rearranged to read

$$y = \frac{[k_1A - \omega^2(k_1B + \varepsilon_1E/4\pi)] + j\omega(k_1E + \varepsilon_1A/4\pi - \omega^2\varepsilon_1B/4\pi)}{C(1 - \omega^2D/C + j\omega F/C)}.$$

$$(6.69)$$

As pointed out previously, a sphere which is surrounded by a dielectric shell requires two relaxation times for a complete description. Therefore, the admittance of the suspension, y, must be transformed into the following form:

$$y = \frac{M}{1 + j\omega T_1} + \frac{N}{1 + j\omega T_2} \qquad\qquad (6.70)$$

where

$$(1 + j\omega T_1)(1 + j\omega T_2) = 1 - \omega^2 T_1 T_2 + j\omega(T_1 + T_2)$$

$$= 1 - \omega^2 D/C + j\omega F/C. \qquad (6.71)$$

Equating the real and imaginary terms of the last two equations, we obtain

$$T_1 T_2 = D/C \qquad \text{and} \qquad T_1 + T_2 = F/C. \qquad (6.72)$$

Solving these equations, expressions for T_1 and T_2 can be found:

$$T_1 = (F/2C)(1 + \sqrt{1 - 4DC/F^2}) \qquad\qquad (6.73)$$

$$T_2 = (F/2C)(1 - \sqrt{1 - 4DC/F^2}). \qquad\qquad (6.74)$$

Now turning to the numerators M and N in (6.70)

$$M = m_1 + \omega^2 m_2 \qquad\qquad (6.75)$$

where

$$m_1 = [(\varepsilon_1/4\pi - k_1 T_1)A + k_1 E]/C(T_2 - T_1) \qquad (6.76)$$

$$m_2 = -[(\varepsilon_1/4\pi - k_1 T_1)B - \varepsilon_1 T_1 E/4\pi]/C(T_2 - T_1) \qquad (6.77)$$

and

$$N = n_1 + \omega^2 n_2 \qquad (6.78)$$

$$n_1 = [-(\varepsilon_1/4\pi - k_1 T_2)A - k_1 E]/C(T_2 - T_1) \qquad (6.79)$$

$$n_2 = -[-(\varepsilon_1/4\pi - k_1 T_2)B + \varepsilon_1 T_2 E/4\pi]/C(T_2 - T_1). \qquad (6.80)$$

Therefore, the complex conductivity of the suspension is characterized by two dispersion terms:

$$k - \frac{j\omega\varepsilon}{4\pi} = \frac{(m_1 + \omega^2 m_2)(1 - j\omega T_1)}{1 + (\omega T_1)^2} + \frac{(n_1 + \omega^2 n_2)(1 - j\omega T_2)}{1 + (\omega T_2)^2}.$$

$$(6.81)$$

Separating real and imaginary parts on the RHS and equating these terms to k and $j\omega\varepsilon/4\pi$:

$$\varepsilon = \frac{4\pi(m_1 + \omega^2 m_2)T_1}{1 + (\omega T_1)^2} + \frac{4\pi(n_1 + \omega^2 n_2)T_2}{1 + (\omega T_2)^2} \qquad (6.82)$$

$$k = \frac{(m_1 + \omega^2 m_2)}{1 + (\omega T_1)^2} + \frac{(n_1 + \omega^2 n_2)}{1 + (\omega T_2)^2}. \qquad (6.83)$$

In general, the dispersion equation for two discretely different relaxation times can be written as

$$\varepsilon = \varepsilon_\infty + \frac{\varepsilon_0 - \varepsilon_m}{1 + (\omega T_1)^2} + \frac{\varepsilon_m - \varepsilon_\infty}{1 + (\omega T_2)^2} \qquad (6.84)$$

or

$$= \frac{\varepsilon_0 - \varepsilon_m}{1 + (\omega T_1)^2} + \frac{\varepsilon_m + \varepsilon_\infty(\omega T_2)^2}{1 + (\omega T_2)^2}. \qquad (6.85)$$

Using (6.82)–(6.85), we can confirm the following equality:

$$\frac{\varepsilon_0 - \varepsilon_m}{1 + (\omega T_1)^2} = \frac{-m_2/T_1 + m_1 T_1}{(1/4\pi)[1 + (\omega T_1)^2]} \qquad (6.86)$$

and

$$\frac{\varepsilon_m + \varepsilon_\infty(\omega T_2)^2}{1 + (\omega T_2)^2} = \frac{(n_1 T_2 + m_2/T_1) + (n_2/T_2 + m_2/T_1)(\omega T_2)^2}{(1/4\pi)[1 + (\omega T_2)^2]}. \qquad (6.87)$$

Equating both sides, term by term, we can now identify

$$\varepsilon_0 = 4\pi(m_1 T_1 + n_1 T_2) \qquad (6.88)$$

$$\varepsilon_m = 4\pi(m_2/T_1 + n_1 T_2) \qquad (6.89)$$

$$\varepsilon_\infty = 4\pi(m_2/T_1 + n_2/T_2). \tag{6.90}$$

Likewise, we can find the equations for k_0, k_m and k_∞. Using the equations we have derived so far, every parameter which is necessary for the characterization of the dielectric properties of shelled spheres can be calculated.

The results which are calculated using this theory are shown schematically in figure 6.11. As noted, the theory predicts two dispersion regions. However, the values of T_1 and T_2 are too far apart and the second dispersion with a relaxation time T_2 can be found only at very high frequencies. Because of this, the second dispersion for biological material such as colloid suspensions and/or cell suspensions has not been detected experimentally. Some of the exceptions observed by Zhang *et al* (1984) with polystyrene microcapsules are shown in figure 6.12.

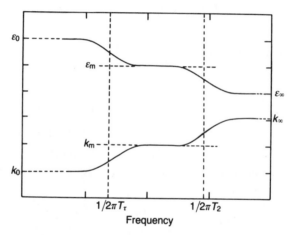

Figure 6.11 Dielectric dispersion curves calculated by Pauly and Schwan's theory. Note the presence of two dispersion regions. (From Pauly and Schwan 1959.)

6.9 Maxwell–Wagner Theory for Non-spherical Particles (Sillars 1936)

So far, Maxwell–Wagner theories have been discussed for spherical particles with and without a thin shell surrounding the core. A spherical model is a good approximation for many colloidal particles, including globular proteins and biological cells. Above all, the simple geometries of these particles make these models amenable to mathematical manipulation. However, in reality, polyions and biological cells often have a complex geometry which cannot be approximated by a simple sphere.

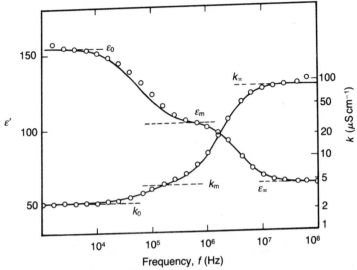

Figure 6.12 Dielectric dispersion curves obtained with a suspension of spherical polystyrene microcapsules. The second dispersion predicted by Pauly and Schwan is clearly seen at frequencies between 1 and 10 MHz. (From Zhang *et al* 1984.)

As discussed in Chapter 5, many protein molecules are either elongated or disc shaped. Under these circumstances, the original Maxwell–Wagner theory must be extended to non-spherical models. Many biological polyions and cells can be approximated by ellipsoids or ellipsoids of revolution (or spheroids). Although the Laplace equation for ellipsoidal coordinates looks formidable, a solution has been obtained and a detailed discussion can be found in Stratton (1941).

Intuitively, the simplest non-spherical particle is a cylinder. However, the calculation of electrical potential for cylinders of finite length is difficult if an electrical field is applied along the major axis. Instead, general ellipsoids or ellipsoids of revolution have been used as a model for non-spherical molecules or particles. The ellipsoidal model used for the calculation of potential is shown in figure 6.13. Obviously ellipsoidal coordinates are the most appropriate choice for this case. The solutions of the Laplace equation for the suspending medium and for inside the ellipsoid are given by the following equations if the direction of applied field is along the x axis:

$$\psi_1 = -E_x x \left(1 - \frac{\varepsilon_2^* - \varepsilon_1^*}{4\pi\varepsilon_1^* + (\varepsilon_2^* - \varepsilon_1^*)l_a} \int_\xi^\infty \frac{d\xi}{(\xi + a^2)\beta(\xi)} \right) \quad (6.91)$$

$$\psi_2 = -E_x x \left(\frac{4\pi\varepsilon_1^*}{4\pi\varepsilon_1^* + (\varepsilon_2^* - \varepsilon_1^*)l_a} \right) \quad (6.92)$$

where ξ represents the surface of the ellipsoid and

$$\beta(\xi) = \frac{\sqrt{(a^2 + \xi)(b^2 + \xi)(c^2 + \xi)}}{abc/2} \tag{6.93}$$

$$l_a = \int_0^\infty \frac{d\xi}{(a^2 + \xi)\beta(\xi)} \qquad l_b = \int_0^\infty \frac{d\xi}{(b^2 + \xi)\beta(\xi)}. \tag{6.94}$$

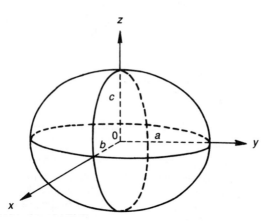

Figure 6.13 The model used for the calculation of dielectric constant for ellipsoidal particles. a, b and c are molecular axes.

If a field is applied along the y axis, a and l_a in (6.91) and (6.92) must be replaced by b and l_b respectively. Analogous to Maxwell's derivation for spherical particles, we surround a group of ellipsoidal particles with a large ellipsoidal boundary and calculate the potential. The potential due to this large ellipsoid with an effective dielectric constant ε^* is given by

$$\psi_p = -E_x x\left[1 - \frac{\varepsilon^* - \varepsilon_1^*}{4\pi\varepsilon_1^* + (\varepsilon^* - \varepsilon_1^*)L_a}\left(\frac{V}{r^3}\right)\right] \tag{6.95}$$

where V is the volume of a large ellipsoid and L_a is equivalent to l_a for small ellipsoids. The potential ψ_p can also be calculated as the sum of the potentials due to small particles:

$$\psi_p = -E_x x\left[1 - N\frac{\varepsilon_2^* - \varepsilon_1^*}{4\pi\varepsilon_1^* + (\varepsilon_2^* - \varepsilon_1^*)l_a}\left(\frac{V}{r^3}\right)\right]. \tag{6.96}$$

Equating (6.95) and (6.96) we obtain

$$\frac{\varepsilon^* - \varepsilon_1^*}{4\pi\varepsilon_1^* + (\varepsilon^* - \varepsilon_1^*)L_a} = p\frac{\varepsilon_2^* - \varepsilon_1^*}{4\pi\varepsilon_1^* + (\varepsilon_2^* - \varepsilon_1^*)l_a} \tag{6.97}$$

where $p = Nv/V$. If we assume that the difference between ε^* and ε_1^* is small, the term $\varepsilon^* - \varepsilon_1^*$ can be eliminated from the denominator and the equation is readily solved for ε^*:

$$\varepsilon^* = \varepsilon_1^* \left(1 + p \frac{4\pi(\varepsilon_2^* - \varepsilon_1^*)}{4\pi\varepsilon_1^* + (\varepsilon_2^* - \varepsilon_1^*)l_a}\right). \tag{6.98}$$

Introducing a new notation such that $n = 4\pi/l_a$, this equation can be rewritten as

$$\varepsilon^* = \varepsilon_1^* \left(1 + p \frac{n(\varepsilon_2^* - \varepsilon_1^*)}{(n - 1)\,\varepsilon_1^* + \varepsilon_2^*}\right). \tag{6.99}$$

In order to simplify the calculation, we assume that the component 1 is an insulator and its conductivity is zero:

$$\varepsilon_1^* = \varepsilon_1 \tag{6.100}$$

$$\varepsilon_2^* = \varepsilon_2 - j4\pi k_2/\omega.$$

Substituting these into (6.99), we obtain the following equations:

$$\varepsilon' = \varepsilon_\infty + \varepsilon_1 \frac{\Delta\varepsilon}{1 + (\omega\tau)^2} \tag{6.101}$$

$$\varepsilon'' = \frac{\varepsilon_1\Delta\varepsilon\omega\tau}{1 + (\omega\tau)^2} \tag{6.102}$$

where

$$\varepsilon_\infty = \varepsilon_1 \left(1 + p\frac{n(\varepsilon_2 - \varepsilon_1)}{(n - 1)\varepsilon_1 + \varepsilon_2}\right) \tag{6.103}$$

$$\Delta\varepsilon = p\frac{n^2\varepsilon_1}{(n - 1)\varepsilon_1 + \varepsilon_2} \tag{6.104}$$

$$\tau = \frac{(n - 1)\varepsilon_1 + \varepsilon_2}{4\pi k_2}. \tag{6.105}$$

As has been discussed, the calculation of ε' and ε'' requires the evaluation of the elliptic integrals l_a and l_b for general ellipsoids. However, practically, the use of a general ellipsoid is not necessary for the analysis of the dielectric properties of biological particles or macromolecules. We can simplify, without the loss of generality, the computation by replacing the general ellipsoid with an ellipsoid of revolution.

Oblate ellipsoid of revolution: a < b = c
In this case, the elliptic integrals l_a and l_b become

$$l_a = 4\pi\left(1/e^2 - \frac{1 - e^2}{e^3}\sin^{-1}e\right) \tag{6.106}$$

$$l_b = \pi/2\left(\frac{1 - e^2}{e^3}\sin^{-1}e - \frac{1 - e^2}{e^2}\right) \tag{6.107}$$

where eccentricity e is defined as

$$e = \sqrt{1 - (a/b)^2}.\qquad(6.108)$$

If the axis b is much larger than a (a very flat disc), the integrals reduce to

$$l_a = 4\pi \qquad l_b = \pi^2(a/b).\qquad(6.109)$$

Figure 6.14 shows the dependence of the dielectric constant on axial ratios for an oblate ellipsoid of revolution calculated using (6.101)–(6.104).

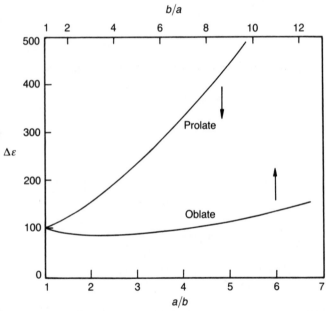

Figure 6.14 The calculated dielectric increment for prolate and oblate ellipsoids. The abscissa shows the axial ratios a/b and b/a. $p = 0.3$, $\varepsilon_1 = 78$ and $\varepsilon_2 = 6$.

Prolate ellipsoid of revolution: $a > b = c$
In this case, the elliptic integrals become

$$l_a = 4\pi\left(\frac{1}{e^2} - 1\right)\left(\frac{1}{2e}\log\frac{1 + e}{1 - e} - 1\right)\qquad(6.110)$$

$$l_b = 2\pi\left(\frac{1}{e^2} - \frac{1 - e^2}{e^2}\log\frac{1 + e}{1 - e}\right).\qquad(6.111)$$

The eccentricity e is defined as

$$e = \sqrt{1 - (b/a)^2}.\qquad(6.112)$$

If $a/b \gg 1$, the integrals reduce to

$$l_a = 4\pi(b/a)^2[\log(2a/b) - 1] \tag{6.113}$$

$$l_b = 2\pi. \tag{6.114}$$

Figure 6.14 also illustrates the dependence of the dielectric constant on axial ratio for an elongated ellipsoid of revolution.

Sphere: $a = b = c$
In this case, the integrals reduce to

$$l_a = l_b = 4\pi/3.$$

However, substitution of this value into (6.103)–(6.105) does not lead to Maxwell's equation for a sphere. This is because of the approximation used for the derivation of (6.98), i.e. $\varepsilon^* \to \varepsilon_1^*$. The frequency profile of the dielectric constant for an ellipsoid of revolution is illustrated in figure 6.15 for three different axial ratios. Note the marked dependence of the dielectric increment on the axial ratio.

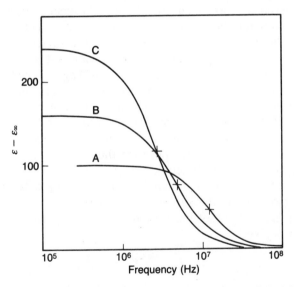

Figure 6.15 The frequency profile of the calculated dielectric constant for prolate ellipsoids. Axial ratios are 1:1 (curve A), 2:1 (curve B) and 3:1 (curve C). $p = 0.3$, $k_2 = 1 \text{ mS cm}^{-1}$, $\varepsilon_1 = 78$ and $\varepsilon_2 = 6$.

We have discussed, in the above, the dielectric properties of spherical particles with and without a thin shell surrounding them. In addition, the theory was generalized to non-spherical particles. The aim was to explain the dielectric behaviour of biological polymers and cells dissolved or suspended in aqueous media. In spite of the differences amongst the models, the physical principles and mathematical techniques are similar except, perhaps, for the mixture theory developed by Hanai. In what follows, theories which are based on different ionic processes will be discussed.

6.10 Surface Conductivity and Counterion Polarization

We have discussed the mechanism for the electric polarization of inhomogeneous systems such as colloid suspensions and biological cells. The underlying physical process which is only implicitly indicated is the creation of apparent dipoles due to charges which are driven by the field and which accumulate at the particle boundary. The separation of the centres of positive and negative charges entails an apparent dipole moment. In this scheme, however, the movement of ions is restricted to the radial direction and tangential movement is not considered. The necessity for tangential movement arises from the observation that the dielectric increment predicted by Maxwell–Wagner theory is far too small to account for the exceedingly large dielectric increment of some biological colloids and cells. For example, the dielectric constant of charged biopolymers, e.g. DNA, reaches a value of several thousand, even at very low concentrations. Maxwell–Wagner theory is simply unable to explain this extraordinary dielectric behaviour of biological macromolecules. Therefore, the introduction of a new concept was necessary and the role of tangential ion movement which had been neglected began to attract the attention of investigators. O'Konski (1960) developed a theory incorporating the concept of surface conductivity into inhomogeneous dielectric theory.

Let us consider a sphere which is surrounded by a layer of mobile ions which move along the surface of the particle as well as in a radial direction. If the mobility of the ith ion is defined by u_i, the total surface conductivity is given by

$$\lambda_s = \sum_i S_i u_i z_i \qquad (6.115)$$

where S_i is the number of charges per unit surface area and z_i is the ion's valence. In addition to this layer there is an ion atmosphere on both sides of the boundary as shown in figure 6.16. The conductivities within these ion atmospheres are distinctly higher than the bulk conductivities k_1 and k_2. If we define the conductivities at arbitrary distances in regions 1 and 2 as k'_1 and k'_2, then the total surface conductivity is given by the following equation:

$$\lambda = \sum_i S_i u_i z_i + \int_{-\infty}^{0} (k_1' - k_1)\, dy + \int_{0}^{\infty} (k_2' - k_2)\, dy. \qquad (6.116)$$

The calculation of surface conductivity has not been carried out. At this stage, however, the value for surface conductivity is not essential to the development of the theory.

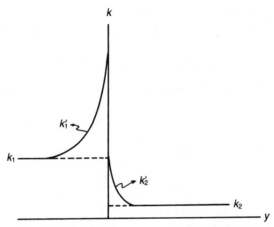

Figure 6.16 The profile of the counterion distribution on both sides of a boundary. Reproduced with permission from O'Konski (1960). Copyright (1960) American Chemical Society.

The next step is to find out how the effective dielectric constant is modified by the presence of surface conductivity. In order to do this, we must find the electrical potential by solving Laplace's equation for the external medium and the interior of particles. The boundary conditions are

$$\psi_1 = \psi_2 \qquad (6.117)$$

$$\varepsilon_2 \frac{\partial \psi_2}{\partial r} - \varepsilon \frac{\partial \psi_1}{\partial r} = 4\pi\rho \qquad (6.118)$$

for $r = R$. ρ is the surface charge density. This term arises from the presence of space charges at the boundary. The transport of ions to and from an element of surface is given by the following continuity equation:

$$\frac{d\sigma_a}{dt} = k \frac{\partial \psi_2}{\partial r} - k_1 \frac{\partial \psi_1}{\partial r} \qquad \text{at } r = R. \qquad (6.119)$$

In addition, the continuity equation of tangential charge movement is given by

$$\frac{d\sigma_b}{dt} = \frac{\lambda}{r^2 \sin\theta} \frac{\partial}{\partial\theta}\left(\sin\theta \left.\frac{\partial\psi_2}{\partial\theta}\right|_{r=R}\right). \tag{6.120}$$

Thus, the total surface charge transfer is given by the sum of these terms, i.e.

$$\frac{d\sigma}{dt} = \frac{d}{dt}(\sigma_a + \sigma_b). \tag{6.121}$$

With this equation combined with the solution of Laplace's equation, the external potential ψ_1 is calculated with the result as shown below:

$$\psi_1 = \left(\frac{R^3}{r^3} \frac{\varepsilon_2^* - \varepsilon_1^*}{\varepsilon_2^* + 2\varepsilon_1^*} - 1\right) Er\cos\theta \tag{6.122}$$

where the quantity ε_2^* is defined by the following expression:

$$\varepsilon_2^* = \varepsilon_2 - j\frac{4\pi}{\omega}\left(k_2 + \frac{2\lambda}{R}\right). \tag{6.123}$$

Comparing this equation with

$$\varepsilon^* = \varepsilon - j\frac{4\pi}{\omega}k$$

we note that the effective conductivity of the particle is modified by a factor of $2\lambda/R$ if the particle is surrounded by an ion atmosphere with a surface conductivity λ. The model used by O'Konski is similar to the one used by Miles and Robertson (see equation (6.66)). In the latter case, however, the tangential movement of ions in the shell was not considered, while O'Konski's solution was obtained using two continuity equations for tangential and radial ion movement. Finally, we discuss how the modified effective conductivity of the particle will affect the dielectric constant of suspension. Using Polder and Van Santen's mixture formula (1946):

$$\frac{\varepsilon^* - \varepsilon_1^*}{3\varepsilon^*} = p\frac{\varepsilon_2^* - \varepsilon_1^*}{\varepsilon_2^* + 2\varepsilon_1^*} \tag{6.124}$$

and substituting into the expression for the complex dielectric constant, the following equation is obtained for the dielectric increment of dilute suspension ($\Delta\varepsilon = \varepsilon - \varepsilon_1$):

$$\frac{\Delta\varepsilon}{p} = \frac{3[(r^2 - 2r - 2)\varepsilon_1 + 3\varepsilon_2]}{(r + 2)^2} \tag{6.125}$$

where $r = k_s/k_1$, and

$$k_s = k_2 + 2\lambda/R. \tag{6.126}$$

As noted, the dielectric increment depends upon the ratio between the bulk and surface conductivities. Figure 6.17 illustrates the dependence of

$\Delta\varepsilon$ upon the ratio k_s/k_1. In this calculation, ε_1 is assumed to be 78.54 and ε_2 to be 3.0. As shown, the value of $\Delta\varepsilon$ increases markedly as the ratio k_s/k_1 increases. For example, $\Delta\varepsilon$ reaches a very high value when the ratio is 100–1000. However, definitive information with regard to the numerical value of the ratio k_s/k_1 is not available at present. Since the value of the dielectric increment critically depends on this ratio, the determination of surface conductivity is the crucial step for the evaluation of this theory. The above discussion was excerpted from O'Konski's article on spherical particles. O'Konski also extended his theory to ellipsoidal particles; however, this problem will not be discussed because of limited space.

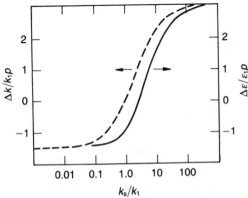

Figure 6.17 The dielectric constant and conductivity increments calculated using O'Konski's theory (1960) as a function of the ratio r ($= k_s/k_1$). Reproduced with permission from O'Konski (1960). Copyright (1960) American Chemical Society.

6.11 Counterion Polarization Theory for a Sphere

As discussed earlier, the surface conductivity which was used by O'Konski was assumed to be purely conductive without a reactive component. Although a consideration of surface conductivity improves the agreement between the observed dielectric increment and the theoretical prediction, nevertheless, a large discrepancy still persists between them. It was suggested by Schwan et al (1962) that the addition of a reactive component to surface admittance would account for the large dielectric increment of biological materials. The counterion polarization theory was formulated by Schwarz for spherical particles (1962). The model used by Schwarz is a charged particle surrounded by a layer of counterions with charge e and mobility u. The counterion layer is assumed to be infinitesimally thin and ions move only tangentially along the surface. The surface charge density in the absence of electrical field is uniformly σ_0 but application of a field produces a charge density $\bar{\sigma}$, in excess of that of constant value σ_0. The surface flux of counterions is given by

$$j_e = e_0 \sigma u E_s \tag{6.127}$$

where the field intensity E_s at the surface is defined by

$$E_s = -(1/R)\partial\psi_s/\partial\theta$$

in spherical coordinates where ψ_s is the surface potential and R is the radius. In addition, there will be a diffusion controlled counter flux

$$j_d = -(ukT/R)\partial\sigma/\partial\theta. \tag{6.128}$$

Thus, the continuity equation is

$$d\sigma/dt = \nabla\cdot(j_e + j_d). \tag{6.129}$$

The full equation is

$$\frac{d\sigma}{dt} = \frac{ukT}{R^2}\frac{1}{\sin\theta}\frac{\partial}{\partial\theta}\left(\sin\theta\frac{\partial\sigma}{\partial\theta} + \frac{e_0}{kT}\frac{\partial\psi_s}{\partial\theta}\sigma\sin\theta\right). \tag{6.130}$$

In general, the excess charge density $\bar{\sigma}$ is negligibly small compared with the mean charge density σ_0. Furthermore, since we apply a periodic field $E = E_0\exp(j\omega t)$, the change of excess charge density $\bar{\sigma}$ will also be periodic. Thus we can write

$$\frac{\partial\sigma}{\partial t} = \frac{\partial\bar{\sigma}}{\partial t} = j\omega\bar{\sigma}. \tag{6.131}$$

Substituting these into (6.130), we can simplify the continuity equation:

$$j\omega\bar{\sigma} = \frac{ukT}{R^2}\frac{1}{\sin\theta}\frac{d}{d\theta}\left(\sin\theta\frac{d\bar{\sigma}}{d\theta} + \frac{e_0\sigma_0}{kT}\sin\theta\frac{d\psi_s}{d\theta}\right). \tag{6.132}$$

The following solutions for $\bar{\sigma}$ and ψ_s are assumed for (6.132):

$$\bar{\sigma} = \sum_n \alpha_n P_n(\cos\theta) \tag{6.133}$$

$$\psi_s = \sum_n \beta_n P_n(\cos\theta) \tag{6.134}$$

where $P_n(\cos\theta)$ is the Legendre polynomial. Substituting these into equation (6.132), an equation relating α_n and β_n is obtained:

$$\alpha_n = -\frac{\beta_n}{1 + j\omega R^2/n(n+1)ukT}\frac{e_0\sigma_0}{kT}. \tag{6.135}$$

As shown below, all the terms with $n \neq 1$ disappear and thus we obtain the following equation relating $\bar{\sigma}$ and ψ_s:

$$\bar{\sigma} = -\frac{1}{1 + j\omega\tau} \frac{e_0\sigma_0}{kT}\psi_s \qquad (6.136)$$

where τ is the relaxation time and is defined as

$$\tau = R^2/2ukT. \qquad (6.137)$$

The surface potential ψ_s is determined as the limiting case of either the potential outside, ψ_a, or that inside, ψ_i, the particle

$$\lim_{r \to R} \psi_a = \lim_{r \to R} \psi_i = \psi_s. \qquad (6.138)$$

As before, the potential functions ψ_a and ψ_i are calculated by solving the Laplace equation with two boundary conditions. One boundary condition is (6.138) and the other is

$$\varepsilon_a^* \frac{\partial \psi}{\partial r} - \varepsilon_i^* \frac{\partial \psi}{\partial r} = 4\pi e_0 \bar{\sigma} \qquad \text{for } r \to R. \qquad (6.139)$$

The solution of the Laplace equation is obtained using (6.138), (6.139) and (6.136):

$$\psi_a = -Er\cos\theta + \frac{\bar{\varepsilon}_i^* - \varepsilon_a^*}{\bar{\varepsilon}_i^* + 2\varepsilon_a^*}R^3E\frac{\cos\theta}{r^2} \qquad (r > R) \quad (6.140)$$

$$\psi_i = -\frac{3\varepsilon_a^*}{2\varepsilon_a^* + \bar{\varepsilon}_i^*}Er\cos\theta \qquad (r < R) \qquad (6.141)$$

where $\bar{\varepsilon}_i^*$ is defined by the following equation:

$$\bar{\varepsilon}_i^* = \varepsilon_i^* + \frac{4\pi}{1 + j\omega\tau} \frac{e_0^2 R\sigma_0}{kT}. \qquad (6.142)$$

This equation clearly demonstrates that the presence of a polarizable counterion layer adds a frequency dependent dielectric increment which is given by

$$\Delta\varepsilon = \frac{4\pi}{1 + j\omega\tau} \frac{e_0^2 R\sigma_0}{kT}. \qquad (6.143)$$

In (6.142), the term $e^2R\sigma_0/kT$ can be written as $\lambda R/D_s$ where λ is the surface conductivity $(= e^2\sigma_0 u)$ and D_s is a diffusion constant $(= ukT)$. Thus an alternate expression for (6.142) is obtained:

$$\bar{\varepsilon}_i = \varepsilon_i + 4\pi\lambda R/D_s. \qquad (6.144)$$

A numerical calculation using (6.143) or (6.144) demonstrates that the dielectric increment of a charged spherical particle suspended in aqueous media rises far above that of the suspending electrolyte solution. Substituting (6.143) into Maxwell's equation we can derive

$$\varepsilon^* = \frac{1-p}{1+p/2}\varepsilon_a^* + \frac{9}{4}\frac{p}{(1+p/2)^2}\,\overline{\varepsilon}_i^* \qquad (6.145)$$

where assumptions are made such as $\varepsilon_i^* \ll \varepsilon_a^*$. p is the volume fraction. From this, we can finally obtain the dielectric increment $\Delta\varepsilon$ at arbitrary volume fraction:

$$\Delta\varepsilon = (9/4)\frac{4\pi p}{(1+p/2)^2}\frac{\sigma_0 e_0^2 R}{kT}. \qquad (6.146)$$

An example of some numerical calculations is shown in figure 6.18, along with the experimental results obtained with a colloidal suspension. The calculation was performed using the following data: $\sigma_0 = 2 \times 10^{13}$ cm^{-2}, $k_a = 0.7$ mS cm^{-1}, $R = 0.585$ μm and $p = 0.22$. As shown, the agreement between theory and the experimental result is exceedingly good. Equations (6.146) and (6.137) predict that the dielectric increment is proportional to R and that the relaxation time is proportional to R^2. These predictions are also confirmed by experiment. It should be pointed out that the theory we derived was with the assumptions that the size of spherical particles is uniform and that there is no distribution of relaxation time. The counterion theory was extended to a non-spherical particle by Takashima (1967a). His calculation will be discussed in Chapter 7 where the dielectric behaviour of DNA is discussed.

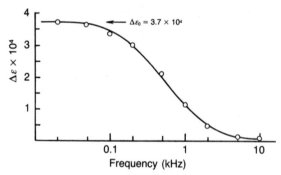

Figure 6.18 The frequency profile of the dielectric constant calculated from (6.146). The circles are the measured points with a polystyrene particle suspension. Reproduced with permission from Schwarz (1962). Copyright (1962) American Chemical Society.

In Schwarz's theory, only the tangential ion flux along the surface was considered and the flux in the radial direction was ignored completely. Also, only the reactive component of surface admittance is considered in this theory. Thus, the in-phase ion fluxes, i.e. the surface conductance, play no role in his mathematical formulation. A hybrid theory which is the composite of O'Konski's model and Schwarz's theory was proposed by Schurr (1964). Instead of the continuity equation used by Schwarz (see equation (6.130)), (6.121) was used by Schurr to derive the following equation:

$$\varepsilon_i^* = \left(\varepsilon_i + \frac{2\varepsilon_s}{R}\right) - j\frac{4\pi}{\omega}\left(k_i + \frac{2\lambda}{R}\right). \tag{6.147}$$

It should be noted that the first term on the RHS of this equation is identical to the expression obtained by Schwarz and that the imaginary part is the same as that obtained by O'Konski. The dielectric increment derived by Schurr is shown below:

$$\varepsilon^* = \varepsilon_\infty^* + \left(\frac{2\lambda}{R}\frac{4\pi}{j\omega} + \frac{4\pi}{1 + j\omega\tau}\frac{e_0^2\sigma_0 R}{kT}\right)\frac{9}{4}\frac{p}{(1 + p/2)^2}. \tag{6.148}$$

As shown, the first term on the RHS is due to surface conductivity. However, this term alters only the imaginary component of the dielectric increment.

As discussed earlier, Maxwell's mixture equation was obtained by solving the Laplace equation, assuming that the charges are driven by applied fields from an infinitesimally thin layer around the surface of the particle. Actually, condensed ions form a layer, of Debye screening length, and this leads to a reduced dielectric increment as shown by Garcia et al (1985). In Schwarz's calculation, the counterion layer was also assumed to be very thin and potential functions were calculated by solving the Laplace equation. As mentioned above, the thickness of the counterion layer is by no means negligible and, under certain conditions, ions spread out and form a diffuse cloud around the particle. Grosse and Foster (1987) pointed out that Schwarz's model can be replaced by a charged particle which is surrounded by a thin ion layer at a distance of $1/\chi_m$ (χ_m is the inverse of Debye length). If we assume the permittivity of a charge layer to be ε_m, then the system can be considered as two capacitors in series, one representing the permittivity of the core particle (equivalent to Schwarz's particle) and the other, the permittivity of the Debye ion layer. Thus, the equation derived by Schwarz can be modified as follows:

$$\Delta\varepsilon = \frac{4\pi\lambda R/D_s}{1 + (4\pi\lambda R/D_s)/\varepsilon_m\chi_m R} \tag{6.149}$$

where D_s is the diffusion constant of a counterion on the surface of the particle, λ is the surface conductivity and R is the radius. Using Schwarz's definition of surface conductivity, $\lambda = e_0^2\sigma_0 u$ and $D_s = ukT$, it is evident that the numerator of this equation is exactly the dielectric increment calculated by Schwarz (see equation (6.144)). The presence of a correction term in the denominator (the second term) reduces the dielectric increment estimated by Schwarz by a certain amount. In fact, Grosse and Foster (1987) found that this term is large enough to reduce the dielectric increment considerably. Thus, the large dielectric increment calculated by Schwarz may be subject to certain error because of the neglect of the thickness of the counterion layer.

6.12 The Counterion Polarization of Rod-like Polymers

The counterion polarization of charged spherical particles of polyions has been discussed in previous subsections. In the model used in these calculations, particles are surrounded by a counterion cloud, without, however, any consideration of the binding of the ions to the charged sites. In what follows, we will discuss a different approach to counterion polarization with rod-like polyelectrolyte molecules as a model.

Mandel's theory (1961)
The model used by Mandel (1961) for rod-like polyelectrolyte molecules is illustrated by figure 6.19. Charged sites are represented as a linear array of square potential wells of height h. Counterions are not allowed

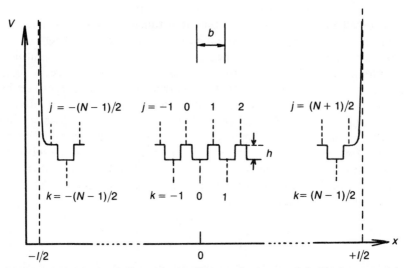

Figure 6.19 The one-dimensional discrete lattice model. Each potential well represents a charged site where counterions are bound. The sharp rise of the potential at the ends indicates that counterions cannot leave the molecule. (From Mandel 1961.) Reproduced by permission of Taylor & Francis Ltd.

to move in radial directions but are allowed to slide along the fibre. A sudden increase in the potential energy at both ends of the polymer indicates that ions cannot move away at the end of the molecule. The origin of the x-coordinate is located at the centre of the rod and the length of the molecule is defined by

$$l = (N + 1)b \tag{6.150}$$

where N is the number of charged sites and b is the distance between them. The dipole moment of the polymer along the x axis is given by the following equation:

$$m_l = \sum_i ex_i - \sum_{r=1}^{n} ze\left(\sum_i p_r(x_i)x_i + \sum_j p_r(x_j)x_j\right) \tag{6.151}$$

where $p_r(x_i)$ represents the probability that a mobile charge r is located at a site x_i. Likewise, $p_r(x_j)$ represents the same probability at an uncharged site x_j. In this equation, the centre of gravity of the charges is assumed to be at $x = 0$. Because of the symmetry involved, the first term on the RHS of (6.151), i.e. the product of charge and the coordinate of the charged site, will vanish.

In the absence of applied field, the probability $p_r(x)$ is defined by

$$p_r(x) = \exp(-V(x)/kT)\left(\sum_{\text{sites}} \exp(-V(x)/kT)\right)^{-1}. \tag{6.152}$$

Noting that there are two types of sites, i.e. charged sites (x_i) with a potential energy $-V_0$ and uncharged sites (x_j) with an energy $-V_0 + h$, we can calculate $p_r(x_i)$ and $p_r(x_j)$ as follows:

$$p_r(x_i) = \frac{\exp(-V_0/kT)}{N\exp(-V_0/kT) + (N+1)\exp(-V_0 + h)/kT}$$

$$\simeq \frac{1}{N[1 + \exp(-h/kT)]} \tag{6.153}$$

$$p_r(x_j) = \frac{\exp[(-V_0 + h)/kT]}{N\exp(-V_0/kT) + (N+1)\exp(-V_0 + h)/kT}$$

$$\simeq \frac{\exp(-h/kT)}{N[1 + \exp(-h/kT)]}. \tag{6.154}$$

Because of the symmetric distribution of ions in the absence of electrical field, the net dipole moment vanishes, thus

$$m_l = -\frac{nze}{N[1 + \exp(-h/kT)]}\left(\sum_i x_i + \exp(-h/kT)\sum_j x_j\right) = 0 \tag{6.155}$$

where n is the number of ions. If a static electrical field is applied along the x axis, the probability functions $p_r(x_i)$ and $p_r(x_j)$ will change to

$$p_r(x_i) = \frac{\exp(zeEx_i/kT)}{S} = \frac{\exp(qi/kT)}{S} \tag{6.156}$$

$$p_r(x_j) = \frac{\exp[(-h + zeEx_j)/kT]}{S} = \frac{\exp(-h/kT)\exp[q(j - \frac{1}{2})/kT]}{S} \tag{6.157}$$

where S is defined by

$$S = \sum_{-(1/2)(N-1)}^{(1/2)(N-1)} \exp(zeEx_i/kT) + \sum_{-(1/2)(N-1)}^{(1/2)(N+1)} \exp[(-h + zeEx_j)/kT] \tag{6.158}$$

$q = zebE$ and $b = l/N$. Substituting these values into (6.151), we obtain an expression for the dipole moment due to mobile charges in the presence of an electric field:

$$m_l = -\frac{nzeb}{S}\left(\sum_i i \exp(qi/kT)\right.$$

$$\left. + \exp(-h/kT)\sum_j (j - \tfrac{1}{2})\exp[q(j - \tfrac{1}{2})/kT]\right). \tag{6.159}$$

The summations can be performed by first assuming that N is very large and, consequently, b $(= l/N)$ is numerically small and by replacing the exponentials by the first term of the series expansion. Considering the symmetrical distribution of the sites of i and j with respect to the origin, m_l may be written as

$$m_l = -\frac{nzeqbN^2/kT}{12} = -\frac{z^2ne^2b^2N^2}{12kT}E. \tag{6.160}$$

Dividing m_l by E, we obtain the polarizability α. According to (6.160), the mobile ion fluctuation moment of linear polymers is proportional to the number of ions and to the square of length.

The relaxation time of mobile ion fluctuation can be calculated as follows. For the sake of simplicity, we assume that the dipole moment is determined by the distribution of bound ions, i.e. we consider the i site only. If we define ξ_r^0 as the mean position of a bound ion in the absence of a field and ξ_r as the mean position when an electrical field is applied, the dipole moment can be defined by

$$m_l = -ze\sum_{r=1}^{n}(\xi_r - \xi_r^0) = -nze\overline{\Delta\xi}. \tag{6.161}$$

Namely, the displacement of bound ions produces a moment of m_l. Thus, the displacement can be defined by combining this equation with (6.160). The relaxation time is calculated by dividing the displacement by its mean velocity, i.e. $v = uzeE$ (where u is the mobility of ions), hence we obtain

$$\tau = \frac{\overline{\Delta\xi}}{v} = \frac{b^2N^2}{12ukT} = \frac{l^2}{12ukT}. \tag{6.162}$$

Since $ukT = D$, the diffusion coefficient, the relaxation time due to mobile charges is proportional to the inverse of the diffusion coefficient for ions which are trapped in potential wells. This diffusion coefficient is different from the one for free ions in bulk media. The calculation of the numerical value for this parameter is not a simple task. This equation shows that the relaxation time of counterion fluctuation is proportional to the square of the length of polyions.

Oosawa's theory (1970)

The underlying assumption used in Mandel's theory is that the field induced displacement of ions along the fibre axis is statistically uniform. In the theory proposed by Oosawa, the ion dispacement is considered to have various amplitudes, i.e. $l, l/2, l/4, \ldots$. Therefore, the fluctuation of ion concentration from the mean value, $\delta C_+(x)$, can be expressed in terms of harmonic components of the Fourier series:

$$\delta C_+(x) = \sum_i [c_i \cos(2\pi i x/l) + c_i' \sin(2\pi i x/l)] \tag{6.163}$$

where the coefficients c_i amd c_i' are defined by

$$c_i = \left(\frac{2}{l}\right)\int_0^l \delta c_+(x) \cos(2\pi i x/l)\,dx \tag{6.164}$$

$$c_i' = \left(\frac{2}{l}\right)\int_0^l \delta c_+(x) \sin(2\pi i x/l)\,dx \tag{6.165}$$

namely, these are the amplitudes of fluctuation of the ith mode. The mean square fluctuation of c_i or c_i' is given by

$$\langle c_i^2 \rangle = \langle c_i'^2 \rangle = \frac{1}{(l/2)(1/c_+) + (l/2)^2(\phi_i/kT)} \tag{6.166}$$

where ϕ_i is the Fourier component of the interaction energy $\phi(r)$ which is defined by

$$\phi_i = \left(\frac{2}{l}\right)\int_0^l \phi(r) \cos(2\pi i r/l)\,dr. \tag{6.167}$$

The dipole moment due to ion fluctuation $\delta c_+(x)$ is

$$\mu = e_0 \int_0^l (x - l/2)\delta c_+(x)\,dx \tag{6.168}$$

or using the Fourier series expansion of δc_+, the mean square dipole moment is given by

$$\langle \mu^2 \rangle = e_0^2 l^2 \sum_i (l/2\pi i)^2 \langle c_i'^2 \rangle. \tag{6.169}$$

Using (6.166), we obtain

$$\langle \mu^2 \rangle = n_+ e_0^2 l^2 (2\pi^2)^{-1} \sum_i (1/i^2)/(1 + n_+\varphi_i/2kT). \tag{6.170}$$

The polarizability α is calculated, using $\alpha_i = \langle \mu^2 \rangle_{E=0}/kT$, as

$$\alpha_i = (n_+ e_0^2/kT)l^2(2\pi^2)^{-1}(1/i^2)/(1 + n_+\phi_i/2kT). \tag{6.171}$$

This equation indicates that the total polarizability of charged linear polyions can be computed as the sum of the contribution of ion displacements which consist of the fundamental mode and harmonic components. As shown, the magnitude of the ith component of polarizability is proportional to the inverse of i^2, the square of wave number, and the fundamental mode is the dominant component. The polarizabilities diminish rapidly for higher harmonic components. Numerical calculation shows that the contributions of harmonic components with i-values higher than 3–4 are negligible.

In order to calculate the frequency profile of polarizability curves using (6.171), we still need to compute the relaxation time of each Fourier component. Analogous to the amplitude of polarizability, the relaxation time consists also of fundamental and harmonic components, as shown by the following:

$$\tau_i = (\zeta/kT)(l/2\pi i)^2/(1 + n_+\varphi_i/2kT) \tag{6.172}$$

where ζ is the frictional coefficient of bound counterion movement. The transformation of polarizability α_1 into the dielectric constants ε' and ε'' can be performed using the usual procedure. The theoretical curves calculated using these equations are shown in figure 6.20. The broken curves are ε' and ε'' due to individual harmonic components. For example, the $i = 1$ term contributes a large dielectric increment at low frequencies while the contributions of harmonic components diminish rapidly because of the inverse i^2-dependency of $\Delta\varepsilon_i$. Close examination of this figure reveals that the slope of the dispersion curve (ε') is steeper at lower frequencies than at higher frequencies. The inset of figure 6.20 shows calculated Cole–Cole plots. The circular plots reveal even more clearly the skewness of the dispersion curve. This is due to the dependence of dielectric increment and relaxation time on i^2. Skewed dispersion curves have, in fact, been observed for DNA by Takashima (1966), as will be discussed in Chapter 7. The dielectric relaxation of DNA is believed to be due to the fluctuation of counterions along the axis of the double helix and it seems to be an ideal sample for the test of Oosawa's theory.

Unlike Mandel's theory, Oosawa's model does not consider the binding of counterions to charged sites. Although the binding is known to have profound effects on the dielectric behaviour of polyions, Oosawa's theory, because of its use of the continuum model, is unable to explain the experimental results. In many ways, discrete lattice models represent the dielectric properties of polyelectrolytes more realistically because the binding of counterions does occur in reality and its effects on the relaxation process cannot be ignored.

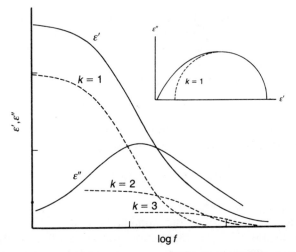

Figure 6.20 The dielectric constant calculated by Oosawa's theory. The full curve is the composite of the harmonic components and the broken curves show the first three modes. The inset shows the calculated Cole–Cole plot. The broken circle is the arc using the lowest mode only. (From Oosawa 1970.) Reproduced by permission of John Wiley & Sons Inc. © 1970.

A theory by Warashina and Minakata (1973) which is based on a discrete lattice model will be discussed in the following. The inset of figure 6.21 illustrates a potential well of charged site with three different

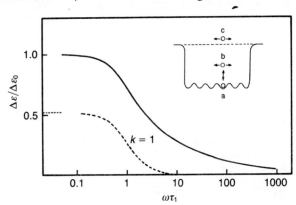

Figure 6.21 The dielectric constants of polyelectrolytes calculated by Warashina and Minakata (1973). The broken curve indicates the contribution of the lowest mode, $n = 1000$, $l = 4 \times 10^{-5}$, $n_0 = 0.6$ and $\varepsilon_0 = 80$. The inset shows a schematic representation of counter-ions: a, bound to charged sites, b, bound but mobile within the potential well and c, a free ion. (From Warashina and Minakata 1973.) Reproduced by permission of the American Institute of Physics.

types of counterions: (i) counterions tightly bound to charged sites: (ii) ions which are bound but mobile within the potential well and (iii) a free counterion. If we assume that the number of counterions in each lattice point is limited to 0 or 1, the partition function for a single lattice point is defined by the following equation:

$$Z_i = 1 + \exp[(g_c - E_i)/kT] \tag{6.173}$$

where g_c is the chemical potential of the counterion and E_i is the interaction energy at a site i. E_i is defined by

$$E_i = E^0 + \sum_{r \neq i} \phi(|i - r|)n_r \tag{6.174}$$

where E^0 is the binding energy and $\phi(|i - r|)$ represents repulsion between sites i and r. n_r is the number of bound ions at site r. This equation indicates that the interaction energy is affected indirectly by ion binding at other lattice points. If E_i is fixed, then the mean value of n_i, the number of ions at the ith site, is defined by the following equation:

$$\langle n_i \rangle = (kT/Z_i)(\partial Z_i / \partial g_c) \tag{6.175}$$

$$= \{1 + \exp[(E_i - g_c)/kT]\}^{-1} \tag{6.176}$$

and its fluctuation is defined by

$$\delta n_i = n_i - n_0 \tag{6.177}$$

where n_0 is the number of ions averaged over all lattice sites at equilibrium, having a range between 0 and 1. The Fourier transform δn_q of δn_i is given by

$$\delta n_q = (1/n)\sum_i \delta n_i \exp(jqi) \tag{6.178}$$

where $j = \sqrt{-1}$ and

$$q = 2\pi m/n \qquad (m = 0, \pm 1, \pm 2, \ldots, \pm n/2) \tag{6.179}$$

and n is the total number of charged sites. The mean square fluctuation of the q-mode is expressed by

$$\langle |\delta n_q|^2 \rangle = (1/n^2)\sum_i \sum_r \langle \delta n_i \delta n_r \rangle \exp[jq(i - r)]. \tag{6.180}$$

After evaluating the sum, this equation becomes

$$\langle |\delta n_q|^2 \rangle = [n_0(1 - n_0)/n]/[1 + nn_0(1 - n_0)\phi(q)/kT] \tag{6.181}$$

where $\phi(q)$ is the interaction energy of the qth Fourier component. The dipole moment induced by the fluctuation of n_i is

$$\mu = eb\sum_i [i - (n + 1)/2]\delta n_i \tag{6.182}$$

where b is the interval between charged sites and e is charge. The mean square moment is given by

$$\langle \mu^2 \rangle = (ebn)^2 \sum_q \langle |\delta n_q|^2 \rangle /(1 - \cos q) \tag{6.183}$$

and the polarizability α is then defined, as $\langle \alpha \rangle = \langle \mu^2 \rangle_{E=0}/3kT$, by

$$\langle \alpha \rangle = (3kT)^{-1}(eb)^2 \sum_q \frac{nn_0(1 - n_0)}{[1 + nn_0(1 - nn_0)\phi(q)/kT](1 - \cos q)}. \tag{6.184}$$

The expression for relaxation time τ_q is

$$\tau_q = \frac{1}{2w(1 - \cos q)[1 + nn_0(1 - n_0)\phi(q)/kT]} \tag{6.185}$$

where w is the rate constant for the transition of a counterion from the ith to the rth site.

If we impose no restriction on the number of counterions at each lattice point, the partition function can be defined in terms of a Boltzmann distribution:

$$Z_i = (n_i!)^{-1} \exp[(g_c - E_i)n_i/kT] \tag{6.186}$$

and the mean value of n_i is

$$\langle n_i \rangle = \exp[(g_c - E_i)/kT]. \tag{6.187}$$

The division of the RHS of (6.186) by $n_i!$ is necessary to count correctly the number of different ways that indistinguishable ions can be bound. Polarizability, in terms of the mean square fluctuation of mode q can be expressed as

$$\langle \alpha \rangle = \frac{(eb)^2}{3kT} \sum_q \frac{nn_0}{(1 + nn_0\phi(q)/kT)(1 - \cos q)} \tag{6.188}$$

and the relaxation time τ_q can be expressed as

$$\tau_q = \frac{1}{2w(1 - \cos q)(1 + nn_0\phi(q)/kT)}. \tag{6.189}$$

A comparison of these equations with (6.184) and (6.185) demonstrates that the only difference between them is nn_0 being replaced by $nn_0(1 - n_0)$. These equations reduce to Oosawa's equations ((6.171) and (6.172)) by replacing $1 - \cos q$ with $q^2/2$ and $w = kT/bz$. Also, by assuming that the interaction energy $\phi(q)$ is zero, (6.188) is virtually the same as Mandel's equation, i.e. (6.160). Some of the results of their calculations are illustrated in figure 6.21.

Matrix method
Counterion polarization has elicited much interest from many investigators. As has been discussed, there are many theories on this problem based on various models and using different mathematical methods. The

reason for this widespread interest may be found in the fact that the problem of counterion distribution around polyelectrolyte molecules has been one of central interest among polymer scientists. Counterion polarization is usually treated as the perturbation of ion distribution by external fields. Therefore, this problem is a subset of an even more fundamental question about polyelectrolytes and counterions.

As has been discussed, early counterion polarization theories by O'Konski (1960) and by Schwarz (1962) are based on the use of macroscopic models. Counterions are treated as a diffuse ion atmosphere with no specific consideration of their binding to charged sites. However, the dielectric increment of polyions is known to depend strongly on the binding of mono- and/or di-valent ions. These observations cannot be explained by the continuum models discussed above. In contrast, the discrete one-dimensional potential well model is better suited to analyze the effect of ion binding. In the following, we will discuss more theories which use matrix methods for the evaluation of partition functions, an essential step to calculating the polarizability or dielectric increment.

The theory by McTague and Gibbs (1966) uses a one-dimensional discrete potential well model which is similar to the one used by Mandel. The partition function is evaluated using the matrix method of Kramer and Wannier (1941) and Montroll (1941) and the mean dipole moment due to counterion perturbation is calculated. Let us find the relationship between mean dipole moment and partition function. The induced mean dipole moment is defined by the following derivative:

$$\langle m \rangle = (-\partial A / \partial E)_{T, V, \lambda} \tag{6.190}$$

where A is the Helmholtz free energy and λ is the activity of the counterion and is given by (6.192). Since Helmholtz free energy is related to the log of the partition function Γ, $\langle m \rangle$ is also defined as

$$\langle m \rangle = -kT(\partial \ln \Gamma / \partial E)_{T, V, \lambda}. \tag{6.191}$$

The probability that a counterion is bound at a particular site is given, without consideration of the ion–ion interaction, by the following equation in the absence of electrical field:

$$P_i = K\lambda' \exp(-ze\psi/kT) = \lambda \tag{6.192}$$

where K is a constant, λ' is the absolute activity of the counterion and ψ is the electrostatic potential. The probability of counterion occupation on a site immediately preceded by an occupied site is

$$P_{ii} = \zeta P_i \qquad \text{or} \qquad = \zeta\lambda \tag{6.193}$$

where ζ is introduced to account for the interaction between nearest neighbours which are occupied by counterions. ζ is of the following form:

$$\zeta = \exp(-z^2 e^2 / \sigma k T r_{i,i+1}) \tag{6.194}$$

where σ is a screening parameter and $r_{i,i+1}$ is the distance between adjacent binding sites. The partition function in the absence of an electrical field is

$$\Gamma^+(N) = \Gamma^+(N) + \Gamma^-(N) \tag{6.195}$$

where $\Gamma^+(N)$ represents all the terms in which an Nth site has a bound counterion and $\Gamma^-(N)$ represents those of an unoccupied Nth site. Each of these can in turn be reduced into contributions from $(N - 1)$ sites:

$$\Gamma^+(N) = \zeta \lambda \Gamma^+(N - 1) + \lambda \Gamma^-(N - 1) \tag{6.196}$$

and

$$\Gamma^-(N) = \Gamma^+(N - 1) + \Gamma^-(N - 1). \tag{6.197}$$

The term $\zeta \lambda \Gamma^+(N - 1)$ represents the contribution of an $(N - 1)$ site which is occupied by an ion while $\Gamma^-(N - 1)$ represents the contribution of an unoccupied $(N - 1)$ site. If an electrical field is applied, ion binding will be perturbed and we have to alter the partition functions by introducing a parameter η as shown below:

$$P_{i+1} = P_i \eta = P_i \exp(-zer l_\theta E / kT) \tag{6.198}$$

where l_θ is the direction cosine between the molecular axis and the electrical field vector. r is the distance between neighbouring sites. We obtain a similar equation for two sites separated by a distance of $2r$ (two sites which are separated by an intervening site):

$$P_{i+2} = P_i \eta^2. \tag{6.199}$$

Introducing the factor η, the partition functions are now written as follows:

$$\Gamma^+(N) = \lambda \zeta \eta_0 \eta^{N-1} \Gamma^+(N - 1) + \lambda \eta_0 \eta^{N-1} \Gamma^-(N - 1) \tag{6.200}$$

$$\Gamma^-(N) = \Gamma^+(N - 1) + \Gamma^-(N - 1) \tag{6.201}$$

where η_0 is inserted to place the zero potential energy at the centre of the molecule. This factor ensures that the moment in the absence of field reduces to zero. The value of η_0 is $\eta^{-(N+1)/2}$ for odd N. The partition function can be given as the product of these terms. In matrix form, it becomes

$$\Gamma(N, \eta, \lambda, \zeta) = [1 \ 1] \begin{pmatrix} \lambda \zeta \eta_0 \eta^{N-1} & \lambda \eta_0 \eta^{N-1} \\ 1 & 1 \end{pmatrix} \begin{pmatrix} \lambda \zeta \eta_0 \eta^{N-2} & \lambda \eta_0 \eta^{N-2} \\ 1 & 1 \end{pmatrix}$$

$$\times \ldots \times \begin{pmatrix} \lambda \zeta \eta_0 \eta & \lambda \eta_0 \eta \\ 1 & 1 \end{pmatrix} \begin{pmatrix} \lambda \eta_0 \\ 1 \end{pmatrix}. \tag{6.202}$$

The row vector $[1 \ 1]$ is an operator, to sum the components of $\Gamma(N)$. If

$\zeta = 1$, the nearest neighbour repulsion is zero and the partition function can be simplified as

$$\Gamma(N) = \prod_{n=0}^{N-1}(1 + a\eta^n) \qquad \text{where } a = \lambda\eta^{-(N+1)/2}. \qquad (6.203)$$

Inserting the partition function in (6.191), we can now calculate the mean dipole moment created by counterion fluctuation:

$$\langle m \rangle = zerl_\theta(\partial \ln \Gamma/\partial \ln E)_{T, V, \lambda}. \qquad (6.204)$$

The polarizability α can be calculated using $\langle \alpha \rangle = \langle m \rangle/E$. In addition, the fraction of binding sites occupied by an ion is given by

$$\frac{\langle n \rangle}{N} = \frac{\lambda}{N\Gamma} \frac{\partial \Gamma}{\partial \lambda} = \frac{\partial \ln \Gamma}{\partial \ln \lambda} \frac{1}{N}. \qquad (6.205)$$

This equation is of particular interest because it has been shown that the induced moment of a polyelectrolyte molecule depends strongly on the fraction of occupied sites.

Figure 6.22 shows the dependence of induced moment on the number of charged sites N. As shown, $\langle m \rangle$ is proportional to N^2 above 10^3.

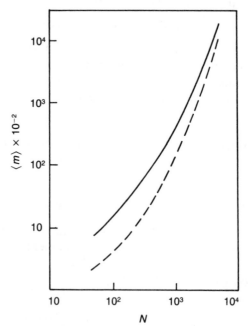

Figure 6.22 The dependence of the induced dipole moment on the degree of polymerization. $\zeta = 0.1$, $\eta = 1.00001$, $\lambda = 0.1$ for the broken curve and $\lambda = 1.0$ for the full curve. (From McTague and Gibbs 1966.) Reproduced by permission of the American Institute of Physics.

However, the slope of this plot changes and becomes proportional to N below 100. This result may indicate the importance of the end effects in short rods. Figure 6.23 shows the dependence of induced moment on the fraction of occupied sites. The parabolic form of the theoretical prediction is in agreement with experimental results. The parabolic dependence of $\langle m \rangle$ on the fraction of occupied sites has been observed, so far, only with synthetic polyelectrolytes. Biological polymers such as DNA become increasingly unstable at low ionic strengths and therefore this limits the measurements to a range where the binding sites are already extensively occupied. Thus, further addition of ions such as Ca or Mg only causes a decrease in induced moment (see figure 6.24).

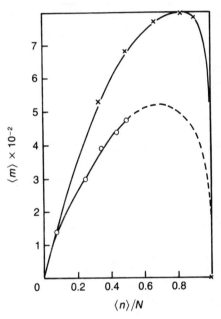

Figure 6.23 The dependence of induced moment on the fraction of occupied sites. $N = 1001$. The crosses are calculated with $\zeta = 1.0$ and the circles with $\zeta = 0.1$. (From McTague and Gibbs 1966.) Reproduced by permission of the American Institute of Physics.

The final topic of this chapter is another matrix method discussed by Minakata et al (1972). The model uses a charged rod having three types of binding sites: (i) where no counterion is bound; (ii) a site occupied by a monovalent ion and (iii) a site occupied by a divalent ion. The interaction energies of the site i with bound ions in states (i), (ii) and (iii) are assumed to be 0, $-eV_1$ and $-2eV_2$ respectively. Moreover, mutual repulsion is considered when two divalent ions or monovalent

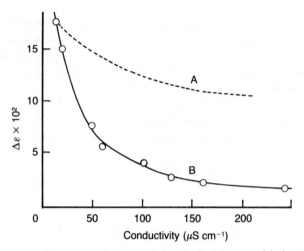

Figure 6.24 The change of the dielectric increment with increasing conductivities of DNA solution. Curve A with NaCl and curve B with $MgCl_2$. (From Takashima 1967b.) Reproduced by permission of John Wiley & Sons, Inc. © 1967.

and divalent ions occupy adjacent sites with interaction energies W_{33} and W_{23}.

The mean square dipole moment $\langle \mu^2 \rangle$ is non-vanishing and is given by

$$\langle \mu^2 \rangle = \sum_i \sum_j \langle (n_i - \langle n \rangle)(n_j - \langle n \rangle) \rangle e^2 r_i r_j \tag{6.206}$$

$$= \sum_i (\langle n_i^2 \rangle - \langle n \rangle^2) e^2 r_i^2 + \sum_i \sum_{j \neq i} (\langle n_i n_j \rangle - \langle n \rangle^2) e^2 r_i r_j \tag{6.207}$$

$$= e^2 \sum_i \Delta n^2 r_i^2 + 2e^2 \sum_i \sum_{m>i} (\Delta_m n) r_i r_{i+m} \tag{6.208}$$

where $\Delta n^2 = \langle n_i^2 \rangle - \langle n \rangle^2$, $\Delta_m n = \langle n_i n_{i+m} \rangle - \langle n \rangle^2$ and n_i is the number of bound ions at the ith site. With some approximation and by introducing $r_i = ib$ and $l = nb$ (where b is the distance between adjacent sites), we obtain

$$\langle \mu^2 \rangle = \tfrac{1}{12}(ne^2l^2\Delta n^2) + \tfrac{1}{6}\left(ne^2l^2\sum_m \Delta_m n\right) \tag{6.209}$$

where $\Delta n^2 = \langle n_i^2 \rangle - \langle n \rangle^2$ and $\Delta_m n = \langle n_i n_{i+m} \rangle - \langle n \rangle^2$. In order to compute $\langle \mu \rangle$, Δn and $\Delta_m n$ must be evaluated.

A matrix **M** is formulated which indicates the statistical weights connecting the sites i and $i + 1$:

$$\mathbf{M} = \begin{pmatrix} 1 & 1 & 1 \\ \lambda_1 u & \lambda_1 u & \lambda_1 \zeta u \\ \lambda_2 v & \lambda_2 \zeta v & \lambda_2 \delta v \end{pmatrix} \qquad (6.210)$$

where λ_1 and λ_2 are the activities of the mono- and di-valent counterions. u, v, ζ and δ are defined as follows:

$$u = \exp(eV_1/kT) \qquad v = \exp(2eV_2/kT)$$
$$\zeta = \exp(-W_{23}/kT) \qquad \delta = \exp(-W_{33}/kT). \qquad (6.211)$$

The partition function Z is given by the trace of matrix $\mathbf{M}$ raised to the nth power:

$$Z = \mathrm{Tr}(\mathbf{M}^n) \qquad (6.212)$$

where n is the number of charged sites. The computation of Δn^2 and $\Delta_m n$ in (6.208) can be performed using the partition function Z. The calculation is quite complex and, therefore, only the results of the calculation are shown, without the detail.

The dielectric increment of randomly oriented rod-like polyions is obtained from (6.208) and $\alpha_E = \langle \mu^2 \rangle_{E=0}/kT$:

$$\Delta \varepsilon = \tfrac{4}{3}\pi NB \frac{ne^2 l^2}{12kT}\left(\Delta n^2 + 2\sum_m \Delta_k n\right) \qquad (6.213)$$

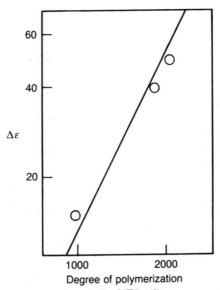

Figure 6.25 The dependence of the dielectric increment on the degree of polymerization. The circles were obtained with polyacrylic acid. (From Minakata *et al* 1972.) Reproduced by permission of John Wiley & Sons, Inc. © 1972.

where N is the number of polyions per unit volume, l is the length of the polyion, and B is the ratio of internal and external fields. Figure 6.25 shows the dependence of the dielectric increment on the degree of polymerization or length l. As shown, $\Delta\varepsilon$ is proportional to l^2. This prediction is verified by experiment (open circles). Figure 6.26 illustrates the dielectric increment as a function of the ratio of divalent (Ca^{++}) and monovalent (Na^+) ions. As shown, the dielectric increment increases with increasing Ca ion concentration and has a maximum value at a ratio of 0.5. This curve, however, decreases again for higher Ca ion concentration. Experimental results show the same tendency although the agreement is semi-quantitative.

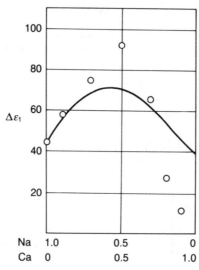

Figure 6.26 The calculated dielectric increment as a function of counterion ratio, Na : Ca. Concentration of polyion, 1 mN; degree of polymerization, 1800; $\alpha = 1.9$, $B = 1.5$ and $l = 1040$ Å. (From Minakata *et al* 1972.) Reproduced by permission of John Wiley & Sons, Inc. © 1972.

As mentioned earlier, a complete discussion of this theory is difficult because of its complexity. Only under certain conditions can the theory be presented by a simple closed form.

(i) No divalent ion is present. In this case, the parameter $v\lambda_2$ vanishes and it is proven that the term $\Delta_{mn}n$ reduces to zero. Δn^2 and $\langle n \rangle$ simplify to

$$\Delta n^2 = \frac{u\lambda_1}{(1 + u\lambda_1)^2} \qquad \langle n \rangle = \frac{u\lambda_1}{1 + u\lambda_1}. \tag{6.214}$$

These equations show that Δn^2 increases as the value of $u\lambda_1$ increases and has a maximum value at $u\lambda_1 = 1$. A further increase in the monovalent ion concentration decreases the value of Δn^2.

(ii) $\zeta = \delta = 1$. This condition means that there is no interaction between counterions (see equation (6.211)), therefore $\Delta_m n$ must vanish. In this case, the mean charge density $\langle n \rangle$ and Δn^2 are given by

$$\langle n \rangle = \frac{u\lambda_1 + 2v\lambda_2}{1 + u\lambda_1 + v\lambda_2} \tag{6.215}$$

$$\Delta n^2 = \frac{u\lambda_1 + 4v\lambda_2 + u\lambda_1 v\lambda_2}{(1 + u\lambda_1 + v\lambda_2)^2}. \tag{6.216}$$

Calculations show that the value of Δn^2 increases with increasing divalent ion concentration.

As discussed previously, ionic processes play a crucial role as an underlying physical mechanism of the electric polarization of polyelectrolytes. Those polymers may have permanent dipole moments, but they play only a secondary role in polyions. The dominant mechanism for the origin of the large dipole moment of polyelectrolytes is the polarization of counterions.

Because of the vast amount of research on the problem of counterion polarization, it is difficult to cover all the theoretical and experimental investigations in full. Two additional references are listed (Chew and Sen 1982, Fixman 1983).

Suggested Reading

Kruyt H R 1949 *Colloid Science* vol I, II (New York–Amsterdam: Elsevier)
Pethig R 1979 *Dielectric and Electronic Properties of Biological Materials* (New York: Wiley)
Rice S A and Nagasawa M 1961 *Polyelectrolyte Solutions* (New York: Academic)

References

Bonincontro A, Cametti C and Di Biasio A 1980 Effect of volume ion polarization on Maxwell-Wagner dielectric dispersion *J. Phys. D: Appl. Phys* **13** 1529–39
Bruggeman D A G 1935 Berechnung verschiedener physikalischen Konstanten von heterogenen Substanzen *Ann. Phys., Lpz* **24** 636–64
Chew W C and Sen P N 1982 Dielectric enhancement due to electrochemical double layer. Thin double layer approximation *J. Chem. Phys.* **77** 4683–93
Daenzer H 1938 *Ergebnisse der Biophysikalischen Forschung* ed. E B Rajewsky (Leipzig: Georg Thieme) pp193–231

Fixman M 1983 Thin double layer approximation for electrophoresis and dielectric response *J. Chem. Phys.* **78** 1483–91

Garcia A, Grosse C and Brito P 1985 On the effect of volume charge distribution on the Maxwell-Wagner relaxation *J. Phys. D: Appl. Phys.* **18** 738–45

Grosse C and Foster K R 1987 Influence of bulk diffusion on the counterion polarization in a condensed counterion model *J. Phys. Chem.* **91** 6415–17

Hanai T 1960 Theory of the dielectric dispersion due to the interfacial polarization and its application to emulsion *Kolloid Z.* **171** 23–31

Hanai T and Koizumi N 1975 Dielectric relaxation of W/O emulsions in particular reference to theories of interfacial polarization *Bull. Inst. Chem. Res.* **53** 153

—— 1976 Numerical estimation in a theory of interfacial polarization developed for disperse systems in higher concentrations *Bull. Inst. Chem. Res.* **54** 248

Kramer H A and Wannier G H 1941 Statistics of the two-dimensional ferromagnet. Part I *Phys. Rev.* **60** 252–62

Mandel M 1961 The electric polarization of rod-like, charged macromolecules *Mol. Phys.* **4** 489–96

Mandel M and Jenard A 1963 Dielectric behavior of aqueous polyelectrolyte solution *Trans. Faraday. Soc.* **59** 2158–77

Maxwell J C 1892 *Treatise on Electricity and Magnetism* (Oxford: Clarendon)

McTague J P and Gibbs J H 1966 Electric polarization of solutions of rodlike polyelectrolytes *J. Chem. Phys.* **44** 4295–301

Miles J B Jr and Robertson H P 1932 The dielectric behavior of colloidal particles with an electric double layer *Phys. Rev.* **40** 583–91

Minakata A, Imai N and Oosawa F 1972 Dielectric properties of polyelectrolytes. II. A theory of dielectric increment due to ion fluctuation by a matrix method *Biopolymers* **11** 347–59

Montroll E W 1941 Statistical mechanics of nearest neighbor systems *J. Chem. Phys.* **9** 706–21

O'Konski C T 1960 Electric properties of macromolecules. V. Theory of ionic polarization in polyelectrolytes *J. Phys. Chem.* **64** 605–19

Oosawa F 1970 Counterion fluctuation and dielectric dispersion in linear polyelectrolytes *Biopolymers* **9** 677–88

Pauly H and Schwan H P 1959 Uber die Impedanz einer Suspension von Kugelformigen Teilchen mit einer Schale *Z. Naturf.* b **14** 125–31

Polder D and Van Santen J H 1946 *Physica* **12** 257

Schurr J M 1964 On the theory of the dielectric dispersion of spherical colloidal particles in electrolyte solution *J. Phys. Chem.* **68** 2407–13

Schwan H P, Schwarz G, Maczuk J and Pauly H 1962 On the low frequency dielectric dispersion of colloidal particles in electrolyte solution *J. Phys. Chem.* **66** 2626–36

Schwarz G 1962 A theory of the low-frequency dielectric dispersion of colloidal particles in electrolyte solution *J. Phys. Chem.* **66** 2636–42

Sillars R W 1936 The properties of dielectric containing semi-conducting particles of various shapes *J. Inst. Electr. Eng.* **80** 378–94

Stratton A 1941 in *Electromagnetic Theory* (New York: McGraw-Hill) pp207–17

Takashima S 1966 Dielectric dispersion of deoxyribonucleic acid. II *J. Phys. Chem.* **70** 1372–80

—— 1967a Mechanism of dielectric relaxation of deoxyribonucleic acid *Adv. Chem. Ser.* **63** 232–52

—— 1967b Effect of ions on the dielectric relaxation of DNA *Biopolymers* **5** 899–913

Trukhan E M 1967 Dispersion of the dielectric constant of heterogeneous systems *Sov. Phys.–Solid State* **4** 2560–70

Wagner K W 1914 Erklarung der dielektrischen Nachwirkungsvorgange auf grund Maxwellscher Forstellung *Ann. Phys., Lpz* **40** 817

Warashina A and Minakata A 1973 Dielectric properties of polyelectrolytes. IV. Calculation of dielectric dispersion by a stochastic model *J. Chem. Phys.* **58** 4743–9

Zhang H Z, Sekine K, Hanai T and Koizumi N 1984 Dielectric approach to polystyrene microcapsule. Analysis and the application to the capsule permeability to potassium chloride *Colloid Polym. Sci.* **262** 513–20

7   Dielectric Properties
of Nucleic Acids

7.1 Introduction

Nucleic acids are one of the most important and exciting biological macromolecules. Not only their genetic functions but also their physical and chemical properties have been investigated exhaustively for many years. In particular, the DNA molecule has an elegant structure and its physical properties remain fascinating to many investigators. Its configuration is now well understood because of the pioneeering work by x-ray crystallographers and the subsequent proposal of the now famous Watson–Crick model.

The dielectric properties of DNA have been investigated since the late 1940s, even before the finding of the double helical structure. The development of research on the dielectric properties of DNA and its homologues will be discussed in this chapter. The discussions are limited to dielectric properties and related problems. The genetic functions of DNA are not within the scope of this book.

7.2 The Structure and Configuration of DNA

A DNA molecule consists of three basic chemical components: deoxyribose, phosphate groups and base molecules such as thymine, adenine, guanine and cytosine. The question has been how to piece them together and construct a model which is compatible with the results obtained from x-ray crystallographic analyses by Wilkins *et al* (1953). This monumental work was accomplished by Watson and Crick in (1953). The model constructed by them is now called the Watson–Crick model. The structure of DNA is shown in figure 7.1. According to this model, a DNA molecule consists of two polynucleotide chains. Each strand is made up of a nucleotide chain in which deoxyribose and phosphate groups alternate. When these two chains form a double stranded configuration, they intertwine around each other with a pitch of 34 Å (see figure 7.1). The crucial part of the DNA structure is the base pairs which bridge two helical strands, as shown in this figure.

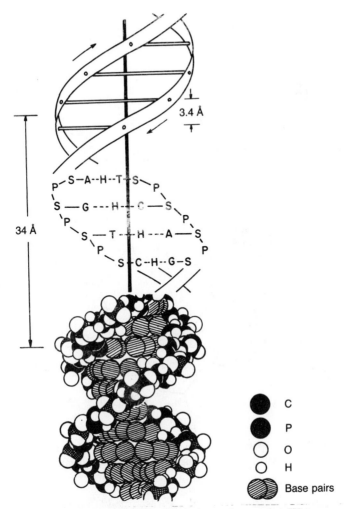

Figure 7.1 The double helical structure of DNA by Watson and Crick. See the text for detail. Taken from Watson and Crick (1953).

Geometric restrictions prevent random base pairing and, accordingly, we find only two types of base pairs in DNA molecules, i.e. adenine–thymine pairs and guanine–cytosine pairs. These pairs are stabilized by hydrogen bonds, namely, two hydrogen bonds in the adenine–thymine (AT) pair and three hydrogen bonds in the guanine–cytosine (GC) pair (see figure 7.2). The vertical distance between base pairs is uniformly 3.4 Å regardless of the type of base pairs. The base planes are tilted relative to the helical axis when DNA fibre is dry (the A form). However, in the presence of moisture, base planes become perpendicular to the helical axis (the B form).

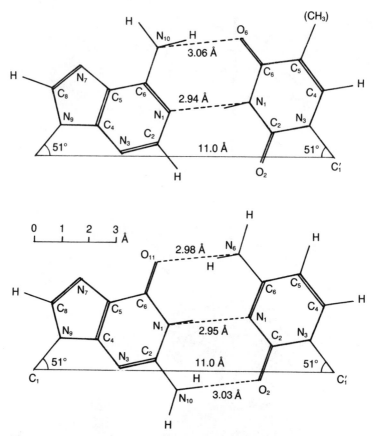

Figure 7.2 The spatial arrangement of the adenine–thymine pair and the guanine–cytosine pair. The broken lines indicate hydrogen bonds.

Since two strands of DNA are intertwined with each other, double helical DNA demonstrates considerable rigidity, even in solution. This is analogous to two threads twisted together. Each thread is quite flexible but when they are twisted around each other, the double threads gain mechanical rigidity. Electron micrographs indicate that double helical DNA has a configuration which is quite different from the random-coil structure. Viscosity measurements and also flow birefringence show that the conformation of DNA molecules can be approximated as a rigid rod if their molecular weights are not excessively large. In addition, as mentioned before, the phosphate groups in nucleotide chains carry negative charges when they are dissolved in neutral aqueous media. Therefore, relatively short DNA can be considered as a charged rigid rod as a first approximation. Because of these features, DNA exhibits behaviour which is characteristic of polyelectrolytes. This fact has an important bearing on the origin of dipole moment of a double helical DNA molecule.

Roughly, a DNA molecule has two distinctly different conformations. One is the double helical configuration and the other is a random-coil structure. The transition from the helical form to the random coil occurs when DNA solution is heated to 80–95 °C. At high temperatures, the two strands of DNA molecule unwind (Doty *et al* 1959). When they separate, each strand transforms itself into a random-coil conformation. The reversal of this process (for some DNA molecules) can be achieved by cooling the solution very slowly after heating. In contrast to double helical DNA, the single strand DNA molecule exhibits no birefringence and has a viscosity which is lower than that of helical DNA by perhaps two orders of magnitude (Geiduschek and Holtzer 1958). Upon a helix–coil transition, the absorption of helical DNA near 260 nm, which is due to the conjugated double bonds of the base molecules, increases by about 40%. In other words, the absorption of helical DNA near 260 nm is lower than that of random-coil DNA (Beaven *et al* 1955). This is called hypochromism. Often hypochromism is used to follow the transition from helical to random-coil configurations. The helix–coil transition of DNA has been investigated exhaustively, experimentally as well as theoretically (see Steiner and Beer 1961).

7.3 Dielectric Properties. Early Research

The investigation of the dielectric properties of DNA began between 1945 and 1950 (Jungner 1945, Jungner and Allgen 1949, Allgen 1950), several years before Watson and Crick published the paper on the double helical structure of DNA. These investigators observed a well defined dielectric dispersion between 30 kHz and 30 MHz using dilute DNA solutions. As will be discussed later, the relaxation frequency observed by them is much higher than those observed by others in recent years. The molecular weight of the DNA samples used by these investigators ranged only from 2×10^5 to 6×10^5. This is very small compared with the samples used for present day dielectric measurements of DNA. One of the data obtained by Allgen is shown in figure 7.3. Note that dielectric constants are plotted against wavelength in this figure. This convention is hardly used now. The centre of these dispersion curves is located between 4.4 and 4.9×10^4 cm which is translated to approximately 1 MHz and a relaxation time of 1.58×10^{-7} s. As discussed later in this chapter, the dispersion of highly polymerized DNA samples is found between 10 and 100 Hz. In spite of the fact that these investigators used small DNA samples, still the difference between these relaxation frequencies cannot be explained on the basis of molecular weight differences alone. The relaxation was attributed by these workers to the dipole moment in the transverse axis of the DNA fibre. This conclusion seemed quite reasonable at that time, based on the magnitude of the dipole moment and the value of relaxation time. The interpretation of the dielectric behaviour of a DNA molecule, however, was drastically modified later as new experimental results began to appear in the literature.

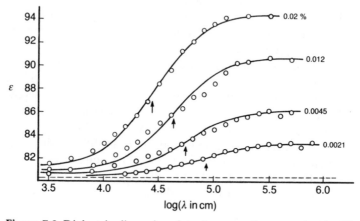

Figure 7.3 Dielectric dispersion data for a small DNA molecule. The abscissa is the wavelength, instead of frequency. (From Allgen 1950.) Reproduced by permission of *Acta Physiologica Scandinavica*.

The dielectric behaviour of nucleohistone (DNHi) was studied by the same authors. Histone is a basic protein with which DNA forms a tight complex *in vivo*. The complexation of DNA with histone is expected to alter the charge distribution and, consequently, the value of the dipole moment. The dielectric dispersion curves of nucleohistone are shown in figure 7.4. The dielectric increment of DNHi is, as expected, smaller than that of DNA. This result proves that the presence of histone molecules shields the charges of DNA and decreases its dipole moment. The dielectric measurements performed by Allgen and co-workers are the only source of information as to the dielectric behaviour of DNHi. These data were, however, obtained many years ago. Since then the technique of purification of nucleic acids has been improved greatly. While the dielectric properties of purified DNA have been investigated repeatedly by many other researchers and new information on the nature of the dipole moment has been available, the dielectric properties of nucleohistone have never been reinvestigated.

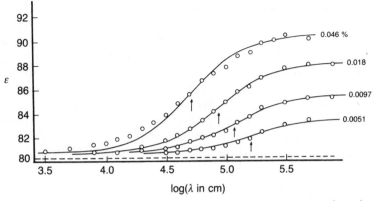

Figure 7.4 Dielectric dispersion data for nucleohistone (DNHi). (From Allgen 1950.) Reproduced by permission of *Acta Physiologica Scandinavica*.

The research by Jungner and co-workers was performed, as already mentioned, even before the double helical structure of DNA became known. These early dielectric measurements of DNA were handicapped for two reasons.

(i) The technique of DNA purification was not adequate at that time. In fact, early investigators used very small DNA samples which might have been fragmented during purification. Recent measurements are all performed using much larger and/or more uniform DNA samples. The conclusion that DNA has a dipole moment in the direction of the transverse axis may have been due, at least partly, to their use of small DNA preparations.

(ii) The structure of the DNA molecule was not well documented at that time. Therefore, the interpretation of experimental results was made difficult because of the lack of sufficient knowledge about the structure.

In spite of these shortcomings, however, the research done by these early workers is highly significant as a landmark for this field.

Subsequently, interesting measurements of the dielectric constant of DNA solution were performed using rotating electrodes by Jacobson (1953) and Jerrard and Simmons (1959). These investigators measured the change in the dielectric constant of DNA solution due to the alignment caused by velocity gradients. For example, Jacobsen observed an increase in the dielectric constant of DNA solution when an electrical field is applied along the laminar flow. On the other hand, a decrease in dielectric constant was noted when electrical fields were applied across the flow. The results obtained by Jerrard and Simmons are similar to those obtained by Jacobson. These results are a clear indication that DNA has a dipole moment in the direction of the longitudinal axis. This conclusion is in contrast to the one reached by Jungner and co-workers some years earlier. As will be discussed shortly, the conclusion that DNA has a longitudinal dipole moment has been confirmed by more recent investigations. Both Jungner and co-workers and Jacobson postulated that DNA had a permanent dipole moment, a concept which was contested later by other investigators. Detailed discussions on rotating electrodes are also found in the following references: Peterlin and Reinhold (1965), Saito and Kato (1957), Takashima (1970) and Barisas (1974).

7.4 Dielectric Properties of DNA at Low Frequencies

Starting in the early 1960s, Takashima (1963, 1966, 1967) performed a series of measurements on the dielectric constant of highly polymerized DNA samples (with a molecular weight of about $2-3 \times 10^6$) in a wide frequency range, which had never been attempted earlier. The range he covered was between 20 Hz and 200 kHz. The lowest frequency used by Jungner and co-workers was 100 kHz. Therefore, the overlap of frequency ranges used by these investigators is only slight. Takashima observed that the dielectric constant of DNA solution increased markedly below 1 kHz and reached a value of 1500 or more at a concentration of only 0.01%. One of the results obtained by Takashima is illustrated in figure 7.5. Note the rise of the dispersion curve below 1 kHz. Also note that the dispersion curve is still rising even at 20–30 Hz. The low-frequency plateau was obtained using the Cole–Cole plot, as shown by the inset of this figure. The dipole moment can be roughly estimated using (5.23) and a value of as large as 100 000 D was found. Based on this observation, it was concluded that the primary dispersion of DNA is located in the 50–100 Hz region and that the small dispersion observed by Jungner and co-workers may have been a secondary dispersion of unknown origin.

It has been mentioned previously that DNA undergoes a transition from the helical configuration to a random-coil structure upon heating at 80–95 °C. Heating, acidic pH and low ionic strength cause unwinding of the double strands and trigger a phase transition. Double helical DNA has a highly ordered internal structure with a remarkable regularity. In other words, helical DNA is in a crystalline state. On the other hand, single strand DNA is a random-coil polymer with no internal regularity, that is to say, single strand DNA is an amorphous substance. Therefore, the helix–coil transition is analogous to the melting of crystals.

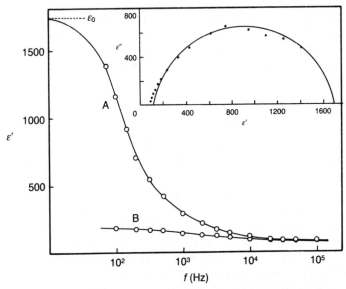

Figure 7.5 The dielectric dispersion of a highly polymerized DNA molecule. Curves: A, double helical DNA; B, single strand random-coil DNA. The inset shows the Cole–Cole plot of the same sample. Reproduced with permission from Takashima (1966). Copyright (1966) American Chemical Society.

The dielectric properties of random-coil DNA have also been investigated by Takashima (1966) and one of his results is shown in figure 7.5, along with that of helical DNA. Even a cursory examination is sufficient to recognize the difference between these two curves. The extraordinarily large dielectric increment of helical DNA is replaced by a small increment upon the helix–coil transition, i.e. from an increment of 1500 to about 20 dielectric units at a concentration of 0.01%. The main feature of the helix–coil transition is the melting of the helix backbone. Therefore, the drastic decrease of dielectric increment following the phase transition is clearly due to the collapse of the secondary structure of helical DNA. In other words, the exceedingly large dielectric increment indicates that the dipole moment of helical DNA is in the direction of the major axis. Application of an electrical field to DNA solution will cause either the orientation of the entire molecule or the fluctuation of mobile ions along the major helical axis. This explanation is consistent with the large relaxation time observed by Takashima.

Let us recall the Watson–Crick model shown in figure 7.1. According to this model, DNA consists of two nucleotide strands. However, geometric constriction requires these two strands to be anti-parallel.

Only an anti-parallel configuration makes these strands compatible to each other without steric hindrance. Nucleotides are basically a polar polymer chain. Although there is no experimental evidence, it is quite reasonable to assume that a single nucleotide chain has a large permanent dipole moment because of the repeating polar structural units, with the polarity in one direction. The DNA samples which were used for the measurements have a molecular weight of about 3×10^6. This can be translated into an extended length of approximately $2 \mu m$. Since the distance between base planes is 3.4Å, there are about 5900 such repeating units in one DNA molecule. If the total dipole moment of DNA is divided by this number, we find the net dipole moment per each 3.4Å unit to be 20 D. Since base pairs are stacked with their molecular plane perpendicular to the helical axis, the dipole moment of base molecules in the direction of the helical axis (the z-direction) is nearly zero except for a slight distortion of the base planes. Therefore, the dipole moment of 20 D must arise from phosphate groups. As mentioned before, nucleotide chains in helical DNA are anti-parallel to one another, in other words, the polarity or dipole moment of one chain opposes the moment of the other chain. Under these circumstances, it is difficult to attribute the large dipole moment to particular groups or structures in the DNA molecule. The origin of the dipole moment of DNA must be sought from different sources such as an induced dipole moment by ion fluctuation. The dipole moment due to ion fluctuation was discussed in Chapter 6. In small globular proteins, the magnitude of the ion fluctuation moment is negligibly small compared with the permanent dipole moment. However, the same argument will not hold for large fibrous polymers such as DNA where the effective distance of ion fluctuation is very long.

In the above, we found that helical DNA has an exceedingly large dipole moment and that the dipole moment disappears upon a helix–coil transition, an indication that the secondary structure has an important bearing on the large dipole moment. Other evidence which supports this statement was found from the length dependence of the dipole moment of DNA. DNA fibres were fragmented using sonic vibration and the mean lengths of DNA samples were estimated from the rotary diffusion constant. The rotary diffusion constant of an elongated polymer is related to its length by the following equation (Perrin 1934):

$$\Theta = \frac{3kT}{16\pi\eta a^3} [2\, ln(2a/b) - 1] \qquad (7.1)$$

where a and b are the major and minor axes of the rod and η is the viscosity. Assuming that the transverse axis of double helical DNA is constant at about 20Å, we can solve this equation numerically. For short DNA, the rigid rod approximation is justified and (7.1) can be used without serious error. Figure 7.6 shows the results of these experiments.

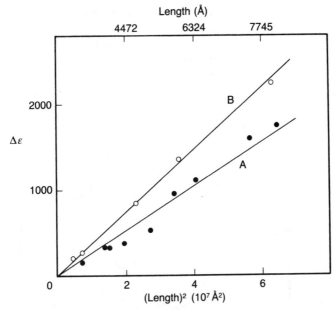

Figure 7.6 A plot of the dielectric increment against hydrodynamic length for salmon sperm (curve A) and calf thymus (curve B) DNA. Note the linear relationship between $\Delta\varepsilon$ and L^2. Reproduced with permission from Takashima (1966). Copyright (1966) American Chemical Society.

Clearly, the dielectric increment is proportional to the square of the effective length of DNA, an indication that helical DNA has a large longitudinal dipole moment. In addition, as shown in figure 7.7, the relaxation time is proportional to the square of the effective length. As we recall, the polarization of polar molecules is in general due to the orientation of the entire molecule if a permanent dipole is a dominant component. Thus, the relaxation time, under this condition, would be proportional to the cube of the effective length of the molecule (see figures 5.9 or 5.18). The relaxation time of DNA was, unexpectedly, found to be proportional to the square of length. This observation raises a serious question as to the nature or the origin of the dipole moment of DNA molecules. It was shown in Chapter 6 that the relaxation time due to ion fluctuation has a quadratic size dependence. In view of the theoretical considerations and the experimental observations, there is a strong indication that the dielectric relaxation of DNA is due to some process other than the orientation of a permanent dipole. DNA is a highly charged rod-like polymer surrounded by counterion clouds. Thus, counterion polarization may be the possible mechanism for the relaxation of DNA. Let us look at (6.137) which was derived by Schwarz (1962) for the counterion polarization of spherical particles. This theory demonstrates the quadratic dependence of the relaxation time on particle radius. Thus, this theory is likely to explain the relaxation of DNA quite nicely, but, since the shape of DNA is best described by a long rigid rod, Schwarz's theory, in its present form, cannot be applied to DNA. Although other counterion polarization theories for rod-like polymers are available (see Chapter 6), a discussion on the generalization of Schwarz's theory may prove useful.

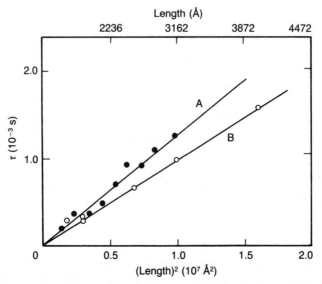

Figure 7.7 A plot of the relaxation time against the square of hydrodynamic length for salmon sperm (curve A) and calf thymus (curve B) DNA. Reproduced with permission from Takashima (1966). Copyright (1966) American Chemical Society.

7.5 The Dipole Moment of DNA due to Counterion Polarization

Since DNA can be approximated as a cylinder, cylindrical coordinates seem to be the most appropriate coordinate system. However, a solution of the potential problem of a cylinder is difficult to obtain when an electrical field is applied along the major axis. This problem was solved by Takashima (1967) using ellipsoidal coordinates. If an ellipsoidal particle is suspended in an aqueous medium and an electrical field is applied along the major axis, the field causes a flux of counterions J_E:

$$J_E = -eu\sigma\nabla\psi_s \qquad (7.2)$$

where u is the mobility of the counterions and ψ_s is the surface potential. The flux of ions driven by the field will be opposed by a diffusion controlled ion flow such as

$$J_D = -ukT\nabla\sigma. \tag{7.3}$$

Therefore, the continuity equation is given by

$$\frac{\partial\sigma}{\partial t} = eu\sigma\nabla^2\psi_s + ukT\nabla^2\sigma. \tag{7.4}$$

So far these formulations are similar to that of Schwarz's derivation. However, instead of spherical coordinates, ellipsoidal coordinates are employed to solve the potential problem for non-spherical particles. Ellipsoidal coordinates consist of three variables: ξ, η and ζ. Of these ξ is the surface of confocal ellipsoids, η represents a hyperboloid of one sheet and ζ is a hyperboloid of two sheets. The Laplace equation in ellipsoidal coordinates is given by

$$\nabla^2\psi = \frac{4}{(\zeta - \xi)(\xi - \eta)(\eta - \zeta)}\Bigg[(\zeta - \xi)R_\eta \frac{\partial}{\partial\eta}\bigg(R_\eta\frac{\partial\psi}{\partial\eta}\bigg)$$

$$+ (\xi - \eta)\, R_\zeta \frac{\partial}{\partial\zeta}\bigg(R_\zeta \frac{\partial\psi}{\partial\zeta}\bigg) + (\eta - \zeta)\, R_\xi \frac{\partial}{\partial\xi}\bigg(R_\xi \frac{\partial\psi}{\partial\xi}\bigg)\Bigg] = 0 \tag{7.5}$$

where R_ξ, R_η, and R_ζ are defined by the following equations:

$$R_\zeta = [(\zeta + a^2)(\zeta + b^2)(\zeta + c^2)]^{1/2} \tag{7.6}$$

$$R_\eta = [(\eta + a^2)(\eta + b^2)(\eta + c^2)]^{1/2} \tag{7.7}$$

$$R_\xi = [(\xi + a^2)(\xi + b^2)(\xi + c^2)]^{1/2}. \tag{7.8}$$

The surface of the ellipsoid represents an equipotential plane. Therefore, the surface potential ψ_s does not depend on ξ, as shown by the following equation:

$$\psi_s = \beta F_2(\eta)F_3(\zeta) \tag{7.9}$$

where β is a constant. Likewise, the counterion density should be of the same form, hence

$$\sigma = \alpha F_2(\eta)F_3(\zeta) \tag{7.10}$$

where α is another constant. Following Schwarz's formulation σ is an excess counterion density which fluctuates with periodic fields, $\sigma = \sigma_0\exp(j\omega t)$, where σ_0 is the average ion density in the absence of the field. If we define $\bar{\sigma}$ as $\bar{\sigma} = \sigma - \sigma_0$, we can replace the derivative $\partial\sigma/\partial t$ in (7.4) with $j\omega\bar{\sigma}$. After rearranging, we obtain an equation relating α to β:

$$\alpha = -\frac{1}{1 + j\omega\tau}\frac{e_0\sigma_0}{kT}\beta. \tag{7.11}$$

Although this equation is the same as (6.135), the relaxation time τ is defined by a quite different expression:

$$\tau = \frac{(\xi - \eta)(\xi - \zeta)}{u(b^2 + c^2 + 2\xi)kT}. \tag{7.12}$$

Since $\xi = 0$ on the surface of an ellipsoid, this reduces to

$$\tau = \frac{\eta\zeta}{u(b^2 + c^2)kT}. \tag{7.13}$$

This equation contains the coordinate variables η and ζ, therefore, it does not define relaxation time uniquely. However, it will be shown later that we can obtain an explicit expression for relaxation time for limiting cases such as very long ellipsoids or spheres. The potentials inside and outside the ellipsoid, ψ_1 and ψ_2, can be determined by solving the Laplace equation for each domain:

$$\nabla^2 \psi_1 = 0 \tag{7.14}$$

$$\nabla^2 \psi_2 = 0. \tag{7.15}$$

The boundary conditions used in this calculation are

$$\psi_1 = \psi_2 = \psi_s \qquad \text{for } \xi = 0 \tag{7.16}$$

where ψ_s is the surface potential. By definition, $\xi = 0$ at the surface of an ellipsoid. The second boundary condition is

$$\psi_2 = \psi_0 = -E_0 x \qquad \text{for } \xi \to \infty \tag{7.17}$$

where ψ_0 is the primary potential due to applied field E_0. The distance x is given in ellipsoidal coordinates by the following equation:

$$x = \left(\frac{(\zeta + a^2)(\eta + a^2)(\xi + a^2)}{(b^2 - a^2)(c^2 - a^2)}\right)^{1/2}. \tag{7.18}$$

Another boundary condition is a continuity of the normal components of displacement:

$$\varepsilon_1\left(\frac{1}{h_1}\frac{\partial\psi_1}{\partial\xi}\right)_{\xi=0} - \varepsilon_2\left(\frac{1}{h_1}\frac{\partial\psi_2}{\partial\xi}\right)_{\xi=0} = -4\pi e\sigma \tag{7.19}$$

where h_1 is defined by

$$h_1 = -\frac{1}{2}\left(\frac{(\xi - \eta)(\xi - \zeta)}{(\xi + a^2)(\xi + b^2)(\xi + c^2)}\right)^{1/2}. \tag{7.20}$$

With these equations, the potential functions outside and inside the ellipsoid are determined:

$$\psi_1 = -\frac{E_{ox}x}{1 + (abc/2)(\bar{\varepsilon}_1 - \varepsilon_2)A_1} \tag{7.21}$$

$$\psi_2 = \psi_0 \frac{1 + (abc)(\bar{\varepsilon}_1 - \varepsilon_2)A_2/2\varepsilon_2}{1 + (abc)(\bar{\varepsilon}_1 - \varepsilon_2)A_1/2\varepsilon_2} \tag{7.22}$$

where A_1 and A_2 are elliptic integrals defined by

$$A_1 = \int_0^\infty \frac{d\xi}{(\xi + a^2)R_\xi} \tag{7.23}$$

$$A_2 = \int_0^\xi \frac{d\xi}{(\xi + a^2)R_\xi}. \tag{7.24}$$

These solutions are similar to the ones given by Stratton except for the presence of the complex dielectric constant $\bar{\varepsilon}_1$ which is defined by

$$\bar{\varepsilon}_1 = \varepsilon_1 + \frac{4\pi a^2}{1 + j\omega\tau} \frac{\sigma e_0^2}{kT} \frac{(\eta\zeta)^{1/2}}{2abc} \tag{7.25}$$

where τ is the relaxation time defined by (7.13). Thus, the presence of mobile counterions on the surface of an ellipsoid entails an increase in the dielectric constant of

$$\Delta\varepsilon = \frac{4\pi a^2}{1 + j\omega\tau} \frac{\sigma e_0^2}{kT} \frac{(\eta\zeta)^{1/2}}{2abc}. \tag{7.26}$$

Again, the RHS of this equation is not unique because of the presence of coordinate variables. However, we can eliminate these for limiting cases, as shown below.

Limiting case I. A long cylinder
In general, the following relation exists on the surface of an ellipsoid because $\xi = 0$:

$$\frac{x^2}{a^4} + \frac{y^2}{b^4} + \frac{z^2}{c^4} = \frac{\eta\zeta}{a^2 b^2 c^2} \tag{7.27}$$

where x, y and z are rectangular coordinates. So far, the derivation has been based on the assumption that DNA is a general ellipsoid. DNA is, however, a very long fibrous polymer and can also be considered as a very long cylinder. Hence, x, y and z in (7.27) can be expressed in cylindrical coordinates:

$$x = a \qquad y = b\cos\theta \qquad z = b\sin\theta. \tag{7.28}$$

Note that x is assumed to be along the major axis of the cylinder. Using these expressions in (7.27), we obtain a simple relation between η and ζ and the axes of an ellipsoid:

$$\eta\zeta = b^2(a^2 + b^2). \tag{7.29}$$

We can substitute this expression into (7.13) and (7.26) to obtain the following equations, which enable us to calculate the dielectric increment and relaxation time uniquely:

$$\tau = \frac{a^2 + b^2}{2ukT} \tag{7.30}$$

$$\Delta\varepsilon = \frac{4\pi}{1 + j\omega\tau} \frac{a^2}{2b} \frac{e^2\sigma_0}{kT} \qquad \text{if } a \gg b. \tag{7.31}$$

Inserting (7.31) in Maxwell's mixture equation:

$$\Delta\varepsilon = \frac{e^2\sigma_0 a^2}{bkT} \frac{9\pi p}{2(1 + p)} \frac{1}{1 + j\omega\tau}. \tag{7.32}$$

In (7.30) and (7.32), the minor axis b is much smaller than the major axis a and is nearly constant at about $20\text{--}30 \times 10^{-8}$ cm. Therefore, from (7.30), we can conclude that the relaxation time is proportional to the square of the major axis a. This is exactly the conclusion reached by experiment. On the other hand, it was also observed that the dielectric increment of DNA increases as a quadratic function of length a. Therefore, (7.32) is in agreement with experimental results. With this observation, we can reach a conclusion that counterion polarization theory can account for the size dependence of the dielectric increment and relaxation time of a DNA molecule. The dominant component of the dipole moment of DNA, therefore, seems to be an induced moment due to counterion fluctuation. The contribution of a permanent dipole in a DNA molecule can be considered negligible in view of these results. Equation (7.32) contains b in the denominator, a very small number of about 20×10^{-8} cm. Thus, the presence of b in this equation increases the value of the dielectric increment enormously. This may explain the reason why DNA solution has an astronomical dielectric increment even at very low concentration.

Limiting case II. A sphere
Another limiting case is a sphere where $a = b = c$. Therefore, (7.27) reduces to

$$a^2 = (\eta\zeta)^{1/2}. \tag{7.33}$$

Substituting this equation into (7.13) and (7.26), we obtain

$$\tau = \frac{a^2}{2ukT} \tag{7.34}$$

$$\Delta\varepsilon = \frac{4\pi}{1 + j\omega\tau} \frac{e^2 a\sigma_0}{kT}. \tag{7.35}$$

Inserting this in Maxwell's mixture formula, the following equation for $\Delta\varepsilon$ is found:

$$\Delta\varepsilon = \frac{e^2\sigma_0 a}{kT} \frac{9\pi p}{(1 + p/2)^2} \frac{1}{1 + j\omega\tau}. \tag{7.36}$$

This is exactly the same expression as obtained by Schwarz for a spherical particle. Thus, the derivation which is based on an ellipsoidal model is consistent with the one using a spherical model.

7.6 Dielectric Properties of DNA at Very Low Frequencies (Four-electrode Measurements)

Although Takashima's measurements covered a wide frequency range, inspection of the experimental data shown in figure 7.5 demonstrates that Takashima's measurements did not uncover the entire dispersion curve of DNA solution. The dielectric constant of solution is still increasing even at 50 Hz and a low-frequency plateau (ε_0) can only be determined by extrapolation using a Cole–Cole plot. This is due to the frequency limitation of the instrument used by this investigator and also because of the error due to electrode polarization (see Appendix E). Obviously, a different measuring technique was desired which enabled the measurements of dielectric constant to be extended to a lower-frequency region without electrode polarization. A four-terminal system which uses an entirely different principle from that of an ordinary two-electrode system, was used by Hanss (1966) for the first time for the dielectric constant measurement of dissipative DNA solutions. Unfortunately, Hanss' measurements were limited to the imaginary part of the complex dielectric constant (ε'') and no attempt was made to measure the real part (ε'). Nevertheless, the results obtained by Hanss indicate the presence of a peak of dielectric loss at 5–10 Hz. This result suggests that the relaxation of highly polymerized DNA would occur at frequencies even lower than that observed by Takashima.

The work by Hanss was followed by extensive measurement of the low-frequency dielectric constant of DNA in a range between 0.3 Hz and 30 kHz by Hayakawa et al (1975) and Sakamoto et al (1976, 1978, 1979). The four-electrode system used by these investigators is discussed in Appendix E. The use of high-quality operational amplifiers which have a very high input impedance, and the use of a sophisticated automatic switching system enabled them to extend the frequency range even below 1 Hz without the error due to electrode polarization. Some exemplary results of these four-electrode measurements are shown in figure 7.8. The molecular weight of the sample used for these measurements was 4×10^6. This diagram demonstrates that the relaxation frequency of DNA solution is located at 3 Hz. This is the lowest relaxation frequency ever observed with DNA solution. However, the most amazing finding is that the low-frequency dielectric constant of 0.05% DNA solution reaches a value of as high as 40 000. Takashima observed a low-frequency limiting dielectric constant of about 2000 using 0.01% solution. Therefore, if normalized, this value is translated into $2 \times 10^4 \, g^{-1} l^{-1}$, while the result of the four-electrode measurement is $8 \times 10^4 \, g^{-1} l^{-1}$. Thus, the difference is fourfold. One of the possible reasons for this discrepancy may be the difference in the degree of polymerization of the samples used by these investigators. As discussed before, the dielectric increment of DNA is proportional to the square of the length of the double helix. Therefore, a slight difference in the length of the sample would be amplified and results in a large difference in dielectric increment.

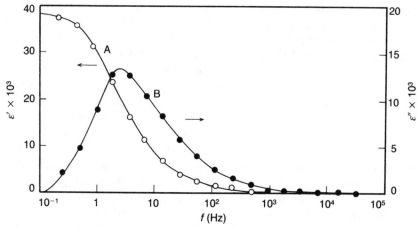

Figure 7.8 The frequency dependence of the dielectric constant of dilute DNA solution. A four-electrode system was used for this measurement. Curve A is the dielectric constant and curve B is the dielectric loss. (From Hayakawa *et al* 1975.) Reproduced by permission of the *Japanese Journal of Applied Physics*.

The greatest advantage of a four-electrode system is its ability to measure the dielectric constant of conductive solutions even in the presence of added salts to the solution. Figure 7.9(*a*) shows the results of dielectric measurements with DNA solution in the presence of NaCl at various concentrations. These figures demonstrate that the presence of salts in solution causes a marked decrease in the dielectric increment. For example, the dielectric increment of 0.05% DNA solution without added salt is 40×10^3 whereas the value in the presence of 2×10^{-3} mol NaCl is only 1000. Also the addition of salts causes a shift of relaxation frequency to a higher region. Note that the peak of the dielectric loss ε''

gradually shifts to higher frequencies as the concentration of added salt increases. It is indeed remarkable to be able to measure the entire dielectric dispersion of DNA solution at frequencies as low as 0.3 Hz without any noticeable effect from electrode polarization, even in the presence of added electrolytes.

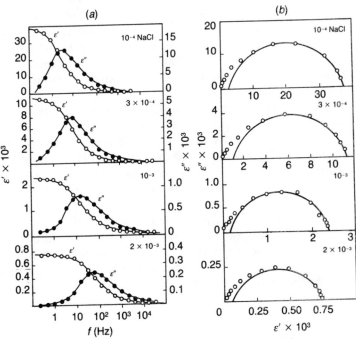

Figure 7.9 (a) The effect of added salt on the dielectric constant of DNA solution. The numbers indicate the concentration of NaCl. Note the drastic decrease in dielectric constant with increasing concentration of NaCl. (b) A Cole–Cole plot representation of the results shown in (a). Note the deviation of measured points from a symmetric arc. (From Hayakawa *et al* 1975.) Reproduced by permission of the *Japanese Journal of Applied Physics*.

Figure 7.9(b) illustrates the Cole–Cole representation of the same results. Inspection of these diagrams reveals the Cole–Cole plot of DNA consistently skewed. This means that the dielectric dispersion of DNA solution is asymmetric, i.e. the slope of the dispersion curves is steeper at lower frequencies than at higher frequencies. Inspection of figure 7.9(a), although not immediately evident, reveals that the dispersion curves are, likewise, steeper at lower frequencies. The curves obtained by Takashima using a two-terminal method also indicate similar asymmetry (see figure 7.5). There are two possible reasons for the asymmetry of dispersion curves. One is the distribution of molecular weights of the DNA samples. As discussed before, the size distribution of fibrous polymers usually follows a logarithmic–normal distribution function. It has been proven that this distribution function gives rise to a symmetric dispersion curve when plotted against the log of frequency. If, however, the size distribution of DNA samples has an asymmetric distribution function, this will give rise to a skewed dispersion curve. A test of this postulate is to use a DNA sample of uniform molecular weight. At present, an enzyme called restriction endonuclease is available which enables one to prepare a DNA sample of uniform molecular weight. Thus, a reinvestigation of the dielectric properties of DNA treated with restriction enzyme will be of interest.

Another cause for the skewed dispersion curve may be due to the nature of the counterion polarization itself. As discussed earlier, the origin of the dipole moment of DNA is most likely to be the polarization of counterions. The theory proposed by Oosawa (1970) is of importance with regard to the asymmetry of the dispersion curve of DNA. As discussed already in Chapter 6, Oosawa's calculation expresses the fluctuation of counterions as various modes of Fourier components. The amplitudes of these modes are, therefore, a, $a/2$, $a/4$, . . ., and they give rise to different relaxation times. Because of the quadratic dependence of ε' and ε'' on the length of DNA, relaxation times will be unevenly distributed on the frequency axis, giving rise to an asymmetric dispersion curve, as shown by figure 7.9.

7.7 An Electro-optical Investigation of the Dipole Moment of DNA

Instead of the frequency domain measurement of the dielectric constant, various time domain electro-optical techniques can be used for the study of the dipole moment of elongated polymers. Electric birefringence (see Yoshioka and Watanabe 1969) which is based on anisotropic refractive indices is the most commonly used technique. In addition, electrical dichroism, which is based on the anisotropy of optical absorptions (Ding et al 1972) and also light scattering in electrical fields (Stoylov 1971), is available. While dielectric measurements are performed using a small field of approximately $1\,\mathrm{V\,cm^{-1}}$ or less, electro-optical techniques require large electrical fields of several $\mathrm{kV\,cm^{-1}}$. Because of technical improvements, where the field strength which is required for these techniques has been lowered considerably, still these methods need higher field strengths compared with frequency domain dielectric measurements where some systems require only 50–100 mV or even less. The requirement of large fields may cause partial denaturation of fragile macromolecules such as DNA (O'Konski and Stellwagen 1965). Hence, the interpretation of experimental results requires consideration of the possible non-linear effects due to applied fields. Nevertheless, these methods, particularly the electric birefringence technique, have been used very effectively for the study of the dipole moment of biological polymers. The comparison of dielectric data with those obtained by either electric birefringence or electric dichroism is useful for gaining insights to the origin and nature of the dipole moment of DNA.

As has been discussed, the origin of dipole moment may be the induced moment due to counterion fluctuation. However, using the electric dichroism technique, Ding et al (1972) reached a different conclusion. There are two experimental results which led them to believe that DNA has a permanent dipole moment under certain conditions. One of these results is the theoretical prediction that the rise of electric dichroism is proportional to the square of field strength, i.e. E^2, if DNA has a permanent dipole moment. Ding et al found that the increase of dichroism followed an E^2-dependence, as predicted by the theory. Other evidence quoted by Ding et al is the observation by Hanss et al (1971) who, using the electric birefringence technique, observed that the reversal of the polarity of applied pulses caused a marked transient decrease of the birefringence. The field reversal technique had been used in order to investigate the nature of the dipole moment of polymers and viruses. If a molecule has a permanent dipole moment, reversal of the applied field will cause the reorientation of the entire molecule. Reorientation, in turn, causes a transient decrease in birefringence. If a molecule has an induced dipole moment, then reversal of the polarity of the field does not cause a transient decrease of birefringence or optical dichroism. This is due to the fact that a direction change of the dipole moment can be implemented by the rearrangement of ion distribution, without reorientation of the molecular axis. Based on these two observations, Ding et al concluded that DNA has a permanent dipole moment. Citing the observation by Neuman and Katchalsky (1972), however, Ding et al postulated that DNA may have behaved as if it has a permanent dipole moment because of the structural alterations and loss of symmetry in high-intensity fields.

7.8 Dielectric Properties of DNA at High Frequencies. Resonance Absorption

The discussion on the dielectric properties of DNA has been focused only on low-frequency relaxation processes. Because of the spectacular dielectric behaviour which DNA solution exhibited at low frequencies, those who were involved in this research at that time were simply preoccupied with the low-frequency behaviour. As has been shown, the dielectric constant of dilute DNA solutions decreases as the frequency of the applied field increases, approaching the value of pure water even at moderate frequencies such as 1 MHz. Under these circumstances, the low-frequency dispersion appeared to be the only problem which was worthy of investigation and there was no interesting problem left at high frequencies. However, this expectation has proven wrong. As discussed at the beginning of this chapter, Allgen and co-workers reported a small dielectric dispersion for DNA at about 1 MHz. Although their finding had been almost forgotten by other investigators, interest, nevertheless, gradually shifted to high-frequency properties. Mandel (1977) and Vreugdenhil et al (1979) reported a small dispersion of DNA samples in the vicinity of 1 MHz which may be the same relaxation process as that observed by Allgen and co-workers. This dispersion is insensitive to the molecular weight and fragmentation of native DNA by sonic vibration did not produce a significant effect on the dielectric increment nor on the relaxation time, in contrast to low-frequency dispersion.

The relaxation of polar molecules usually produces a broad dispersion curve. However, at very high frequencies, these broad band dispersion curves are replaced by sharp resonance-type absorptions. It had been believed that resonance absorption of electromagnetic energy by biological polymers would not be observed unless the measurements are performed at extremely high frequencies. Even in the low GHz region (10^9 Hz), resonance-type absorption was deemed unlikely mainly because of the viscous damping due to the lossy solvent, water. Also small resonance absorption peaks were considered extremely difficult to detect because of the presence of a large dielectric loss of water in the same frequency range. However, there has been a series of theoretical and experimental work in recent years which may alter our view on resonance phenomena in highly dissipative systems such as aqueous solutions of DNA. In the following, a short preliminary discussion on the elementary theory of resonance absorption of electromagnetic energy by a simple harmonic oscillator will be presented.

In general, two types of absorption of electromagnetic energy are expected to occur in dielectric materials. One is due to the fluctuation of charges or the orientation of dipoles between equilibrium positions which are separated by a potential barrier. This type of power absorption gives rise to broad band relaxation phenomena and has been dealt with in the previous and present chapters. The second power absorption is due to the displacement of charges bound elastically to an equilibrium position. Oscillation of these charges leads to a very sharp power absorption. This type of absorption is the topic of the present discussion. The model used for the derivation of the theory is a harmonic oscillator in which charges are bound to the nucleus elastically with a spring (Boettcher and Bordewijk 1978). Application of this basic theory to overdamped biological polymers requires the analysis of the viscous damping due to solvent molecules and it thus becomes extremely complex.

The elastic movement of charges in the absence of external forces is given by the following equation of motion:

$$m\ddot{x} + kx = 0 \tag{7.37}$$

where x is the displacement from the equilibrium position, m is the mass of charge carrier and k is a spring constant. The solution of this equation is

$$x = x_0 \exp(\mathrm{j}\omega_0 t) + x_0' \exp(-\mathrm{j}\omega_0 t) \tag{7.38}$$

where $\mathrm{j} = \sqrt{-1}$, ω_0 is the angular frequency of unforced oscillation, and x_0 and x_0' are constants. Substituting this into (7.37), we obtain

$$k = m\omega_0^2. \tag{7.39}$$

This is the solution for the unforced oscillation of elastically bound charges. Also, the equation of motion will be modified when we apply a periodic electric field such as $E = E_0 e^{\mathrm{j}\omega t}$. Thus the equation of motion becomes

$$m\ddot{x} + m\gamma\dot{x} + m\omega_0^2 x = -e_0 E_0 e^{\mathrm{j}\omega t}. \tag{7.40}$$

Note that the spring constant which was obtained above is inserted on the LHS of this equation (the third term). The parameter γ in the second term on the LHS is a damping constant. By substituting an assumed solution $x = a e^{\mathrm{j}\omega t}$ into (7.40), we can obtain an equation for the constant a:

$$a = -\frac{e_0}{m} \frac{E_0}{\omega_0^2 - \omega^2 + \mathrm{j}\omega\gamma}. \tag{7.41}$$

Hence, the displacement x becomes

$$x = -\frac{e_0}{m} \frac{E_0 e^{\mathrm{j}\omega t}}{\omega_0^2 - \omega^2 + \mathrm{j}\omega\gamma}. \tag{7.42}$$

The induced dipole moment is given by $p = -xe$ and the polarizability $\alpha = p/E$. Therefore, we obtain the following expression for the induced polarizability:

$$\alpha = \frac{e_0^2}{m} \frac{1}{\omega_0^2 - \omega^2 + \mathrm{j}\omega\gamma}. \tag{7.43}$$

For real systems, there are more than one electron and the polarizability is the sum of the terms due to these harmonic oscillators. The numerator of the complex term in (7.43) is now replaced by an oscillator strength f_j and is summed up for the number of electrons:

$$\alpha = \frac{e_0^2}{m} \sum_j \frac{f_j}{\omega_j^2 - \omega^2 + j\omega\gamma}. \tag{7.44}$$

The oscillator strength is a quantum mechanical quantity which is related to the transition dipole moment M_{ij}. M_{ij} represents a hypothetical dipole moment of an electron before and after excitation from the ith ground state to the jth excited state. The transition moment is defined by

$$M_{ij} = e_0 \int \psi_i r \psi_j \, dr \tag{7.45}$$

where e_0 is the electronic charge, r is distance, and ψ_i and ψ_j are the wave functions of the ith and jth states respectively. The oscillator strength f_j is related to the transition moment by the following equation (Kauzman 1964):

$$f_j = \frac{2m\omega_{ij}|M_{ij}|^2}{3he_0^2} \tag{7.46}$$

where ω_{ij} is the transition frequency, m is the mass of an electron and h is Planck's constant. Inserting this in (7.44), we obtain

$$\alpha = \frac{2}{3h} \sum \frac{\omega_{ij}|M_{ij}|^2}{\omega_{ij}^2 - \omega^2 + j\omega\gamma}. \tag{7.47}$$

This is a quantum mechanical expression for electronic polarizability. However, for the present discussion, we will adhere to the classical expression of equation (7.44).

Polarizability is a complex quantity and can be separated into real and imaginary parts, namely

$$\alpha(\text{Re}) = \frac{e_0^2}{m} \sum_j \frac{f_i(\omega_j^2 - \omega^2)}{(\omega_j^2 - \omega^2)^2 + \omega^2\gamma^2} \tag{7.48}$$

$$\alpha(\text{Im}) = \frac{e_0^2}{m} \sum_j \frac{f_i\omega\gamma}{(\omega_j^2 - \omega^2)^2 + \omega^2\gamma^2}. \tag{7.49}$$

The remaining task is to relate polarizability to a quantity which is experimentally observable. Oscillation of charges occurs at very high frequencies, at which the dielectric constant approaches the square of refractive index, i.e. $\varepsilon = n^2$, as stated previously. Therefore, we can relate polarizability to refractive index using the Clausius–Mosotti equation:

$$\alpha = \frac{4\pi N}{3} \frac{n^2 - 1}{n^2 + 2}. \tag{7.50}$$

Since n is usually very close to 1, we can simplify this as

$$n = 1 + 2\pi N\alpha. \tag{7.51}$$

Using this expression, we can now obtain the final solution which relates refractive index to polarizability:

$$n(\text{Re}) = 1 + \frac{2\pi N e_0^2}{m} \sum_j \frac{f_j(\omega_j^2 - \omega^2)}{(\omega_j^2 - \omega^2)^2 + \gamma^2 \omega^2} \qquad (7.52)$$

$$n(\text{Im}) = \frac{2\pi N^2 e_0^2}{m} \sum_j \frac{f_j \omega \gamma}{(\omega_j^2 - \omega_2)^2 + \gamma^2 \omega^2}. \qquad (7.53)$$

Figure 7.10 illustrates the dependence of $n(\text{Re})$ and $n(\text{Im})$ on the circular frequency. As shown by these figures, the real part of the refractive index undergoes a dispersion whenever the frequency of the applied field approaches the characteristic frequency ω_j; first an increase and then a decrease through zero. On the other hand, the imaginary part, which is absorption by definition, reaches a maximum value at $\omega = \omega_j$. Thus, the shape of the dispersion of refractive index is quite different from that of dipole relaxation. In addition, the frequency range of resonance absorption is much narrower than that of dielectric relaxation.

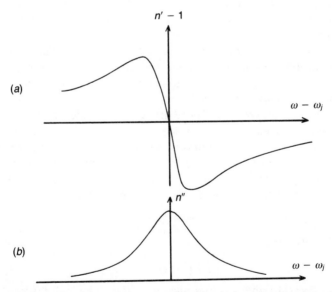

Figure 7.10 The dispersion of refractive index due to resonance power absorption. The curves are calculated using a simple harmonic oscillator model. (a) The real part of refractive index and (b) the imaginary part.

The resonance absorption of electromagnetic energy in biological polymers is not expected to occur in the RF or even in the lower microwave range. This expectation was born out of the belief that the viscous damping of water, particularly bound water, is so severe that possible resonances will be either frequency shifted or their intensity greatly attenuated. However, Lu *et al* (1977), Kohli and Van Zandt (1982) and Van Zandt *et al* (1977, 1982) demonstrated theoretically that the vibrational modes in the double helical DNA molecule can couple with microwave fields and produce resonance absorption in the GHz range. Their early calculations did not consider the viscous damping of the surrounding water molecules. Later work, however, shows that the viscous damping due to water causes a frequency shift and line broadening but the oscillator strength, or the intensity of resonance absorption, is unaffected.

The measurements of power absorption by DNA in the low GHz range were performed by Edwards *et al* (1984, 1985). Although their results underwent various stages of evolution, only the latest results and their interpretation will be discussed. They observed that only the DNA samples which were treated with restriction endonuclease and which were extremely well characterized with a uniform length enhanced the power absorption in the low GHz region and exhibited a series of sharp resonance peaks at 2.55, 4.00, 6.6 and 8.75 GHz. The solution of linear DNA, on the other hand, exhibited three peaks at slightly different frequencies.

These resonance absorption data clearly indicate the presence of certain vibrational modes in the DNA molecule. The vibrational modes in microwave frequencies are longitudinal acoustic modes. The length of DNA, which shows enhanced absorption coefficients, is a few wavelengths of sound waves. There are a few considerations with regard to the field induced acoustic modes. The acoustic wavelength must satisfy certain boundary conditions so that microwave fields and the acoustic modes are able to couple in order to allow energy transfer. The resonant frequencies v_n observed with circular as well as linear DNA, having a length of l, must satisfy the following equations:

$$v_{cir}(n + 1) = v_n l \qquad (7.54)$$

for circular DNA and

$$v_{lin}(n + \tfrac{1}{2}) = v_n l \qquad (7.55)$$

for linear DNA. In these equations, v is the velocity of sound in the molecule. n indicates the nth harmonic, the fundamental mode being $n = 0$. The acoustic modes for linear and circular DNA molecules are illustrated in figure 7.11. The velocity of sound calculated using (7.54) and (7.55) is 1.69 km s^{-1}. This value is in good agreement with the value of 1.67 km s^{-1} obtained by Hakim *et al* using Brillouin scattering measurements (1984).

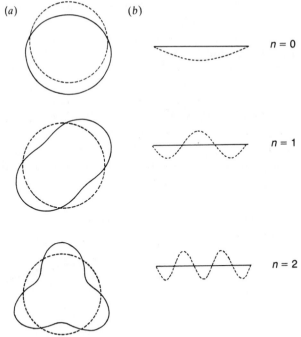

Figure 7.11 Fundamental and harmonic components of the vibration of (*a*) circular and (*b*) linear DNA molecules. (From Edwards *et al* 1984.)

Although these results are some of the most exciting developments in the dielectric research of DNA in recent years, the results obtained by Edwards *et al* are not universally accepted as yet. One of the reasons for the controversy is the most stringent conditions with which the experiments must be performed in order to detect these resonance absorptions. The size of DNA must be uniform and its length must be within the right range. Highly polymerized DNA which is often used for dielectric measurements does not show any noticeable absorption in the frequency range of interest. In addition, the amplitudes of the resonance absorptions are very small. Because of these reasons, the results obtained by Edwards *et al* have not been reproduced by other investigators.

A paper published by Gabriel *et al* (1987) reports two sets of data independently obtained in two different laboratories (London and Uppsala) using different techniques, yet using, however, the same DNA preparation—cyclic plasmid DNA, having 5480 base pairs. One of the data presented in their article is shown in figure 7.12. This figure illustrates the dielectric constant and loss factor of DNA solution between

1 and 10 GHz. The abscissa is on a linear scale, in order to conform to the plot used by Edwards *et al*. It can be noted that the expected small resonances are conspicuously missing in these diagrams. Foster *et al* (1987) also performed similar experiments but were unable to confirm the presence of resonance peaks. They point out that the amplitudes of the resonance signals are comparable to or even smaller than the oscillatory noises arising from the connector used to attach the probe to the DNA test set. Other work by Maleev *et al* (1987), using *E. Coli* DNA, fails to confirm the presence of resonance peaks. Because of the importance of the theoretical prediction by Van Zandt, Lu and their collaborators and Edwards *et al*'s experimental observations, discussions on this problem will continue for some time to come.

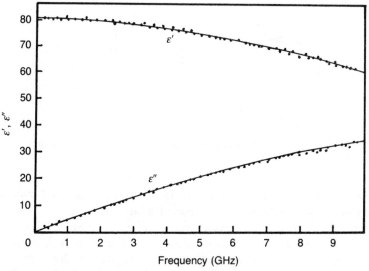

Figure 7.12 The dielectric constant and dielectric loss of DNA solution between 1 and 10 GHz. (From Gabriel *et al* 1987.) Note the absence of resonance peaks. Reprinted by permission from *Nature*, vol 328, p145. Copyright © 1987 Macmillan Journals Limited.

The dielectric properties of unfractionated and highly polymerized DNA were studied in the microwave frequency range by Foster *et al* (1984) and by Takashima *et al* (1984). Only the results obtained by Takashima *et al* are presented. Figure 7.13 illustrates the dielectric constant (ε') and loss factor (ε'') of calf thymus DNA between 50 MHz and 8 GHz. (The actual measurement was carried out between 5 MHz and 70 GHz.) As will be discussed in Chapter 8, the dielectric constant of water undergoes an anomalous dispersion in the same frequency range. In this diagram, the dielectric constant and loss factor of NaCl solution were subtracted from those of the DNA solution. This manipulation reveals more clearly the existence of a dispersion of dielectric constant and the power absorption of DNA solution between 100 MHz and 1 GHz. The broad band absorption does not resemble those observed by Edwards *et al* and is typical of dielectric relaxation. Power absorption was observed with double helical and single stranded DNA. Therefore, the double helical structure is not essential for absorption. Under these circumstances, it is reasonable to conclude that this broad band power absorption is unrelated to the results by Edward *et al*.

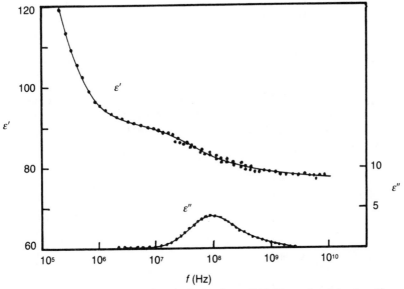

Figure 7.13 The broad band absorption of highly polymerized calf thymus DNA solution (1% by weight). The dielectric constant and dielectric loss of dilute NaCl solution have been subtracted. (From Takashima *et al* 1984.) Reproduced from the *Biophysical Journal*, 1984, vol 46, pp 29–34 by copyright permission of the Biophysical Society.

The results by Takashima *et al* were confirmed recently by Mashimo *et al* (1987) using a time domain refractometer (TDR). Only the real part of the complex dielectric constant is reproduced in figure 7.14. As illustrated, measurements were performed below and above the helix–coil transition temperature (70–90 °C). At low temperatures, when DNA is in the helical form, the amplitude of the dielectric dispersion is relatively small. However, upon the helix–coil transition, the amplitude increased considerably as shown in figure 7.15. The origin of this relaxation process is still unknown. However, the observation that a helix–coil transition increases the amplitude of the dielectric dispersion

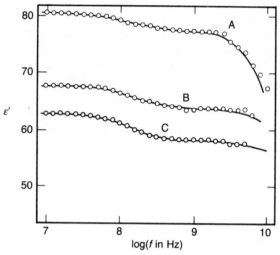

Figure 7.14 The dielectric constant of DNA solution at various temperatures. Curves A, B and C are obtained at 30, 70 and 90 °C respectively. (From Mashimo *et al* 1987.) Reproduced by permission of Elsevier Science Publishers.

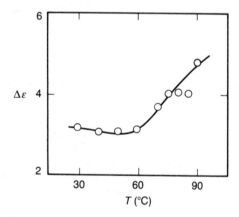

Figure 7.15 The amplitude of the dielectric dispersion of DNA below and above the transition temperature. (From Mashimo *et al* 1987.) Reproduced by permission of Elsevier Science Publishers.

is indicative of the participation of internal groups in this relaxation process. Whereas the internal motion of helical DNA is hindered because of its highly ordered, almost crystalline structure, melting of the helix gives rise to an amorphous liquid-like state in the DNA molecule. An understanding of this relaxation requires the identity of this unknown group. The use of synthetic polynucleotides such as poly-adenine (Poly A), which consists of only one type of nucleotide, will help identify the origin of this relaxation. Takashima *et al* (1986), using poly-A, observed a similar but more distinct dispersion at nearly the same frequency. Moreover, Takashima (1985, unpublished) found, with a dilute solution of adnosine, a similar dispersion at the same frequency. In view of these observations, the relaxation process in question is most likely to be due to the hindered motion of base molecules. Since bases are hydrogen bonded to their complementary base in the helical form, their motion is limited and this gives rise to a small dielectric increment. On the other hand, the motion of bases in coil DNA, because of the absence of inter-chain linkages, is much less hindered, thus giving rise to a larger increment. If this postulate is proven correct, this result may be one of the few cases where a dielectric relaxation technique has uncovered the internal molecular motion of biological macromolecules. As has been discussed, research on the high-frequency dielectric properties of DNA is relatively new. Although there is some controversy about the power absorption due to field driven acoustic modes, these results, no doubt, have invigorated the field and stirred considerable discussion.

7.9 The Effect of Dye Binding on the Dielectric Properties of DNA

The binding of dye molecules by DNA has an important biological implication (Orgel and Brenner 1961, Chan and Ball 1971). Some dyes are intercalated between base pairs, causing conformation changes, and other dyes bind externally, neutralizing the charges of phosphate groups. The effects of amino acridine dyes were investigated by Goswami and co-workers (1973, 1974). Proflavine which is one of the amino acridine dyes used by these workers was found to increase the relaxation time, indicating an increase in the length of the DNA molecule. On the other hand, the binding of proflavine caused a marked decrease in the dielectric increment, which is an indication of the neutralization of surface charges. The genetic implication of these results is yet to be found by further investigation.

Suggested Reading

Steiner R F and Beer R F 1961 *Polynucleotides* (Amsterdam–New York: Elsevier)

References

Allgen L G 1950 A dielectric study of nucleohistone from calf thymus *Acta Physiol. Scand. Suppl.* **22** 32

Barisas B H 1974 Effect of shearing laminar flow on dielectric polarization of suspension of rigid particles *Macromolecules* **7** 930–3

Beaven G, Holiday E and Johnson E 1955 *The Nucleic Acids* vol 1 (New York: Academic) p493

Boettcher C T F and Bordewijk P 1978 *Theory of Electric Polarisation* (Amsterdam: Elsevier) p240

Chan E W and Ball J K 1971 Interaction of DNA with three dimethyl derivatives of ben(c)acridines *Biochim. Biophys. Acta* **238** 31–45

Cole R H 1977 Dielectric theory and properties of DNA in solution *Ann. New York Acad. Sci.* **303** 59–73

Ding D W, Rill R and Van Holde K E 1972 The dichroism of DNA in electric fields *Biopolymers* **11** 2109–24

Doty P, Boedtker H, Fresco J, Haselkorn R and Litt M N 1959 Secondary structure in ribonucleic acid *Proc. Natl. Acad. Sci. USA* **45** 482–92

Edwards G S, Davis C C, Saffer J D and Swicord M L 1984 Resonant microwave absorption of selected DNA molecules *Phys. Rev. Lett.* **53** 1284–7

Edwards G S, Davis C C, Swicord M L and Saffer J D 1985 Microwave-field driven acoustic modes in DNA *Biophys. J.* **47** 799–807

Foster K R, Epstein B R and Gealt M A 1987 'Resonances' in the dielectric absorption of DNA? *Biophys. J.* **52** 421–5

Foster K R, Stuchly M A, Kraszewski A and Stuchly S S 1984 Microwave dielectric absorption of DNA in aqueous solution *Biopolymers* **23** 593–9

Gabriel C, Grant E H, Tata R, Brown P R, Gestblom B and Noreland E 1987 Microwave absorption in aqueous solution of DNA *Nature* **328** 145–6

Geiduschek E and Holtzer A 1958 Application of light scattering to biological systems: deoxyribonucleic acid and muscle proteins *Adv. Biol. Med. Phys.* **6** 431–551

Goswami D N, Das J and Das Gupta N N 1973 Dielectric behavior of DNA-proflavine complex *Biopolymers* **12** 1047–52

Goswami D N and Das Gupta N N 1974 Dielectric behavior of DNA-dye complex-II *Biopolymers* **13** 391–400

Hakim H B, Lindsay S M and Powell J 1984 Speed of sound in DNA *Biopolymers* **23** 1185–92

Hanss M 1966 Mesure de la dispersion dielectrique en tres basse frequence de solutions de DNA *Biopolymers* **4** 1035–41

Hanss M, Bernengo J C and Roux B 1971 *Results presented at XXIII Int. Cong. Pure Appl. Chem., Boston*

Hayakawa R, Kanda H, Sakamoto M and Wada Y 1975 New apparatus for measuring the complex dielectric constant of a highly conductive material *Japan. J. Appl. Phys.* **14** 2039–52

Jacobson B 1953 Method for obtaining streaming orientation and simultaneous determination of dielectric properties in macromolecular solutions *Rev. Sci. Instrum.* **24** 949–54

Jerrard H D and Simmons B A W 1959 Dielectric studies on deoxynucleic acid *Nature* **184** 1715–16

Jungner G 1945 Dielectric study of polynucleotides *Acta Physiol. Scand. Suppl.* **10** 32

Jungner G and Allgen L G 1949 Molecular weight determination on thymonucleic acid components by dielectric measurements *Nature* **163** 849–51

Kauzman W 1964 *Quantum Chemistry* (New York: Academic) ch 16

Kohli M and Van Zandt L L 1982 Microwave absorption by folded DNA chains *Biopolymers* **21** 1399–410

Lu K C, Prohofsky E W and Van Zandt L L 1977 Vibrational modes of A-DNA, B-DNA and A-RNA backbones: an application of a Green-function refinement procedure *Biopolymers* **16** 2491–506

Maleev V T, Kashpur V A, Glibitsky G M, Krasnitskaya A A and Veretelnik Y V 1987 Does DNA absorb microwave energy? *Biopolymers* **26** 1965–70

Mandel M 1977 Dielectric properties of charged linear macromolecules with particular reference to DNA *Ann. New York Acad. Sci.* **303** 74–87

Mashimo S, Umehara T, Ota T, Kuwabara S, Shinayashiki N and Yagihara S 1987 Evaluation of complex permittivity of aqueous solution by time domain reflectrometry *J. Mol. Liq.* **36** 135–51

Neuman E and Katchalsky A 1972 Long lived conformation changes induced by electric impulses in biopolymers *Proc. Natl. Acad. Sci. USA* **69** 993–7

O'Konski C T and Stellwagen N C 1965 Structural transition produced by electric fields in aqueous sodium deoxyribonucleate *Biophys. J.* **5** 607–13

Oosawa F 1970 Counterion fluctuation and dielectric dispersion in linear polyelectrolytes *Biopolymers* **9** 677–88

Orgel A and Brenner S 1961 Mutagenesis of bactriophage T4 by acridines *J. Mol. Biol.* **3** 762–8

Perrin F 1934 Mouvement Brownien d'un ellipsoide. Dispersion dielectrique pour des molecules ellipsoidales *J. Phys. Radium* **5** 497–511

Peterlin A and Reinhold C 1965 The influence of laminar flow on the dielectric constant of dilute polymer solutions *Kolloid Z.* **204** 23–8

Saito N and Kato T 1957 On the viscoelasticity and complex dielectric constant in the presence of an electric field and shearing laminar flow in a solution of macromolecules *J. Phys. Soc. Japan* **12** 1393–402

Sakamoto M, Hayakawa R and Wada Y 1978 Dielectric relaxation of DNA solution. II *Biopolymers* **17** 1507–12

—— 1979 Dielectric relaxation of DNA solution. III. Effects of DNA concentration, protein contamination and mixed solvents *Biopolymers* **18** 2769–82

Sakamoto M, Kanda H, Hayakawa R and Wada Y 1976 Dielectric relaxation of DNA solution *Biopolymers* **15** 879–92

Schwarz G 1962 A theory of the low-frequency dielectric dispersion of colloidal particles in electrolyte solution *J. Phys. Chem.* **66** 2636–42

Stoylov S P 1971 Colloid electro-optics. Electrically induced optical phenomena in dispersion systems *Adv. Colloid Interface Sci.* **3** 45–110

Takashima S 1963 Dielectric dispersion of DNA *J. Mol. Biol.* **7** 455–67

—— 1966 Dielectric dispersion of deoxyribonucleic acid. II. *J. Phys. Chem.* **70** 1372–80

—— 1967 Mechanism of dielectric relaxation of deoxyribonucleic acid *Adv. Chem. Ser.* **63** 232–52

—— 1970 Dielectric behavior of helical polyamino acids in shear gradients *J. Phys. Chem.* **74** 4446–52

Takashima S, Casaleggio A, Giuliano F, Morando M, Arrigo P and Ridella S 1986 Study of bound water of poly-adenine using high frequency dielectric measurement *Biophys. J.* **49** 1003–8

Takashima S, Gabriel C, Sheppard R J and Grant E H 1984 Dielectric behavior of DNA solution at radio and microwave frequencies (at 20 °C) *Biophys. J.* **46** 29–34

Van Zandt L L, Kohli M and Prohofsky E W 1982 Absorption of microwave radiation by DNA double helix in aquo *Biopolymers* **21** 1465–8

Van Zandt L L, Lu K C and Prohofsky E W 1977 A new procedure for refining force constants in normal coordinate calculations on large molecules *Biopolymers* **16** 2481–90

Vreugdenhil T, Van der Touw F and Mandel M 1979 Electric permittivity and dielectric dispersion of low-molecular weight DNA at low ionic strength *Biophys. Chem.* **10** 67–80

Watson J D and Crick F H C 1953 Molecular structure of nucleic acids *Nature* **171** 737

Wilkins M, Stokes A and Wilson H 1953 Molecular structure of deoxypentose nucleic acids *Nature* **171** 738–40

Yoshioka K and Watanabe H 1969 in *Physical Principles and Techniques of Protein Chemistry* ed. S J Leach (New York: Academic) p335

8   Water in Biological Systems

8.1 Introduction

Water is an essential substance for the origin, stability and function of biological systems, yet, it is one of the most enigmatic and elusive substances. After so many years of intense research, we do not yet have a satisfactory model which explains thermodynamic and fluid mechanical properties of liquid water. On the other hand, surprisingly, liquid water exhibits simple dielectric properties which can be described by Debye one relaxation time theory. However, this deceptively simple dielectric behaviour is actually very difficult to explain. The dielectric behaviour of water in biological systems is even harder to understand. This is expected in view of the puzzling nature of pure water.

8.2 The Structure of Water and Ice

It is well known that the structure of liquid water is not much different from that of ice. For example, figure 8.1 shows the radial distribution function of liquid water obtained by x-ray analysis (Morgan and Warren 1938, Narten *et al* 1967). The first peak at 3 Å is due to the first nearest neighbours and the second and somewhat broader peak is due to the second nearest neighbours. Other peaks are due to water molecules further away from the central molecule. If we compare the radial distribution function at 4 °C with that of ice, we find a striking resemblance between them. This similarity indicates that the structure of ice remains essentially unchanged upon melting. However, in spite of the apparent structural rigidity of liquid water, the fluid dynamical properties change drastically upon the melting of ice, as is well known to everyone. Although some other liquids show a more or less similar behaviour, the solid–liquid transition of water is still puzzling.

Let us look at the structure of ice Ih (hexagonal ice) which is the most common form of ice not far below the freezing point (Whalley *et al*

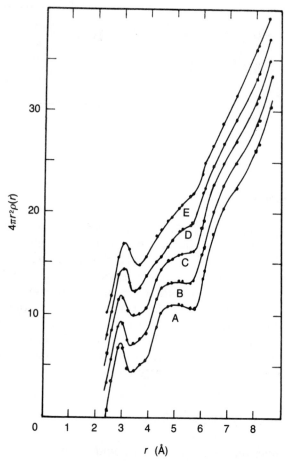

Figure 8.1 The radial distribution function of liquid water in the temperature range 1.5–80 °C. Curves: A, at 1.5 °C; B, 13 °C; C, 30 °C; D, 62 °C and E, at 80 °C. The curves are vertically shifted by 2.5 units at each temperature. (From Morgan and Warren 1938.)

1968) (see figure 8.2). As we can see, one water molecule is surrounded by four other water molecules and they form a tetrahedral configuration (Pauling 1940; see also Gabler 1978). These nearest neighbours are 2.76 Å away from the central molecule. The sharp peak in the x-ray diffraction pattern is due to these nearest neighbours. If we look at the second nearest neighbours, we begin to see the outline of a hexagonal configuration. These molecules are 4–5 Å away from the oxygen atom of the central molecule. The broad second peak in the radial distribution function is due to these water molecules. The implication of figure 8.1 is that the tetrahedral structure is essentially intact, even above the freezing point. The broken lines in figure 8.2 which connect the oxygen and hydrogen atoms between neighbouring molecules are hydrogen bonds. If the temperature reaches the melting point, some of these hydrogen bonds will break and this, in turn, introduces the phase transition from the solid to the liquid state. Hydrogen bonds seem to be the dominant factor for the stability of the tetrahedral structure of ice. As mentioned before, the number of hydrogen bonds which are broken upon a phase change seems to be small in view of the evidence provided by x-ray analysis. Thus, the rigidity of the water structure is maintained by extensive hydrogen bonding throughout its liquid phase. It may be appropriate to discuss, at this juncture, the origin and nature of the hydrogen bond in water (see Hadzi and Thompson 1957).

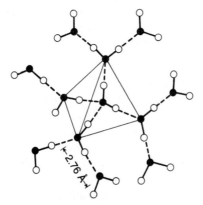

Figure 8.2 The structure of hexagonal ice. Note the tetrahedral configuration between nearest neighbours. Also note the hexagonal structure with second nearest neighbours. (Taken from Gabler 1978.) Reproduced by permission of Academic Press, Inc.

As shown in figure 8.3, hydrogen bonds are formed between the lone pair electrons of oxygen and the 1s electron of hydrogen. Oxygen has an electronic structure of $(1s)^2(2s)^2(2p_x^2 2p_y 2p_z)$. Using 2s and 2p electrons, oxygen forms sp^3 hybridized orbitals. The wave functions of these orbitals are given by the following:

$$\psi_1 = 1/2(s + \sqrt{3}z) \tag{8.1a}$$

$$\psi_\sigma' = s/2 + \sqrt{2}/3x - z/2\sqrt{3} \tag{8.1b}$$

$$\psi_1 = -s/2 + x/\sqrt{6} + z/\sqrt{12} - y/\sqrt{2} \tag{8.1c}$$

$$\psi_1' = -s/2 + x/\sqrt{6} + z/\sqrt{12} + y/\sqrt{2} \tag{8.1d}$$

where s, x, y, z represent 2s, $2p_x$, $2p_y$ and $2p_z$ atomic orbitals of

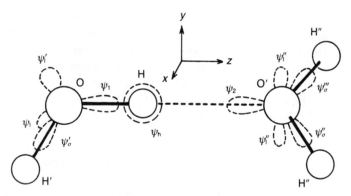

Figure 8.3 A schematic illustration of a hydrogen bond. The circle around the hydrogen atom is the 1s electron shell and the broken line between the 1s and ψ_2 orbitals represents a hydrogen bond. (From Tsubomura 1954.) Reproduced by permission of the *Bulletin of the Chemical Society of Japan*.

oxygen. Mapping of these orbitals reveals that they form a tetrahedral configuration in 3D space. Of these, ψ_1 and ψ'_s are directed toward adjacent hydrogen atoms and form a covalent bond with the 1s electron of hydrogen. The other orbitals which remain unbonded are called 'lone pair' electrons. The presence of lone pair electrons increases the density of negative charges around oxygen. This entails an increase in the magnitude of the dipole moment of water. Lone pair electrons form a weak bond with the 1s electron of a hydrogen which is already bonded covalently to another oxygen. This bond is commonly called a 'hydrogen bond'. The nature of a hydrogen bond has been debated for many years. It is generally agreed that a hydrogen bond has ionic character in addition to covalent character. The types of linkages which may contribute to a hydrogen bond are shown by the following arrangements (Tsubomura 1954):

$$O_x - H \quad \ldots \ldots O_y \qquad (8.2a)$$

$$O_x^- \quad H^+ \ldots \ldots O_y \qquad (8.2b)$$

$$O_x^+ \quad H^- \ldots \ldots O_y \qquad (8.2c)$$

$$O_x^- \quad H \quad \ldots \ldots O_y^+ \qquad (8.2d)$$

$$O_x \quad H^- \qquad \quad O_y^+. \qquad (8.2e)$$

Of these, the first structure represents the covalent character of a hydrogen bond. The energy of a hydrogen bond in water has been calculated using the valence bond method (Tsubomura 1954) and also by using the molecular orbital method (Morokuma 1970). Both calculations indicate that the hydrogen bond energy is approximately 5–12 kcal mol^{-1}. The energy–distance diagram of dimeric water molecules is shown in figure 8.4. The minimum at 1.75 Å from O_1 indicates the equilibrium distance of the hydrogen atom on the $O \cdots O$ axis. The second trough which has been inferred to exist due to a proton jump is not visible in this diagram.

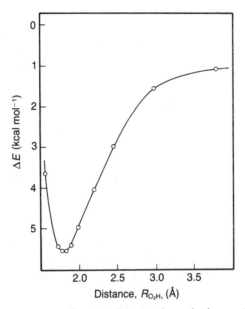

Figure 8.4 The energy–distance diagram for a hydrogen bond. The abscissa represents the distance between two oxygen atoms of neighbouring water molecules. (From Morokuma 1970.) Reproduced by permission of the American Institute of Physics.

The proton tunnelling effect which has been postulated for water and ice occurs in the presence of (a) 'excess' protons and (b) 'defect' protons, as will be discussed in more detail later. The migration of an 'excess' and a 'defect' proton in a repeating long chain of OH groups is depicted schematically in figure 8.5 (Eigen and De Maeyer 1958). This mechanism has been used by some investigators as an explanation for the unusually high mobility of a proton in ice.

A number of models for liquid water have been proposed over many years. However, these models can be classified into two categories. One is to treat water as a homogeneous system (see Pethig 1979). In this model, the flexibility of liquid water is considered to arise from the bending and stretching of hydrogen bonds. The second model is a cluster model in which water molecules form clusters of various sizes. Clusters are not stable entities and they are considered to form and break within a very short time constant. This model is called a flickering cluster model (Frank and Wen 1957) and it was the basis of a statistical thermodynamic calculation by Nemethy and Scheraga (1962a, b). This chapter is not designed to discuss these models in detail but suffice it to say that we still do not have a satisfactory model for liquid water which is universally accepted.

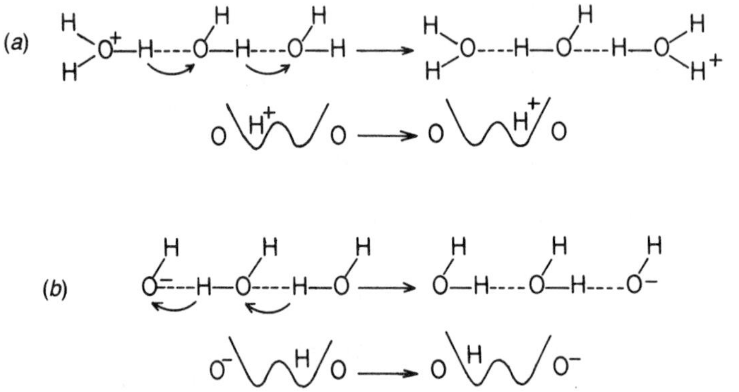

Figure 8.5 A chain of repeating O–H groups which are connected with hydrogen bonds. (*a*) The jump of an 'excess proton'. (*b*) The jump mechanism for a 'defect' proton. (After Eigen and De Maeyer 1958.)

The dielectric properties of water and ice have been investigated extensively for many years. The dipole moment of a water molecule is known to be 1.84 ± 0.02 D (see Smyth 1955). The vector of a dipole moment bisects the H–O–H angle and is directed from oxygen toward the region of positive charges of hydrogen. A theoretical calculation of the dipole moment of water is rendered difficult because of the contribution of lone pair electrons, which tend to increase the separation of positive and negative charges and shift the charge centres away from hydrogen atoms toward oxygen.

The low-frequency dielectric constant has been measured by a number of investigators at various temperatures (see table 8.1). No frequency dependence of the dielectric constant is observed below 100 MHz. The static dielectric constant at various temperatures can be fitted to the following empirical equation (Malmberg and Maryotte 1956) with an uncertainty of ± 0.05 dielectric units (see the calculated values in table

8.1). There are other empirical equations proposed by other investigators. However, for the purpose of this chapter, the following equation seems to be sufficiently accurate:

$$\varepsilon = 87.740 - 0.4008t + 9.398 \times 10^{-4}t^2 - 1.410 \times 10^{-6}t^3 \quad (8.3)$$

where t is the temperature in °C.

Table 8.1 Static and high-frequency dielectric constants for liquid water.

Temperature	g-factor	ε_s	ε_∞	$\tau \ (10^{-11} \text{ s})$
0	2.87	88.3	4.46	1.79
10		84.1	4.10	1.26
20		80.4	4.23	0.93
30		76.8	4.20	0.72
40		73.2	4.16	0.58
50		70.0	4.13	0.48
60		66.6	4.21	0.39
75		62.1	4.49	0.32
100	2.56			
200	2.24			
300	1.64			

8.3 The Theory of the Static Dielectric Constant of Water

We discussed Kirkwood's equation for the static dielectric constant in Chapter 2. The equation is reproduced once again for the sake of convenience:

$$\frac{(\varepsilon - 1)(2\varepsilon + 1)}{9\varepsilon} = \frac{4\pi N_A}{3VkT} g\mu_0^2 \quad (2.66)$$

where g is a correlation parameter which plays a central role in the analysis of the dielectric properties of liquid water. This factor is given by (2.67), as reproduced below:

$$g = 1 + N \langle \cos \theta \rangle \quad (2.67)$$

where $\langle \cos \theta \rangle$ is the mean of the cosine of the angle between two adjacent dipoles and N is the number of nearest neighbours. It is generally agreed that Kirkwood's equation does not represent the dielectric constant of water correctly. Froehlich (1949) derived an equation, using Onsager's reaction field and Kirkwood's g-factor, as shown below:

$$\frac{(\varepsilon - n^2)(2\varepsilon + n^2)}{\varepsilon(n^2 + 2)^2} = \frac{4\pi N_A}{9VkT} g\mu_0^2 \tag{8.4}$$

where n is the refractive index and N_A is Avogadro's number. This equation was further modified by replacing n^2 with ε_∞:

$$\frac{(\varepsilon - \varepsilon_\infty)(2\varepsilon + \varepsilon_\infty)}{\varepsilon(\varepsilon_\infty + 2)^2} = \frac{4\pi N_A}{9VkT} g\mu_0^2 \tag{8.5}$$

where ε_∞ refers to the dielectric constant at sufficiently high frequencies where orientational polarization has no contribution, but yet not high enough to dampen out the induced polarizations due to bond vibrations, rotation and electronic polarizations. In (8.4) or (8.5), the parameter g is the crucial quantity and is the most difficult to determine. Let us assume μ_0 to be 1.84 D. Since ε and ε_∞ or n can be measured and other parameters are known, we can calculate the value of the g-factor backwards using either (8.4) or (8.5). The result of this calculation is shown in table 8.1. Note that the value of the g-factor is large at 0 °C, indicating a high degree of correlation between neighbouring dipoles. As the temperature increases, the g-factor decreases and has a value of 2.56 at 100 °C. At even higher temperatures, the correlation parameter has a value of 1.64, indicating that there is still a considerable correlation, even in the vapour phase.

A theoretical calculation of the static dielectric constant depends critically on the value of the g-factor. As a first approximation, the correlation parameter is defined by (2.67). In this equation, N is the number of nearest neighbours. This means that the correlation between the central molecule and the second and third nearest neighbours is ignored in the original derivation by Oster and Kirkwood (1943). Using the radial distribution data obtained by x-ray analyses, Oster and Kirkwood calculated the value of the g-factor. They obtained a value of 2.63 at 0 °C and 2.82 at 83 °C. Obviously, these results are unacceptable because they indicate an enhancement of the correlation with increasing temperature. The static dielectric constant of water calculated by Oster and Kirkwood is shown in figure 8.6. The poor agreement between the theoretical and observed values is not surprising. This outcome is presumed to be due to the neglect of the long-range correlation between the central molecule and second and third nearest neighbours.

The theoretical study on the g-factor was improved substantially by Pople (1951). The model used by Pople is shown in figure 8.7. Although the structure of water was assumed to be tetrahedral, bending of the O–H bond and of the lone pair to which it is hydrogen bonded are allowed as shown by the kinked lines at angles ϕ_A, ϕ_B and ϕ_C, etc. If the bonding force constant is designated by k_d, the correlation parameter g is given by the following equation:

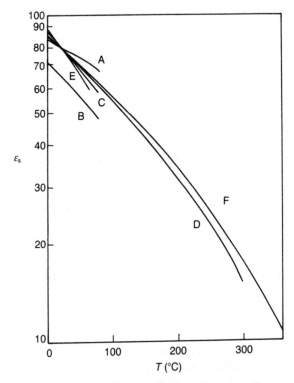

Figure 8.6 Calculated dielectric constants for water. Curves: A, by Oster and Kirkwood (1943); B, by Pople (1951); C, Pople's calculation improved by Coulson and Eisenberg (1966); D, by Haggis *et al* (1952); E, by Harris (1955); F, experimental results by Akerlof and Oshery (1950). (From Hasted 1973.) Reproduced by permission of Chapman & Hall.

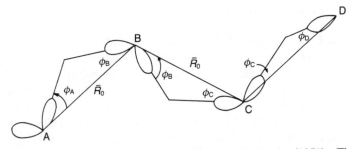

Figure 8.7 The model for water proposed by Pople (1951). The bending of the hydrogen bonds is characterized by the angles ϕ_A, ϕ_B etc. (From Hasted 1973.) Reproduced by permission of Chapman & Hall.

$$g = 1 + \sum_{i=1}^{\infty} N_i \, 3^{(1-i)} \cos^{2i} \alpha [1 - (kT/k_d)]^2 \qquad (8.6)$$

where i is the number of coordination shells of the hydrogen bonded neighbours. α is one-half of the H–O–H bond angle. The value of k_d was empirically chosen to be 3.78×10^{-13} erg rad^{-2}. The numbers of second and third shells are assumed to be 11 and 22 respectively. With these substitutions, the value of the g-factor was found to be 2.60 at $0\,°C$ and 2.46 at $83\,°C$. These are much more reasonable results than those obtained by Oster and Kirkwood. The dielectric constant calculated by Pople using these values for the g-factor is shown in figure 8.6 (curve C).

Curve D in figure 8.6 was calculated by Haggis *et al* (1952), considering hydrogen bond breaking instead of bond bending. Haggis *et al* used the following equation for the g-factor for fully bonded water molecules:

$$g = 0.862\{4 + [4 + (2 + 24/9)\cos 71°]\cos 71°\}\cos 54.5° = 2.91. \quad (8.7)$$

On the other hand, the expression for zero bonded molecules becomes

$$g = 0.862(4 + 2\cos 71°)\cos 54.5° = 2.34. \qquad (8.8)$$

Note that the value of the correlation parameter decreases considerably as the extent of bonding diminishes. The dielectric constant calculated by Haggis *et al* agrees very well with experimental results over a wide range of temperature. Thus, the bond breaking model yields a better result than the bond bending model used by Pople and others. The bond bending model yields good agreement with experimental values only in a limited temperature range. Furthermore, this model is not applicable above $100\,°C$ where extensive bond breaking would take place.

8.4 The Frequency Dependence of the Dielectric Constant

The dielectric constant of liquid water is independent of frequency below 100 MHz. The onset of dielectric dispersion is known to occur around 1 GHz. The dielectric constants of liquid water between 0.5769 and 890 GHz, which were obtained by several investigators, are summarized in figure 8.8. The two curves illustrated in this figure were obtained at 0 and 30 °C. As shown, the experimental results from these independent measurements are remarkably consistent. According to this figure, the dielectric constant of liquid water changes from 88.3 to 4.23 (using the data at 0 °C). The relaxation frequency at 0 °C is found at 16 GHz and this is translated into a relaxation time of about 10^{-11} s. The full curve drawn in this figure is calculated using Debye's equation. Although measured points fit the Debye theory closely, there have been

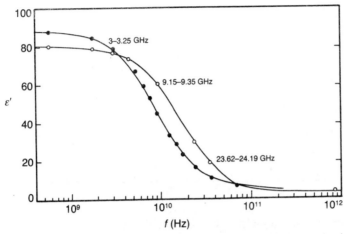

Figure 8.8 The frequency dependence of the dielectric constant ε' of water. This graph is a composite of the results obtained by various authors. (Grant and Shack 1967, Grant *et al* 1957, Chamberlain *et al* 1966, Cook 1952, Buchanan 1952, Sandus and Lubitz 1961, Collie *et al* 1948, Hasted and El-Sabeh 1953, Pottel 1972 (unpublished, see Franks 1973), Saxton 1952, Saxton and Lane 1952.)

some debates about the distribution of relaxation times for liquid water. Some authors claim that the dielectric relaxation of liquid water is better explained by the Cole–Cole equation (equation (3.71)), with a small distribution parameter ranging from 0.023 to 0.015 (Schwan and Grant 1976). The Cole–Cole arc which was constructed using currently available data is shown in figure 8.9 (see Hasted 1973). The depressed arc shows the Cole–Cole plot with an α-value of 0.012–0.013. Although the deviation of measured points from the single time constant arc is very small, nevertheless, measured points seem to be better fitted to the depressed arc than the Debye arc. However, the deviation is extremely small and the relaxation of water in the liquid state can be represented adequately by one time constant. As has been discussed, an analysis of the dielectric constant of water requires either bending or breaking of hydrogen bonds. The deformation and/or rupture of hydrogen bonds are subject to statistical variations and this in turn causes a distribution of relaxation times. However, the expected wide distribution of relaxation time has not been observed experimentally. This is one of many unanswered questions with regard to the relaxation of water dipoles.

It has been mentioned briefly before that the activation enthalpy of the dipole orientation of water is approximately 4.5 kcal mol^{-1}. This value corresponds to the breaking of one hydrogen bond. This observation suggests that bond breaking is the dominant polarization mechanism. Let us now look at the actual experimental data more carefully.

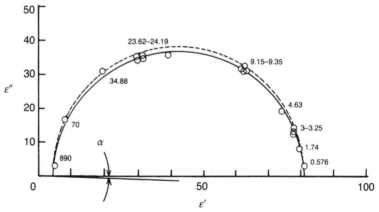

Figure 8.9 A Cole–Cole plot for water. The experimental results were obtained by the same authors as in figure 8.8. The depressed circle gives a distribution parameter α of 0.012–0.013.

Figure 8.10 shows a semi-logarithmic plot of the relaxation frequency of water against $1/T$ (see Hasted 1973). The bold curves represent experimental data. Clearly, the plot is not linear and bending of the curve occurs around 40 °C. This is an indication of structural changes at this temperature. The two thin straight lines represent the limiting slopes in the low and high temperature ranges. Using the low-temperature limiting slope, we obtain, using (3.105), an activation enthalpy of 4.5 kcal mol^{-1}. The implication of this result is that the reorientation of one water molecule requires the breaking of one hydrogen bond only. However, the high-temperature limiting slope in figure 8.10 gives rise to an activation enthalpy of 3 kcal mol^{-1}. This means that enhanced thermal motion causes the breaking of a number of hydrogen bonds even before the application of a field, thus producing more non-bonded as well as singly bonded water molecules. The polarization of non-bonded water does not require the breaking of a hydrogen bond. As the number of non-bonded water molecules increases, the number of hydrogen bonds broken by applied field will diminish and, on average, the activation enthalpy decreases. This explanation seems to account for the decrease of activation enthalpy from 4.5 to 3 kcal mol^{-1} with increasing temperature. The number of broken hydrogen bonds due to the orientation of water depends on the number with which a water molecule is bonded to neighbours. If doubly or triply bonded water is the dominant fraction, then the activation enthalpy would be much larger than 4 kcal mol^{-1}. The observation that the activation enthalpy of liquid water near 0 °C is close to 4.5 kcal mol^{-1} indicates that water molecules are not fully hydrogen bonded to four neighbours at this temperature. We surmise that some of them are singly bonded and perhaps most of them are doubly bonded. This observation does not seem to be compatible with the results of x-ray analyses which indicate that water molecules are extensively bonded to neighbours near the freezing point.

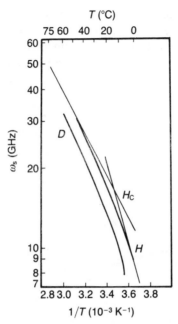

Figure 8.10 A semi-logarithmic plot of the relaxation frequencies of water (H) and D_2O (D) against $1/T$. Note the bending of the plot around 40 °C (H_C), indicating a configuration change. (From Hasted 1973.) Reproduced by permission of Chapman & Hall.

8.5 Dielectric Properties of Ice

The structure of ice Ih (hexagonal ice) is illustrated in figure 8.2. As shown, each water molecule is hydrogen bonded to four nearest neighbours, thus forming a tetrahedral configuration. However, because of the ionization of water, approximately 10^{-7} molar hydronium and hydroxyl ions exist even in the purest ice crystals, creating two types of lattice defect (Bjerrum 1951). The L defects are created by missing protons due to the presence of OH ions and the D defects are produced by the excess protons of hydronium ions. Water dipoles can rotate in ice because of these defects and this seems to be the dominant mechanism for the electric polarization of water in pure ice. Since the number of these defects is small, one might expect the dielectric constant of ice to be much smaller than that of liquid water. Figure 8.11 shows the dielectric constants of water and ice below and above the transition temperature. As shown, the dielectric constant of water continues to increase even after freezing and reaches a value of as high as 130 at $-60\,°C$. In general, the phase transition of polar substances from the liquid to the solid state causes a sudden decrease of dielectric constant because of the loss of rotational freedom in the solid state. However, water is one polar substance which does not follow this general rule.

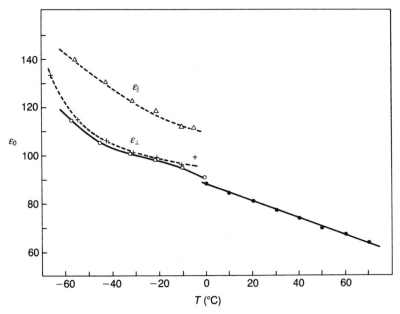

Figure 8.11 The dielectric constants of water and ice above (full circles) and below (open circles) the freezing point. The two broken curves below the freezing point indicate the anisotropy of the electric polarization of a single ice crystal.

In spite of the continuity of the dielectric constant below and above the transition temperature, a substantial shift in the relaxation frequency is observed from 16 GHz to about 5 kHz upon freezing. Hence, a million-fold decrease. The dielectric relaxation of ice had been investigated using polycrystalline ice by early workers (Auty and Cole 1952). However, Granicher *et al* (1957), Steineman (1957) and Steineman and Granicher (1957) performed a series of dielectric measurements using very pure single crystals. By definition, single crystals consist of only one domain and these preparations enabled these investigators to study the dielectric behaviour of ice unequivocally, including anisotropy and the effect of regulated doping with hydrogen fluoride. The dielectric dispersion curve obtained by Granicher *et al* (1957) is shown in figure 8.12. The average low-frequency dielectric constant of single ice crystals, $\varepsilon(\|)$ and $\varepsilon(\perp)$, is 120–130, which is basically in agreement with the value found by Auty and Cole using polycrystalline ice. However, the relaxation of a single crystal is characterized by only one time constant while that of polycrystalline ice shows a small distribution of the relaxation time. As illustrated in figure 8.11, single crystals shows a small but well defined anisotropy of about 15 per cent.

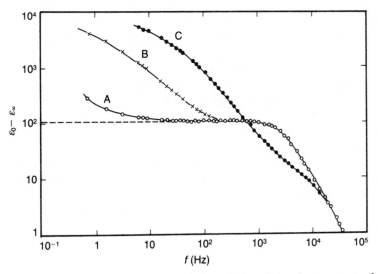

Figure 8.12 The frequency dependence of the dielectric constant of single crystalline ice. The full curve was calculated using Debye's equation. Curves: A, pure ice; B and C, with HF at two different concentrations. (From Granicher *et al* 1957.) Reproduced by permission of the Royal Society of Chemistry.

The activation enthalpy of the electric polarization in pure ice is found to be $13\,\mathrm{kcal\,mol^{-1}}$. This value is equivalent to the energy of three hydrogen bonds. We may be able to conclude from this observation that the orientation of water in ice requires the breaking of three hydrogen bonds. However, the activation enthalpy of the polarization of ice can also be interpreted as the energy required to create D and L defects by applied electrical fields. Hydrogen bonds are broken in order to form D defects and the formation of L defects results in the formation of H_3^+O

ions. If the creation of lattice defects or charge separation is the rate limiting step, the activation enthalpy can be attributed to the formation of lattice defects rather than to the orientation of water molecules.

The last topic on the dielectric properties of ice is the effect of doping with hydrogen fluoride. The size of a fluoride ion is approximately the same as that of oxygen. Therefore, HF can fit in the lattice of ice crystals without any significant structural perturbation. However, unlike the water molecule, HF has only one hydrogen. Thus, the insertion of HF creates D-type defects in ice crystals. The migration of protons utilizing these defects produces a large dielectric constant, as shown in figure 8.12. Judging from the value of the dielectric constant, it can be surmised that the migration of protons may not be limited to short distances. The peak value of the dielectric constant is observed at a low concentration of around 10^{-4} mol. At high concentrations of HF, however, the dielectric constant decreases, indicating that there is some mechanism which prevents proton migration. Perhaps a high concentration of HF causes extensive perturbation of the crystal lattice, which is not favourable for proton fluctuations.

Since excellent reviews are available on this subject, I shall not go into the details of the relaxation mechanism of water and ice any further. In order to understand the structure and physical properties of bound water in biological cells, it is important to have some insight into the dielectric properties of liquid water and ice crystals. It is not easy to understand the dielectric properties of bound or cell water without some knowledge of pure water and ice. It has been postulated that the physical properties of cell water are somewhere between liquid water and ice. Needless to say, this is a very vague expression for describing their properties.

8.6 The Dielectric Constant of Water Surrounding an Ion

When an ion is placed in aqueous media, it will attract a number of water molecules, because of the interaction between the charge of the ion and the dipole moment of water. The interaction energy, attractive in this case, between the ion and water dipole is written as

$$W = -E \cdot \mu = -E\mu \cos \theta \qquad (8.9)$$

where E is the field due to the ion and μ is the electric moment of the dipole. The field E is given by $e/\varepsilon r^2$ and, thus, the interaction energy is defined by

$$W = -(e/\varepsilon r^2)\mu \cos \theta. \qquad (8.10)$$

The field intensity in the vicinity of an ion is very large. For example, if the distance between an ion and the water molecule is 2 Å, the field is as high as 10^6 V cm^{-1}. As indicated by (8.10), the maximum interaction energy is obtained when the angle θ is 0, that is, water dipoles are oriented along the line connecting the centre of the dipole and the ion. As mentioned before, water molecules have a stable structure when they are in a tetrahedral configuration, if external forces are absent. However, the strong field created by a charge is sufficient to perturb the tetrahedral configuration. As the distance increases, the radial configuration diminishes and the tetrahedral structure begins to dominate again.

As discussed previously, tetrahedral water which is extensively hydrogen bonded has a large dielectric constant. This statement holds both for the liquid and the crystal phases. The dielectric behaviour of the water of hydration or the water in close proximity to an ion is not well understood. However, it is reasonable to assume that the orientation of water dipoles along their primary axis is modulated extensively by the overwhelmingly large field due to the ion. Thus figure 8.13 which shows a water molecule in the vicinity of an anion may be an accurate illustration of the solvation water in general. For the water surrounding a cation, the direction of the water molecule should be rotated by 180°. Small external fields are used for the measurement of dielectric constant. These small fields perturb the direction of water dipoles slightly along the transverse and longitudinal axes, depending on the direction of the applied field.

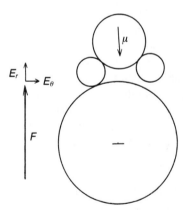

Figure 8.13 A water molecule near an anion. F is an electrical field due to ionic charge. E_r and E_θ are the small fields used for dielectric measurements. (From Takashima *et al* 1986.) Reproduced from the *Biophysical Journal*, 1986, vol 49, pp 1003–8 by copyright permission of the Biophysical Society.

A theoretical calculation of the dielectric constant of hydration water was performed by Debye many years ago (1929). The method used by

Debye for this calculation is summarized as follows. In order to perform this calculation, we need to compute two quantities, the displacement $D(= \varepsilon E)$ and the force F due to the electrical field. The general expressions for these quantities are:

$$D = E + 4\pi P \qquad (8.11)$$

and

$$F = E + 4\pi P/3 \qquad (8.12)$$

where P is the polarization. The force F gives rise to a non-vanishing mean moment μ (see Chapter 2)

$$\mu = \alpha F + \mu_0 L(\mu_0 F/kT) \qquad (8.13)$$

where μ_0 is the dipole moment of an individual molecule. Using this equation and $p = N\mu$ in (8.11) and (8.12), we can eliminate P to obtain

$$D = F + 8\pi/3[N\alpha F + N\mu_0 L(\mu_0 F/kT)] \qquad (8.14)$$

$$E = F - 4\pi/3[N\alpha F + N\mu_0 L(\mu_0 F/kT)]. \qquad (8.15)$$

If we assume an arbitrary value for the force F (a field due to ionic charge), then we can calculate D using (8.14), since all other quantities are known. The displacement D is given by

$$D = e/r^2. \qquad (8.16)$$

Therefore, we can readily calculate the distance r using this equation. Thus, the assumed value for F automatically determines the distance r. It appears that we can calculate the distance r using the well known equation $E = e/\varepsilon r^2$. However, at present, the value of ε, the dielectric constant of water, is an unknown variable and it is apparent that this method cannot be used. A small external field is then superposed on the field of the ion longitudinally or transversely. If the field is applied radially or in a longitudinal direction, then the value of the displacement will change minutely, i.e. by δD, and also the value of E changes by δE. Thus, the dielectric constant ε_r is given by

$$\varepsilon_r = \delta D/\delta E \qquad \text{or} \qquad = dD/dE. \qquad (8.17)$$

The derivative dD/dE can be rewritten as

$$dD/dE = (dD/dF)/(dE/dF). \qquad (8.18)$$

Differentiation of (8.14) and (8.15) gives rise to

$$dD/dF = 1 + 8\pi/3[N\alpha + N\mu_0^2/kT \, L'(\mu_0 F/kT)] \qquad (8.19)$$

$$dE/dF = 1 - 4\pi/3[N\alpha + N\mu_0^2/kT \, L'(\mu_0 F/kT)] \qquad (8.20)$$

where L' is the derivative of the Langevin function with respect to F.

We now have every quantity which is needed to calculate the dielectric constant ε_r. The results of Debye's calculation are shown in figure 8.14. This figure shows the change of the dielectric constant of the hydration water of an ion as a function of distance from the central ion. Obviously, ε_r is very small within 2–3 Å from the ion. However, ε_r gradually increases as the distance increases to 10–15 Å, reaching a value of 78, the normal dielectric constant of water.

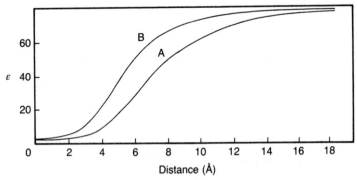

Figure 8.14 The dielectric constant of water near an ion. (From Debye 1929.) Curves: A, radial field (ε_r); B, transverse field (ε_θ).

If we apply a field perpendicularly, the direction of the large field due to the ion will be changed slightly by an angle $\delta\theta$ which gives rise to a small transverse vector δE. This vector is given by

$$\delta E = E \sin \delta\theta = E\delta\theta. \tag{8.21}$$

Likewise, the displacement D will rotate and produce an angular component, thus we obtain

$$\delta D = D\delta\theta = (D/E)\delta E. \tag{8.22}$$

Using these, we derive the following equation:

$$\varepsilon_\theta = D/E. \tag{8.23}$$

Since D and E are already defined by (8.14) and (8.15), we can easily calculate the dielectric constant ε_θ. This result is also shown in figure 8.14. This figure indicates that the presence of an ion exerts a force which is sufficient to alter the rotational characteristics of water molecules within 10–15 Å. This means that the dielectric behaviour of five to six layers of hydration water is affected by the presence of an ion. Since the electrostatic force due to an ion diminishes with the inverse square of distance, this conclusion is rather surprising.

It was pointed out in Chapter 2 that Debye's theory utilizes the Lorentz field as the foundation of his theory. As discussed, the Lorentz field is now deemed inappropriate to represent the internal field for polar molecules. This problem was solved by Onsager using the concept of reaction field in addition to internal field. The dielectric behaviour of hydration water was reinvestigated by Hasted *et al* (1948) using Onsager's formalism and Kirkwood's general theory in the aforementioned Debye method. The equation used by Hasted *et al* is shown below:

$$\varepsilon E = \frac{4\pi}{V} \frac{(n^2 + 2)(2\varepsilon + 1)}{3(2\varepsilon + n^2)} L\left(\frac{\varepsilon(n^2 + 2)}{(2\varepsilon + n^2)} \frac{\mu_0^2}{kT} E\right) + \frac{3\varepsilon(n^2 - 1)E}{2\varepsilon + n^2}$$

(8.24)

where L is the Langevin function and n is refractive index. This equation can be approximated as

$$\varepsilon E = \frac{4\pi}{V} \frac{(n^2 + 2)}{3} L\left(\frac{(n^2 + 2)}{2} \frac{\mu_0^2}{kT} E\right) + \frac{3}{2}(n^2 - 1)E.$$

(8.25)

On substituting $n^2 = 4.13$ and $\mu = 1.84$ D into this equation we obtain

$$\varepsilon E = 5.69E + 1.58 \times 10^6 \, L(1.39 \times 10^{-4}E).$$

(8.26)

Hasted *et al* used Kirkwood's equation as their second method:

$$D = \varepsilon E = \left(\frac{4\pi}{V}\right)[(\alpha E) + \mu L(\mu_0 E/kT)].$$

(8.27)

Comparing this equation with (2.66) we see that while (2.66) is a weak field approximation, (8.27) is valid even for strong fields. Using the value of 2.45 for μ_0, instead of 1.84 D, Hasted *et al* obtained

$$D = \varepsilon E = 4.0E + 1.03 \times 10^6 \, L(2.18 \times 10^{-4}E).$$

(8.28)

This is the second equation used for the calculation of the dielectric constant of hydration water. Note that the LHS of (8.26) and (8.28) is the displacement D which is equal to e/r^2. Equations (8.26) and (8.28) are special cases of the following general equation:

$$AE_r + BL(CE_r) = e^2/r^2.$$

(8.29)

In (8.26), $A = 5.69$, $B = 1.58 \times 10^6$ and $C = 2.18 \times 10^{-4}$, and in (8.28), $A = 4.0$, $B = 1.03 \times 10^6$ and $C = 2.18 \times 10^{-4}$.

The dielectric constants ε_r and ε_θ are calculated using the following equations (see also (8.17) and (8.23)):

$$\varepsilon_r = \partial D/\partial E_r = A + BC[dL(CE_r)/d(CE_r)]$$

(8.30)

and

$$\varepsilon_\theta = D/E_\theta = A + B[L(CE_\theta)/E_\theta].$$

(8.31)

The results of the calculations using Onsager's theory are illustrated in figure 8.15. The results obtained using Kirkwood's equation are very similar to those of Onsager's formalism. The striking feature of these results is the abrupt increase of the dielectric constant of water near a central ion. Namely, ε_r and ε_θ increase to their normal value within a distance of 2–3 Å, indicating that the charge of a central ion affects the rotational characteristics of the water dipole within only 4 Å from the ion. This result is in contrast to that calculated by Debye who concluded that the electrostatic force of an ion reaches more than 10 Å. It is, however, well known, as discussed before, that the Debye theory of electric polarization is inadequate for polar molecules such as water. Because of this, the results obtained by Hasted *et al* using Onsager's and Kirkwood's theories can be considered more reliable and free of internal inconsistency.

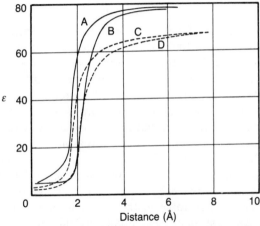

Figure 8.15 The dielectric constant of water near an ion. Curves A and B were calculated by Hasted *et al* (1948) with radial and angular fields. (From Hasted *et al* 1948.) Reproduced by permission of the American Institute of Physics. Curves C and D were calculated by Takashima *et al* (1986) for radial and angular fields. (From Takashima *et al* 1986.) Reproduced from the *Biophysical Journal*, 1986, vol 49, pp 1003–8 by copyright permission of the Biophysical Society.

As noted, (2.84) is a cumbersome equation to solve and, therefore, only approximate solutions were obtained. For example, in (8.24), ε is the unknown to be computed. However, this equation cannot be transformed into an explicit expression for ε. Hasted *et al* replaced

(8.24) with (8.25) in order to eliminate ε from the RHS of this equation. This approximation is, however, valid only if $\varepsilon \gg n^2$. However, the result of this calculation shows that the computed value of ε is actually smaller than or comparable to n^2 when the distance is smaller than 3 Å. This contradicts the original assumption.

Takashima et al (1986) carried out the calculation of the dielectric constant of hydration water using Onsager's theory in Debye's formalism. The computational difficulties which Hasted et al encountered were circumvented by an iteration procedure until the calculated dielectric constant became self-consistent. They used the following equations for the force F and internal field E:

$$D = F + \frac{3\varepsilon(n^2 - 1)F}{2\varepsilon + n^2} + \frac{4\pi(n^2 + 2)(2\varepsilon + 1)\mu_0}{3(2\varepsilon + n^2)} L\left(\frac{\varepsilon(n^2 + 2)}{2\varepsilon + n^2} \frac{\mu_0 F}{kT}\right)$$

(8.32)

$$E = \frac{3\varepsilon}{2\varepsilon + 1} F.$$

(8.33)

In (8.33), the reaction field is ignored with an assumption that the reaction field does not actually exert a torque on the dipoles. Following Debye's method, the dielectric constants ε_r and ε_θ are calculated using (8.17) and (8.23). The procedure is to assume an arbitrary value for F and to calculate the displacement D and the internal field E using (8.32) and (8.33). Starting with an arbitrary but realistic value for ε, the iteration procedure is repeated until self-consistency is reached. Usually 10–15 iterations are sufficient to obtain a reasonable convergence of the calculated dielectric constant. The calculated dielectric constants ε_r and ε_θ are illustrated in figure 8.15. The results of this calculation are quite similar to those of Hasted et al; namely, the charge of the central ion exerts only a short-range force to the surrounding water molecules, i.e. the dielectric constant increases rapidly as the distance between the central ion and the water molecules increases above 3 Å. This indicates that only the first or part of the second layer of hydration water has a dielectric constant which is significantly different from that of normal water. Since the dielectric constant of ice is not different from that of liquid water (although its relaxation frequency is a million-fold lower), the first layer of hydration water, having a very small dielectric constant, does not resemble either that of liquid water or ice.

It must be remembered that the dielectric constant is only a measure of rotational characteristics. Therefore, the magnitude of the dielectric constant has no relevance to other transport phenomena such as self-diffusion, viscosity and even ionic conductivity. Thus, the discussion presented above pertains only to the rotational behaviour of water molecules in the vicinity of ions.

8.7 A Molecular Dynamic Simulation of Water near a Hydrophobic Wall

We have discussed, in the foregoing, the dielectric behaviour of water in the vicinity of an ionic charge. It was concluded that the effect of the charge of an ion does not reach beyond 2–4 Å and the dielectric constant increases to a normal value rapidly as the distance between the ion and the water molecules increases. In biological systems, however, hydrophobic or van der Waals interactions play a role which is as important as that of charge–charge and/or charge–dipole interactions for the formation of bound water. The dynamic properties of water molecules in the proximity of uncharged walls have been investigated using molecular dynamic computer simulation. Although there are a number of papers on this subject, the articles published by Marchesi (1983) and by Barabino et al (1984) seem to be most relevant to our current interests on the behaviour of water near the uncharged surfaces of large molecules and cell walls.

The model used by these authors is of water molecules placed between two planar walls, separated by 20 Å. The Boltzmann potential which is averaged over the surface of the wall is used as the potential between the wall and water molecules:

$$\exp(-\phi_w(z)/kT) = A^{-1} \int_A \exp(-\phi_w(x, y, z)/kT)\,dxdy. \quad (8.34)$$

This potential function can be interpolated to a Lennard-Jones potential, i.e.

$$\phi_w(z) = \gamma[(\sigma/z)^4 - (\sigma/z)^2]. \quad (8.35)$$

The numerical values for the constants γ and σ for oxygen and hydrogen are given below:

$$\gamma_O = 1.2 \times 10^{-20} \quad \text{and} \quad \sigma_O = 2.0 \text{ Å}$$

$$\gamma_H = 0.6 \times 10^{-20} \quad \text{and} \quad \sigma_H = 1.3 \text{ Å}.$$

A potential function which is called a CI potential (configuration interaction potential) was used for the interaction between water molecules (for the detail, see Matsuoka et al 1976). In addition, the potential due to image charge was taken into consideration. If a charge q is placed near a wall, this charge will interact with the wall through the image charge q_i on the other side. The effective image charge is defined by

$$q_i = [(1 - \varepsilon_w)/(1 + \varepsilon_w)]q. \quad (8.36)$$

If the dielectric constant of the wall ε_w is assumed to be 3, q_i reduces to

$-\frac{1}{2}q$. In the following, some of the results which are relevant to our present discussion will be summarized.

Angular profile of water

In the case of water near an ion, the orientation of water dipoles is dictated by charge–dipole interactions which tend to orient water dipoles along the line connecting the charge and the centre of the dipole. In the case of hydrophobic interaction, the orientation of a dipole is more complex. Part (a) of figure 8.16 shows the results of the simulation without the image charge effect. The abscissa is the cosine of the angle between the water dipole and the wall. The zero of this coordinate means that the water dipole is parallel to the wall and +1 or −1 mean that the dipoles are perpendicular to the wall with their vector pointing away from or toward it. The ordinate is the probability density function

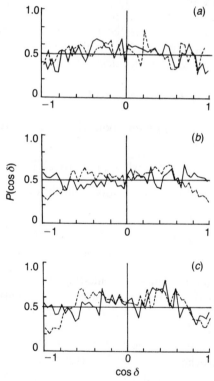

Figure 8.16 The orientation of water molecules near a hydrophobic wall. The ordinate is the direction cosine. The 0 means water is parallel to the wall and +1 and −1 mean 90° and 270° with the wall. (From Barabino et al 1984.) Reproduced by permission of Elsevier Science Publishers.

of having these orientations. The full lines in these diagrams are for bulk water and the broken lines are for wall water. In part (a) where no image charge is considered, the presence of a wall does not seem to have a noticeable effect on the orientation of water dipoles. However, the consideration of image charges (part (b)) causes a preferential orientation of water dipoles parallel to the wall. Lastly, part (c) shows the results of the calculation considering only the repulsive force. In this case, the tendency of orientation parallel to the wall is accelerated.

Dipole–dipole correlation

Interesting information can be obtained by calculating the correlation function of a water dipole with other dipoles. The two-dimensional dipole–dipole correlation function in the x–y plane is defined by

$$g_\mu^{xy}(r) = (1/2\pi\rho r d)\langle \cos\theta_{ij}\rangle_{r,r+dr}/dr \qquad (8.37)$$

where ρ and d are the density and thickness of the layer, and θ_{ij} is the angle between dipoles i and j. The average is taken over all molecules whose O–O distance is between r and $r + dr$. Figure 8.17 shows the value of $g_\mu^{xy}(r)$ calculated for a distance of 8–10 Å from the wall (a), for 2.5–3.5 Å (b) and for 0.5–1.5 Å (c) respectively. As the distance from the wall increases, the correlation function exhibits a sharp peak at an O–O distance of about 2.5–2.7 Å. This is the distance we find in hydrogen bonded tetrahedral water molecules. In other words, the configuration of water molecules returns to the ordinary structure of bulk water when they move away from the wall. However, when the distance from the wall is small (part (b)), the height of the correlation function diminishes markedly. At a distance of 0.5–1.5 Å, the peak of the correlation function nearly vanishes, which is an indication that the correlation among water dipoles near a wall is vanishingly small. Based on this result, we can conclude that there is no coherent interaction between water dipoles near hydrophobic planar walls.

In conclusion, the presence of an uncharged wall does not have a profound effect on the angular distribution of water molecules. However, the wall has a prominent effect on dipole–dipole correlations even though the wall is itself uncharged. The purpose of these molecular dynamic calculations was to investigate theoretically the effect of van der Waal's forces on the structuring of water molecules using a simple model. The underlying assumption of this calculation is that the hydrophobic interaction is one of the dominant factors in polymer–water interactions. This assumption is by and large reasonable because many uncharged polymers attract a number of water molecules. However, molecular dynamic simulations indicate that the presence of an uncharged boundary does not seem to have a marked effect on the spatial configuration, even at short distances, except for dipole–dipole correlations.

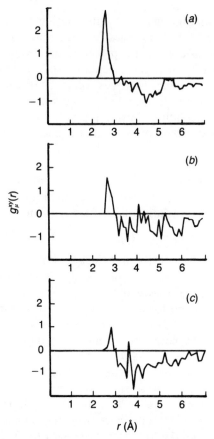

Figure 8.17 The dipole–dipole correlation function $g_\mu^{xy}(r)$ for wall water. (*a*) The distance from the wall is between 8 and 10 Å, (*b*) between 2.5 and 3.5 Å and (*c*) between 0.5 and 1.5 Å. (From Marchesi 1983.) Reproduced by permission of Elsevier Science Publishers.

8.8 Bound Water in Biological Macromolecules

The structure and physical properties of bound water have been investigated for many years, yet we do not have a satisfactory model for the water of hydration due to various reasons. Bound water is a special form of bulk water perturbed by charges or hydrophobic interactions. Water molecules can have a variety of structures and physical states depending strongly on the circumstances. As discussed earlier, some biological water can be nearly identical to normal water. On the other hand, some portions of bound water may be like ice. Furthermore, as discussed, water molecules in the vicinity of a charge have a structure which does not resemble either normal water or ice.

Structural water
First of all, some water molecules are found in biological polymers as an integral part of the structure. A number of x-ray crystallographic studies on proteins are able to detect these water molecules on electron density maps. The investigations by Lipscomb *et al* (1969) and by Dickerson and Geis (1969) on carboxypeptidase A are some of these examples. Figure 8.18 illustrates the spatial arrangement of water molecules in a triple helical collagen molecule (Ramachandran and Chandrasekharan 1968). As shown, a water molecule is hydrogen bonded to a carbonyl group as well as to an NH residue of one of the chains and also to a carbonyl group of another strand. The role of these water molecules is to bridge the helical chains and to stabilize the triple helical structure.

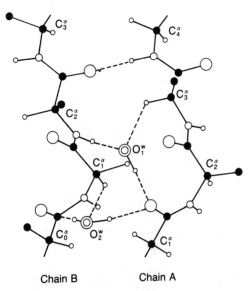

Chain B   Chain A

Figure 8.18 A side view of the chains A and B of collagen molecules. This diagram shows three different types of cross-linking secondary bonds. O_w^1 and O_w^2 represent oxygens of the water molecules. The full and open small circles are carbons and nitrogens. The large open circles represent carbonyl oxygens. (From Ramachandran and Chandrasekharan 1968.) Reproduced by permission of John Wiley & Sons, Inc. © 1968.

Thus, it can be inferred that removal of these water molecules would cause an instability of the helical structure. The presence of structural water of this kind is presumed to be ubiquitous among biopolymers. A recent report by Kopka *et al* (1984) shows a chain of water molecules within a groove of six base pairs in double helical DNA (see figure 8.19). These water molecules are an integral part of the structure of biopolymers and have a very specific configuration which can be identified by x-ray analysis. Clearly, they are a part of the polymer and cannot be removed without disrupting the internal structure. Although they play an important role for the stability of biopolymers, their dielectric properties are unknown because of the difficulty of identifying them dielectrically. The number of these water molecules is relatively small compared with those which are bound to the surface of polymers. However, we can easily surmise that the orientational freedom of these water molecules is very restricted, with their dielectric constant being much smaller than that of normal water.

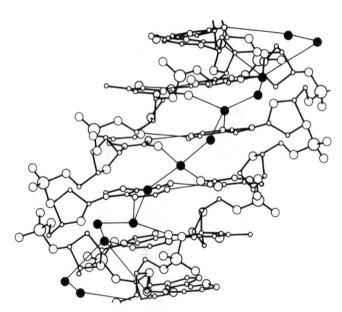

Figure 8.19 Water molecules found in the groove of the double helix of B-DNA. The sequence is CGCGAATTCG with the spine of hydration. Full spheres represent water oxygen. (From Kopka *et al* 1984.) Reproduced by copyright permission of the Biophysical Society.

Dielectric properties of water of sorption
The adsorption or desorption of water on the surface of protein crystals has been of considerable interest for many years. It is not certain

whether the adsorption of water is strictly a surface phenomenon or that adsorbed water penetrates, to some extent, into crystals and becomes bound to individual protein molecules. However, for the sake of simplicity, let us assume that the adsorption of water on protein crystals is strictly a surface phenomenon. When the crystalline powder of a protein is exposed to water vapour of known partial pressure p/p_0, uptake of water vapour on the surface of crystals will occur. This adsorption will continue until an equilibrium state is reached. A plot of the amount of adsorbed water at equilibrium against the partial pressure of water vapour is shown in figure 8.20. The amount of adsorbed water increases, as shown, as the partial pressure of water vapour increases, and the curve reaches a plateau. However, the curve rises again as the partial pressure of H_2O further increases. Intuitively, this curve indicates the formation of multilayers of adsorbed water. Some adsorption isotherms which have been observed with solid materials, however, do not indicate the formation of multilayers. For example, the adsorption of gas molecules on metal surfaces can be represented by curve A in the same figure. This curve is obviously different from curve B and indicates the formation of only one layer of adsorbed gas molecules.

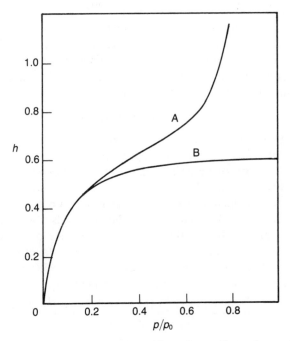

Figure 8.20 A schematic representation of monolayer (curve B) and multilayer (curve A) adsorptions of gaseous molecules on solid surfaces. h is the amount of adsorbed water in grams.

The theory of monolayer adsorption was formulated by Langmuir (1918) who assumed that the adsorption of gas molecules takes place at discrete sites. If N represents the total number of adsorption sites and n is the number of sites which are already occupied by gas molecules, the rate of adsorption at an arbitrary time is given by

$$v_a = k_a(N - n)p \tag{8.38}$$

where k_a is the rate constant of adsorption and p is the partial pressure of the gases. On the other hand, the desorption rate of adsorbed gases is expressed by

$$v_d = k_d n \tag{8.39}$$

where k_d is the rate constant of desorption. At equilibrium, the following equality exists:

$$k_a(N - n)p = k_d n. \tag{8.40}$$

Rearranging this equation, we obtain

$$n/N = k_a p/(k_d + k_a p). \tag{8.41}$$

This equation can be rewritten as follows:

$$h = h_m bp/(1 + bp) \tag{8.42}$$

where h is the amount of adsorbed water in grams and h_m is the amount of water when all the active sites are occupied. $b = k_a/k_d$. This equation, when h is plotted against p, yields a curve which is shown in figure 8.20 (curve A). Another way to display the adsorption is to rearrange the equation as follows:

$$1/h = 1/h_m + 1/h_m bp. \tag{8.43}$$

This equation produces a straight line when $1/h$ is plotted against $1/p$ with a slope of $1/h_m b$ and an intercept with the ordinate of $1/h_m$.

Langmuir's theory of monolayer adsorption is applicable to the cases where the surface of a solid is chemically inert. However, virtually all biological materials have charged or polar groups on or near the surface and they exert forces beyond the first layer of adsorbed water. Under these circumstances, the uptake of water molecules by protein powder results in the formation of multiple adsorption layers. In order to provide a quantitative theory, Brunauer et al (1938) generalized the Langmuir equation as shown below (BET theory):

$$h = \frac{h_m Cx[1 - (n + 1)x^n + nx^{n+1}]}{(1 - x)[1 + (C - 1)x - Cx^{n+1}]} \tag{8.44}$$

where $x = p/p_0$ and n is the number of layers. The quantity C is given by the following approximate formula:

$$C = \exp[-(E_f - E_l)/RT] \tag{8.45}$$

where $E_f - E_1$ is the difference in the heat of adsorption between the first and last layers. It can be assumed that E_1 is equal to the heat of condensation of water vapour. Note that the BET equation reduces to the Langmuir equation when $n = 1$:

$$h = h_m Cx/(1 + Cx).\qquad(8.46)$$

The curve calculated by BET theory is illustrated schematically in figure 8.20 (curve B). As shown, this diagram is similar to the adsorption isotherms obtained experimentally with protein powders. Analogous to Langmuir's equation, we can reduce the BET equation to a linear function as shown below:

$$x/h(1 - x) = 1/h_m C + (C - 1)x/h_m C.\qquad(8.47)$$

The plot of $x/h(1 - x)$ against x will result in a straight line.

Dielectric properties of adsorbed water
The dielectric constant of dry protein powder is very small and a value of 3–5 has been accepted, although an exact value is difficult to obtain. Usually, protein powder is packed to form a disc for the measurement of dielectric constant, however this introduces two sources of error. One is poor contact between electrodes and the solid specimen and the other is air space trapped in the packed powder, which reduces the dielectric constant of samples. The dielectric constants of dry and hydrated protein powders were first measured by Bayley (1951a, b) in a wide frequency range. Bayley found that the adsorption of a small amount of water increases the dielectric constant of protein powder. Moreover, the protein powders he studied exhibited a broad frequency dependent dielectric constant. Bayley attributed this dispersion to the orientation of adsorbed water molecules. The dispersion encompasses a frequency range between 300 kHz and about 3 GHz. This behaviour is substantially different from that of bulk water, particularly in the shift of relaxation frequency by a few orders of magnitude from the GHz region to the MHz range. The frequency shift can be interpreted as a result of the hindrance of orientation due to electrostatic interaction between water and protein surfaces.

Rosen (1963), using a variety of protein powders, investigated the effect of adsorbed water on the dielectric properties. As shown in figure 8.21, the static dielectric constant of protein powders increases very slowly with the increasing partial pressure of water vapour during the initial stage. However, there is a critical point at which the slope of the dielectric constant changes abruptly, indicating a change in the nature of the adsorbed water. Since every protein powder Rosen investigated exhibited a similar critical point, he concluded that this behaviour is the fundamental feature of the sorption water on protein surfaces. Rosen's findings indicated that the dielectric behaviour of adsorbed water is not as simple as the interpretation by Bayley.

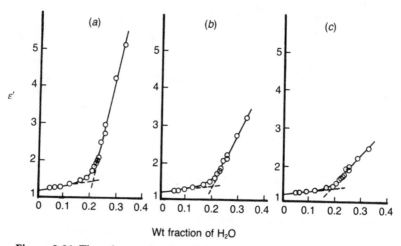

Figure 8.21 The change in the dielectric constant of bovine serum albumin powders as a function of water of adsorption (*a*) at 100 kHz, (*b*) at 1 MHz and (*c*) at 10 MHz. (From Rosen 1963.) Reproduced by permission of the Royal Society of Chemistry.

Takashima and co-workers (1962, 1965) using egg and horse serum albumin measured the dielectric constant of compressed powder with and without water of adsorption. First, they estimated the water content of freeze dried proteins and found that lyophilized protein powders still contain as much water as 12–15% of the total weight. These water molecules can be extracted by drying in vacuum over phosphorus pentoxide. Only when the total amount of sorption water exceeds 20% did the dielectric constant begin to increase rapidly. As the amount of sorption water further increases, the dielectric constant of the powder increases beyond that of bulk water (78 at 25 °C) and reaches a value of 170–220. The dielectric constant of hydrated protein crystals does not increase indefinitely, however, with a further increase of sorbed water. In other words, as shown in figure 8.22, the plot of dielectric constant against water content seems to fit the Langmuir isotherm whereas actual water sorption follows BET theory. This result indicates that the large dielectric constant is due to the formation of the first water layer. The contribution of the second and third layers appears to be negligibly small. From this observation, Takashima and co-workers concluded that the dielectric constant of protein crystals with sorption water is due to the formation of an electrical double layer between the protein surface and the first layer of sorption water. The dielectric constant of adsorbed water is small compared with that of a double layer and is overshadowed by the very large dielectric constant due to surface polarization. So far, no explanation is available as to the origin of the initial phase (region A in figure 8.22) of adsorption where the addition of water does not produce an increase in the dielectric constant. The underlying concept of Takashima's interpretation is that the adsorption of water on protein surfaces proceeds sequentially, forming the first layer and then the second and third layers. In this scheme, liquid-like multilayers do not appear during the formation of the first layer. This concept is implicit in the BET theory.

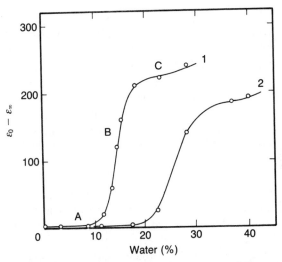

Figure 8.22 The change in the dielectric constant of the powders of bovine serum albumin (curve 1) and egg albumin (curve 2). The percentage of water is determined as $(w - w_0)/g \times 100$. (From Takashima 1962.) Reproduced by permission of John Wiley & Sons, Inc. © 1962.

Whereas the measurements by Rosen and also by Takashima and co-workers were limited to frequencies below 10^7 Hz, the analysis by Kent (1972) was based solely on measurements in the microwave region (approximately 1.5–9 GHz). Using the observation that the dielectric loss of protein powder increases markedly with increasing temperature, Kent was led to a conclusion that the major part of dielectric loss arises from the water adsorbed into liquid-like multilayers, even at very low hydrations. At the frequencies Kent used, the first layer of water is irrotational. However, this water can be relocated, at high temperatures, to outer layers and become polarizable. In order to allow the thermal jump, however, the binding energies of the first and outer layers must not be very different. Kent calculated, using the following equation, the dielectric losses of water in the first layer (ε_1'') and in outer layers (ε_2''):

$$\varepsilon''(1 + V) - \varepsilon_p'' - V\varepsilon_1'' = Vx(\varepsilon_2'' - \varepsilon_1'') \qquad (8.48)$$

where V is the amount of sorbed water expressed as a fraction of the protein dry weight, x is the relative vapour pressure and ε_p'', ε_1'' and ε_2'' are the loss factors of proteins, monolayers and multilayers. This equation can be rearranged as

$$\frac{\varepsilon''(1 + V) - \varepsilon_p''}{V} = \varepsilon_1'' + x(\varepsilon_2'' - \varepsilon_1''). \qquad (8.49)$$

A plot of $[\varepsilon''(1 + V) - \varepsilon_p'']/V$ against x should be linear and ε_1'' can be determined from the intercept of this plot with the ordinate and its slope. The values obtained are listed in table 8.2. This table demonstrates that while the loss factor of protein ε_p'' is nearly independent of temperature, those of monolayers and multilayers increase markedly with increasing temperature. The dielectric loss of monolayer water is by no means negligible compared with that of multilayers. The similarity between ε_1'' and ε_2'' indicates that the rotational hindrance of water in monolayers and in multilayers is not as different as has been postulated. Based on these observations, Kent was led to a conclusion that the formation of BET multilayers does not proceed stepwise, in other words, multilayers can appear at an early stage of adsorption, even during the formation of the first layer. According to this interpretation, the concept of the aforementioned 'critical hydration state' is difficult to interpret.

Table 8.2 The dielectric loss for protein, monolayer and multilayer water molecules.

Sample	Temperature (°C)	$\varepsilon_p''/\varepsilon_0$	$\varepsilon_1''/\varepsilon_0$	$\varepsilon_2''/\varepsilon_0$
Haemoglobin[†] (Denatured)	0	0.011	0.31	0.63
	25	0.011	0.55	1.06
	40	0.011	0.68	1.25
BSA[†]	0	0.012	0.38	0.97
	21	0.015	0.48	1.14
	40	0.018	0.61	1.59
White fish meat[‡]	10	0.018	0.17	0.63
	23.6	0.020	0.19	0.82
	90	0.032	0.48	3.58

† Leslie (1962).
‡ Kent (1972).

The large dielectric constant of hydrated protein powders which was observed at low frequencies by Rosen and Takashima and co-workers is likely to be due to the double layer capacitance rather than to the orientation of individual water molecules. The real rotational behaviour of hydration water molecules must be investigated at much higher frequencies, where the surface polarization is completely damped. Harvey and Hoekstra (1972), using a time domain refractometer (TDR), measured the dielectric relaxation of adsorbed water on the surface of lysozyme. These authors observed that the dielectric dispersion of rotationally bound water molecules fits the Debye equation within experimental error. At high water content, two dispersion regions were found that were presumably due to the water molecules in the Langmuir layer and BET outer layers.

Hydration water of proteins in solution
The dielectric behaviour of sorbed water on the surface of protein powder has already been discussed. Let us now turn to a discussion on the dielectric behaviour of water bound to protein molecules that are dissolved in aqueous media. As illustrated schematically in figure 5.13, the dielectric constant of protein solution usually exhibits three dispersion regions. One is found at around 10 GHz and is due to the orientation of bulk water. This dispersion is usually similar to that of pure water. As has been postulated, because of the dipole–charge interaction between water and the polar groups of proteins or nucleic acids, water molecules within a few angstroms from the surface seem to be irrotational and those which are located outside this range are similar to bulk water. Therefore, the question is whether this irrotational water near the polymer surface is, in spite of the constraints, still polarizable and gives rise to a dispersion which is distinctly different from that of bulk water.

A search of the dispersion due to bound water was first undertaken by Schwan (1965) using haemoglobin solution. The data observed by Schwan are shown in figure 8.23. The dispersion below 30 MHz is obviously due to the orientation of protein dipoles. At high frequencies, protein molecules are no longer rotational and behave as a non-polar particle. The high-frequency limiting value which is designated by ε_∞ represents the dielectric constant of bulk water, including the cavities created by protein molecules. At these frequencies, the dielectric constant of water is independent of frequency as indicated by the broken line. However, Schwan observed that the dielectric constant did not reach a plateau but continued, instead, to decrease with increasing frequency as shown by the full curve. This small dispersion (Schwan called it a δ-dispersion) was attributed to the hindered orientation of bound water molecules, assuming their structure is partially ice-like and partially liquid-like. Since liquid water has a relaxation frequency at around 16–20 GHz, while that of ice is at about 1–5 kHz, the relaxation frequency of bound water should be found at intermediate frequencies. Therefore, Schwan concluded that 300–500 MHz is quite reasonable for the characteristic frequency of the orientation of bound water. Similar but more detailed work was performed by Pennock and Schwan (1969), also using haemoglobin.

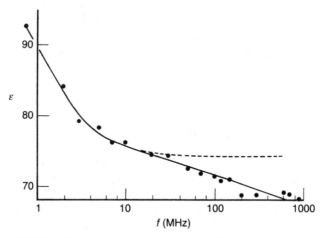

Figure 8.23 The dielectric constant of human haemoglobin between 10 MHz and 1 GHz. The broken line shows the possible high-frequency plateau and the circles with the full curve show the measured points. (From Schwan 1965.) Reproduced by permission of the *Annals* of the New York Academy of Sciences.

Another high-frequency measurement of the dielectric constant of protein solution was carried out by Grant and co-workers (1965, 1968) using serum albumin as the sample. The data obtained by them are essentially the same as those of Schwan and co-workers for haemoglobin. Grant also attributed δ-dispersion to the hindered rotation of bound water. However, Grant and co-workers suggested, in addition, that this dispersion can also be explained by the partial rotation of polar side chains of albumin molecules. The centre frequency of δ-dispersion is similar to the characteristic frequencies of α-amino acids. Thus, if proteins have bulky polar side chains, the dispersion due to the orientation of these groups is likely to be found in the same frequency region.

As shown in figure 8.23, the magnitude of the δ-dispersion is very small and, moreover, this dispersion is located between two major dispersions due to the protein dipole and water dipoles. Thus, an accurate determination of the full amplitude and relaxation frequency of δ-dispersion is not feasible.

8.9 The Dielectric Behaviour of Water Bound to DNA

The amount of water bound to DNA was studied by Tunis and Hearst (1968) using ultracentrifugation and it was found that about five or six molecules of water are bound to a base pair. This amount is equivalent to approximately 0.3–0.4 gram g^{-1} of DNA. As discussed before, there are water molecules which are tightly bonded to base molecules and which help stabilize the double helical structure (Lewin 1967). Such water was shown to exist in DNA by Kopka et al (1984) (see figure 8.19). As stated, these water molecules are tightly bonded and are likely to be irrotational.

What is more relevant from the viewpoint of electric polarization is the water which is bound loosely with DNA molecules. There are two possible sites for binding. One is the base molecules and the other is the sugar phosphate groups. The result of a Monte Carlo simulation of the solvation water surrounding a base pair and sugar phosphate by Clementi and Corongiu (1981) is shown in figure 8.24. This figure shows the sites of water molecules in the vicinity of an AT pair including two sugar phosphate groups. The horizontal bar indicates a distance of 2 Å. This figure shows that clusters of water molecules are found between two concentric circles with a gap width of about 10 Å. It should be noted that the density of bound water is scant lateral to the base pairs but a large number of water molecules are found surrounding phosphate groups, indicating the importance of charge–dipole interactions. Also of importance is the fact that at least five or six water molecules are found within 2–3 Å from a phosphate group. According to the calculations by Hasted et al (1948) and Takashima et al (1986) the dielectric constant of water within 2–4 Å radius from charged sites is extremely small. Thus, we have good reason to believe that at least ten water molecules per two phosphate groups are irrotational. Other water molecules, although they are within the circular region of 10 Å wide, are believed to be rotational. Thus, some of the bound water molecules are likely to produce a frequency dependent dielectric constant in spite of limited orientational freedom. Another interesting observation is that bound water does not have a tetrahedral configuration, which is the dominant form of bulk water. However, close examination of this figure indicates the presence of hydrogen bonds between some of the adjacent water molecules such as the 104–129 pair and the 101–135 pair. We can postulate from this figure that the configuration of bound water is not ice-like although some water may be hydrogen bonded.

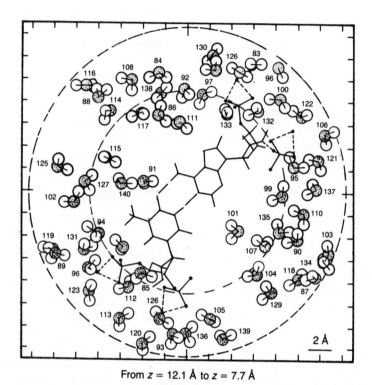

From $z = 12.1$ Å to $z = 7.7$ Å

Figure 8.24 A Monte Carlo simulation of the water molecule configuration near an AT pair. (From Clementi and Corongiu 1981.) Reproduced by permission of the *Annals* of the New York Academy of Sciences.

Experimental studies of the dielectric behaviour of bound water are very difficult and, in fact, there is no straightforward method to determine the dielectric constant of the water surrounding charged biological polymers. The only exception is the δ-dispersion reported by Schwan and Grant. Although this dispersion is believed to be due to bound water, the evidence is not yet conclusive. This uncertainty creates some confusion with regard to the interpretation of the physical state of bound water. The simplest interpretation is the two-state model, i.e. the water which is very close to the surface of polyions is irrotational and has a very small dielectric constant. That which is outside these layers is rotational and behaves as normal water. However, there are many other hypotheses which are based on the concept of long-range ordering forces. It remains to be seen which model represents the physical state of bound water more realistically.

8.10 Dielectric Properties of Living Tissue

The water content in biological tissue or cells is very high and many reseachers postulated that the structure of tissue water is different from normal bulk water (Ling and Negendank 1970, Ling 1970). As has been discussed in this chapter, the structure of water which is bound to proteins and nucleic acids does not appear to be drastically different from that of bulk water except for that which is located within a few angstroms from the polar groups of macromolecules.

The above discussion is based on the observation of the dielectric relaxation behaviour of bound or cell water. However, there are other transport processes which may be affected by the small confines of cells, such as self-diffusion and ionic conductivity. The self-diffusion and conductivity of solvent molecules in the presence of polymers were investigated by Foster et al (1984) and Blum et al (1986) using NMR and conductivity measuring techniques. Only the results of the conductivity experiments will be discussed. The conductivity of the solvent in solution, k, relative to that of pure solvent, k_w, is given by the Fricke equation (1924, 1925):

$$\frac{k_\infty - k_w}{k_\infty + xk_w} = p \frac{k_i - k_w}{k_i + xk_w} \tag{8.50}$$

where k_∞, k_w and k_i are the conductances of suspension, the solvent and the particle respectively. Figure 8.25 shows a plot of the relative conductivity of polyethylene oxide (PEO) solution, k/k_w, against the volume fraction of solid material.

As shown in this figure, the relative conductivity decreases as the total volume fraction p increases. It is striking to note that a reduction of conductivity does not depend significantly on the size nor on the type of polymer. This means that a reduction of ion movement does not involve specific interactions between solvents and solid material. Deviations of the experimental results from the Fricke equation and/or Hanai's mixture theory is another interesting aspect of this work. However, these authors did not analyze the cause of this deviation in detail. The most probable reason for the deviation is the variability of the conductivity of the solvent k_w in the Fricke and Hanai equations. If we solve the Fricke or Hanai equation for k_w, we can calculate the conductivity of the solvent at each volume fraction p. Some of the results of this calculation, using Foster et al's data, are shown in table 8.3. As expected, k_w decreases markedly as the volume fraction p increases.

Conversely, a decrease in the activity coefficient and/or the conductivity of solvent, perhaps due to hydration, may be the cause of the deviation of the suspension conductance from the mixture equations.

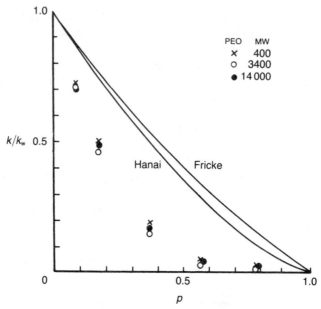

Figure 8.25 The electrical conductivity of PEO solutions in 0.1 KCl solution. The ordinate is the relative conductivity k/k_w and the abscissa is the volume fraction. Theoretical curves by Fricke and by Hanai are shown by the full curves. (From Foster *et al* 1984.) Reproduced from the *Biophysical Journal* 1984, vol 45, pp 975–84 by copyright permission of the Biophysical Society.

The analysis of conductivity data can be applied to the suspension of biological cells. Figure 8.26 shows the conductivity of a human erythrocyte suspension (Takashima 1987, unpublished). The conductivity of the erythrocyte suspension decreases markedly as the volume fraction p increases. However, a deviation of the curve from the Fricke equation or from Hanai's equation is not observed. The conductance of suspending media at various volume fractions are calculated using (8.50) and are tabulated in table 8.3. Clearly, the conductance of the suspending medium does not change noticeably with p this time. Unlike the polymer solution used by Foster *et al*, the size of erythrocyte is so much larger than that of the solvent molecule that this may be the cause for the different results.

Table 8.3 The conductivity of KCl solution in the presence of synthetic polymer molecules.

Volume fraction	k/k_w†	k/k'_w‡	k'_w/k_w§
0.1	0.81	0.70	0.864
0.2	0.72	0.43	0.597
0.3	0.60	0.26	0.433
0.4	0.464	0.14	0.301
0.5	0.353	0.07	0.198
0.6	0.252	0.02	0.079

The conductivity of NaCl solution in the presence of erythrocytes.

Volume fraction	k/k_w†	k/k'_w‡	k'_w/k_w§
0.1	0.81	0.825	1.01
0.2	0.72	0.650	0.902
0.3	0.608	0.557	0.916
0.4	0.529	0.425	0.803
0.5	0.353	0.325	0.920
0.6	0.252	0.240	0.952

† Calculated by Hanai's mixture theory. k is the conductance of suspension and k_w is the conductance of salt solution.
‡ Measured experimentally.
§ k_w is the conductance of salt in the absence of polymers and k'_w is the conductance of salt solution in the presence of polymers.

The transport of ions inside red blood cells was investigated by Pauly and Schwan (1966). The conductivity of the cell interior is typically 4.5 to 5.0 mS cm^{-1}. This conductance is much smaller than that of the suspending medium (about 15 mS cm^{-1}). The low conductivity is partially due to the volume exclusion of haemoglobin molecules (35%). The correction for the volume exclusion by haemoglobin raises the conductivity to about 9 mS cm^{-1}. This value is still lower than the internal ionic conductance (about 14.5 mS cm^{-1}) of erythrocyte, calculated using the ionic composition in the cell and the equivalent conductances of each ion. We can repeat the measurement of internal conductances with lysed and presumably resealed erythrocytes (without intracellular haemoglobin). Pauly and Schwan found that the internal conductance of lysed erythrocytes lies between 13.5 and 14 mS cm^{-1}. This is very close to the theoretical value. Based on these data, we can draw a conclusion that the reduced internal conductance of intact erythrocytes is not due to the confinement of ions within the cell but is due to the reduction of ion movements by a bulky haemoglobin molecule.

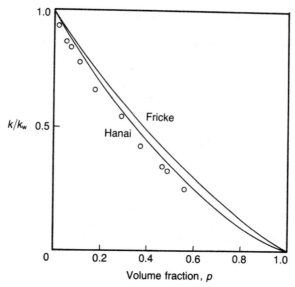

Figure 8.26 The conductivity of human erythrocyte suspension at various volume concentrations (Takashima 1987, unpublished).

As discussed, DC conductance is due to the transport of ions whereas the dielectric constant and AC conductance are due to the polarization of dipolar species. Thus, the origins of these two processes are entirely different. The discussion given above pertains only to DC conductance. The interior dielectric constant of erythrocyte was found to be about 50. If this value is corrected for the volume of haemoglobin, a value which is very similar to the dielectric constant of the external solution, i.e. 78.5, is obtained. This means that the orientation of water dipoles in erythrocytes does not exhibit any anomalies. The conclusion we can draw from these observations is that the dielectric behaviour of water in biological cells is essentially the same as that of bulk water although other transport properties such as ion mobilities may be affected by the presence of bulky polymers.

Some papers have been published reporting that the relaxation time of tissue water is shifted by 20–25% from that of pure water. Moreover, the slope of the dielectric dispersion curve of tissue water is broader than that of bulk water, indicating the presence of a distribution of relaxation times. In this context, the result obtained by Takashima *et al* (1984) for DNA solution may be of some significance. Figure 8.27 shows the dispersion curve for DNA solution between 0.5 and 70 GHz. This figure demonstrates that the dispersion curve of DNA solution above 1 GHz, which is due to bulk water, is slightly broader than that of salt solution with a 10% shift in relaxation frequency. However, the shift is extremely small in spite of the enormous viscosity of DNA solution at this concentration (1%). This result suggests that the macroscopic viscosity in biological systems has only a small effect on the orientation of water dipoles.

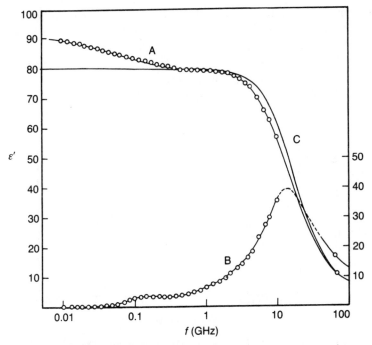

Figure 8.27 The dielectric constant (curve A) and loss factor (curve B) for concentrated calf thymus DNA (1%) between 5 MHz and 70 GHz. Curve C is calculated using Debye's equation. (From Takashima *et al* 1984.) Reproduced from the *Biophysical Journal*, 1984, vol 46, pp 29–34 by copyright permission of the Biophysical Society.

8.11 Ultrahigh-frequency Dielectric Measurements with Protein Powders

As has been discussed, the frequency range of interest for biological polymers has been limited to the RF region. The dielectric relaxation of these substances ceases at low GHz and what is left above this region is the polarization of water molecules, may they be free or bound to polymers. However, this statement is restricted to the macroscopic motion of polymer molecules. When we turn our attention to the microscopic motion of the internal groups of macromolecules, there are abundant sources of information on molecular dynamics, even above 100 GHz, where even water dipoles cease to polarize. The tools which are capable of uncovering this information utilize ultrahigh-frequency dielectric techniques. The range we are going to discuss is from 50 to 150 GHz. This range, though very high from the viewpoint of dielectric measurements, is still considerably lower than those used for infrared spectroscopy. 150 GHz is equivalent to 5 cm^{-1} (2 mm). The absorption coefficient of haemoglobin film between 20 and 470 cm^{-1} is shown in figure 8.28. The frequency range we are currently interested in is located at the low-frequency end of this spectrum.

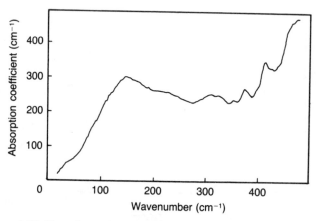

Figure 8.28 The absorption coefficient of haemoglobin film at 6 K, from 20 to 470 cm^{-1}, measured by Fourier transform spectroscopy. (From Genzel *et al* 1983.) Reproduced by permission of John Wiley & Sons, Inc. © 1983.

The technique used by Genzel *et al* (1983) and Poglitsch *et al* (1984) is an untuned cavity method. It involves the creation of approximately isotropic and homogeneous fields inside a large cavity (of diameter 36 cm) by supplying radiation from an approximately tunable source. The quality factor Q of a multimode resonator is only slightly dependent on the frequency of radiation so long as the wavelength remains small compared with the size of the resonator. The sample, which is usually a compacted disc of dried or hydrated proteins and model substances, is inserted into the cavity and the resultant change in Q is measured with precision. Since any scattering or reflection of radiation by the sample is confined within the cavity, the loss of Q is entirely due to absorption by the sample. Using this system, the absorption coefficients of haemoglobin, polyalanine and L-alanine (Genzel *et al* 1983), lysozyme (Poglitsch *et al* 1984), and alkylamides (Haessler *et al* 1986) have been investigated.

The absorption coefficients of egg-white lysozyme are shown in figure 8.29 as an example. The absorption coefficients are strongly dependent upon temperature below 50 K while they increase more slowly above 50 K. Furthermore, in all temperature ranges, absorption coefficients increase with increasing frequency. This indicates that dried lysozyme has an absorption peak at even higher frequencies. As shown in figure 8.28, haemoglobin film has a broad band absorption in the far-infrared region with a maximum at 147 cm^{-1}. This absorption is common for many proteins. Assuming that lysozyme has a similar absorption, the result obtained by Poglitsch *et al* may be, at least in part, the low-frequency tail of this absorption. The results obtained with haemoglobin and polyalanine are similar to those of lysozyme. However, the absorption coefficients are one order of magnitude smaller with simple L-alanine. The difference between proteins and simple amino acids is the absence of intra-chain hydrogen bonds in the latter. Therefore, it is reasonable to assume that the polarization of intra-chain hydrogen bonds may be the source of absorption of proteins in the millimetre wavelength region.

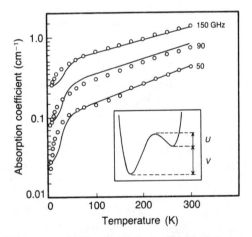

Figure 8.29 The absorption coefficient of dried lysozyme (water content: 0.5%). The full curves represent the values calculated using (8.51). The inset shows the asymmetric double-well potential used to describe the relaxation processes. (From Poglitsch *et al* 1984.) Reproduced by permission of Academic Press Inc.

The following equations were used by Genzel *et al* to explain the absorption of dried proteins at ultrahigh frequencies:

$$\varepsilon(\omega) = \varepsilon_\infty + \frac{S_0 v_0^2}{v_0^2 - v^2 - jv\gamma_0} + \sum_{i=1}^{3} \frac{C_i}{T} \frac{\Delta\varepsilon}{1 - jv/v_i} \qquad (8.51)$$

where

$$\Delta\varepsilon = \frac{\exp(-V_i/T)}{[1 + \exp(-V_i/T)]^2} \qquad (8.52)$$

and

$$v_i = v_\infty \exp(-U_i/T)[1 + \exp(-V_i/T)]. \qquad (8.53)$$

U_i and V_i are the depths of the asymmetric potential wells, and v_0 and v are the resonance frequency and the frequency of measurement. C_i is given by

$$C_i = \frac{N_i g_i \mu_i^2}{k} \frac{4\pi}{3} \frac{(\varepsilon_\infty + 2)^2}{9} \frac{3\varepsilon_s}{\varepsilon_\infty + 2\varepsilon_s} \qquad (8.54)$$

where ε_s is the static dielectric constant, N_i is the density of dipoles, m_i is the microscopic dipole moment, g_i is the Onsager correction for local fields, and k is Boltzmann's constant. Equation (8.51) consists of a harmonic oscillator term in the far-infrared with an oscillator strength S_0 centred around 4400 GHz (147 cm^{-1}). The remaining terms represent relaxation processes of proton jumps between the asymmetric double well potentials indicated in figure 8.29. The conversion from dielectric constant to absorption coefficient is done as follows:

$$\alpha = \frac{1}{nc} \frac{(\varepsilon_s - \varepsilon_\infty)\omega^2\tau}{1 + \omega^2\tau^2} = \frac{\omega\varepsilon''}{nc} \qquad (8.55)$$

where n and c are the refractive index and the velocity of light in vacuum respectively. Note that the absorption coefficient, by definition, is equivalent to the imaginary part of the complex permittivity multiplied by a factor ω/nc. The detail of these derivations is discussed by Genzel *et al* (1983). They found by curve fitting that at least three relaxation processes are necessary to explain the experimental results. The proof that these absorptions are not due to water of hydration was found by repeating the measurements at various water contents and by extrapolating the data to zero water concentration. Figure 8.30 illustrates the absorption coefficient of a hydrated lysozyme sample as well as that of carefully dried protein powder with 0.5% water content. As illustrated, nearly complete drying does not eliminate the aforementioned absorption. The contribution of polar side chains is also ruled out because polyalanine, which has no polar side groups, exhibits the same absorptions.

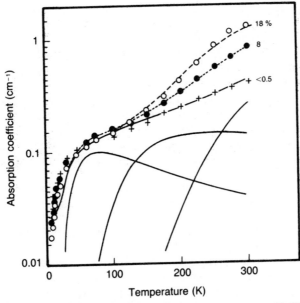

Figure 8.30 The absorption coefficient of lysozyme at 50 GHz for three hydration levels, 18%, 8% and 0.5%. The dielectric constant of the dried material is composed of three relaxation processes as represented by the three full curves. (From Poglitsch *et al* 1984.) Reproduced by permission of Academic Press Inc.

Suggested Reading

Franks F 1973 *Water. A Comprehensive Treatise* (New York: Plenum)
Hadzi D 1959 *IUPAC Symp. Hydrogen Bonding* (New York: Pergamon)
Hasted J 1973 *Aqueous Dielectrics* (London: Chapman–Hall)
Saenger W 1987 Structure and dynamics of water surrounding biomolecules *Ann. Rev. Biophys. Chem.* **16** 93–114

References

Akerlof G C and Oshery H I 1950 The dielectric constant of water at high temperatures and in equilibrium with its vapor *J. Am. Chem. Soc.* **72** 2844–7
Auty R P and Cole R H 1952 Dielectric properties of ice and solid D_2O *J. Chem. Phys.* **20** 1309–14
Barabino G, Gavotti C and Marchesi M 1984 Molecular dynamics simulation of water near walls using an improved wall-water interaction potential *Chem. Phys. Lett.* **104** 478–84
Bayley S T 1951a The dielectric properties of various solid crystalline proteins,

amino acids and peptides *Trans. Faraday Soc.* **47** 509–17

—— 1951b The dielectric properties of adsorbed water layers on inorganic crystals *Trans. Faraday Soc.* **47** 518–22

Bjerrum N 1951 *Klg. Danske Videnskab. Seiskab Mat-Fys. Medd.* **27** 41

Blum F D, Pickup S and Foster K R 1986 Solvent self-diffusion in polymer solutions *J. Colloid Surf. Sci.* **113** 336–41

Brunauer S, Emmett P H and Teller E 1938 Adsorption of gases in multi-molecular layers *J. Am. Chem. Soc.* **60** 309–19

Buchanan T J 1952 *Proc. IEEE* **99** 61

Chamberlain J E, Chantry G W, Gebbie H A, Stone N W B, Taylor T B and Wyllie G 1966 Submillimetre absorption and dispersion of liquid water *Nature* **210** 790

Clementi E and Corongiu G 1981 Simulations of the solvent structure for macromolecules: solvation model for B-DNA and Na-B-DNA double helix at 300 K *Ann. New York Acad. Sci.* **367** 83–107

Collie C H, Hasted J B and Ritson D M 1948 *Proc. Phys. Soc.* **73** 145

Cook H F 1952 *Br. J. Appl. Phys.* **3** 249

Coulson C A and Eisenberg D 1966 Interaction of H_2O molecules in ice. I. The dipole moment of an H_2O molecule in ice *Proc. R. Soc.* A **291** 445–59

Debye P 1929 *Polar Molecules* (New York: Dover) p115

Dickerson R E and Geis I 1969 in *The Structure and Action of Proteins* (New York: Harper and Row) p94

Eigen M and De Maeyer L 1958 Self dissociation and proton charge transport in water and ice *Proc. R. Soc.* A **247** 505–33

Foster K R, Cheever E, Leonard J B and Blum F D 1984 Transport properties of polymer solutions. A comparative approach *Biophys. J.* **45** 975–84

Frank H S and Wen W Y 1957 Structural aspects of ion-solvent interaction in aqueous solutions: a suggested picture of water structure *Faraday Discuss. Chem. Soc.* **24** 133–40

Fricke H 1924 A mathematical treatment of the electric conductivity and capacity of disperse systems. I. The electric conductivity of a suspension of homogeneous spheroids *Phys. Rev.* **24** 575–87

—— 1925 A mathematical treatment of the electric conductivity and capacity of disperse systems *Phys. Rev.* **26** 678–81

Froehlich H 1949 *Theory of Dielectrics* (Oxford: OUP)

Gabler R 1978 *Electrical Interactions in Molecular Biophysics* (New York: Academic)

Genzel L, Kremer F, Poglitsch A and Bechtold G 1983 Relaxation processes on a picosecond time scale in hemoglobin and poly (L-alanine) observed by millimeter wave spectroscopy *Biopolymers* **22** 1715–29

Granicher H, Jaccard C, Scherrer P and Steineman A 1957 Dielectric relaxation and the electrical conductivity of ice crystals *Faraday Discuss. Chem. Soc.* **23** 50–62

Grant E H 1965 The structure of water neighboring proteins, peptides and amino acids as deduced from dielectric measurements *Ann. New York Acad. Sci.* **125** 418–27

Grant E H, Buchanan T J and Cook H F 1957 Dielectric behavior of water at microwave frequencies *J. Chem. Phys.* **26** 156–61

Grant E H, Keefe S E and Takashima S 1968 The dielectric behavior of aqueous solutions of BSA from radiowave to microwave frequencies *J. Phys. Chem.* **72** 4373

Grant E H and Shack R 1967 *Br. J. Appl. Phys.* **18** 1807

Hadzi D and Thompson H W 1957 *Symp. Hydrogen Bonding, Ljunljana* (New York: Pergamon)

Haggis G H, Hasted J B and Buchanan T J 1952 The dielectric properties of water in solutions *J. Chem. Phys.* **20** 1452–65

Harris F E 1955 Contributions of fluctuations and anisotropy to dielectric polarization in polar substances *J. Chem. Phys.* **23** 1663–72

Harvey S C and Hoekstra P 1972 Dielectric relaxation spectra of water adsorbed on lysozyme *J. Phys. Chem.* **76** 2987–94

Haessler S, Poglitsch A, Genzel L and Kremer F 1986 Pico-second relaxations in model substances for proteins. A millimeter-wave investigation on crystalline alkyl amides *Biopolymers* **25** 677–91

Hasted J B and El-Sabeh S H M 1953 The dielectric properties of water in solution *Trans. Faraday Soc.* **49** 1003–11

Hasted J B, Ritson D M and Collie C H 1948 Dielectric properties of aqueous ionic solutions. Parts I and II *J. Chem. Phys.* **16** 1–21

Kent M 1972 Complex permittivity of protein powders at 9.4 GHz as a function of temperature and hydration *J. Phys. D: Appl. Phys.* **5** 394–409

Kopka M L, Pjura P, Yoon C, Goodsell D and Dickerson R E 1984 *Proc. Int. Biophysics Cong., Bristol, IUPAB*

Langmuir I 1918 The adsorption of gases on plane surfaces of glass, mica and platinum *J. Am. Chem. Soc.* **40** 1361–403

Leslie R B 1962 *PhD Thesis* University of Nottingham

Lewin S 1967 Some aspects of hydration and stability of the native state of DNA *J. Theor. Biol.* **17** 181–212

Ling G N 1970 The physical state of water in living cells and its physiological significance *Int. J. Neurosci.* **1** 129–52

Ling G N and Negendank W 1970 The physical state of water in frog muscles *Physiol. Chem. Phys.* **2** 15–33

Lipscomb W N, Hartsuck J A, Quiocho F A and Recke G N 1969 The structure of carboxypeptidase A. IX. The x-ray diffraction results in the light of the chemical sequence *Proc. Natl. Acad. Sci. USA* **64** 28–35

Malmberg C G and Maryotte A A 1956 Dielectric constant of water from 0° to 100° *J. Res. Natl. Bur. Stand.* **56** 1–8

Marchesi M 1983 Molecular dynamics simulation of liquid water between two walls *Chem. Phys. Lett.* **97** 224–30

Matsuoka O, Clementi E and Yoshimine M 1976 CI study of the water dimer potential surface *J. Chem. Phys.* **64** 1351–60

Morgan J and Warren B E 1938 X-ray analysis of the structure of water *J. Chem. Phys.* **6** 666

Morokuma K 1970 Molecular orbital studies of hydrogen bonds: dimeric H_2O with the Slater minimal basis set *J. Chem. Phys.* **52** 1301

Narten A H, Danford M D and Levy H A 1967 X-ray diffraction study of liquid water in the temperature range 4–200 °C *Faraday Discuss. Chem. Soc.* **43** 97–107

Nemethy G and Scheraga H A 1962a Structure of water and hydrophobic bonding in proteins. I. Model for the thermodynamic properties of liquid water *J. Chem. Phys.* **36** 3382–400

—— 1962b Structure of water and hydrophobic bonding in proteins. II. Model for the thermodynamic properties of aqueous solutions of hydrocarbons *J. Chem. Phys.* **36** 3401–17

Oster G and Kirkwood J G 1943 The influence of hindered molecular rotation on the dielectric constant of water, alcohols and other polar liquids *J. Chem. Phys.* **11** 175–8

Pauling L 1940 *The Nature of the Chemical Bond* (Ithaca, NY: Cornell University Press) ch IX

Pauly H and Schwan H P 1966 Dielectric properties and ion mobility in erythrocytes *Biophys. J.* **6** 621–39

Pennock B E and Schwan H P 1969 Further observations on the electrical properties of haemoglobin-bound water *J. Phys. Chem.* **73** 2600–10

Pethig R 1979 *Dielectric and Electronic Properties of Biological Materials* (New York: Wiley) ch 4

Poglitsch A, Kremer F and Genzel L 1984 Picosecond relaxations in hydrated lysozyme observed by mm-wave spectroscopy *J. Mol. Biol.* **173** 137–42

Pople J A 1951 Molecular association in liquids. II. A theory of the structure of water *Proc. R. Soc.* A **205** 163–78

Ramachandran G N and Chandrasekharan R 1968 Interchain hydrogen bonds via bound water molecules in the collagen triple helix *Biopolymers* **6** 1649–58

Rosen D 1963 Dielectric properties of protein powders with adsorbed water *Trans. Faraday Soc.* **59** 2178–91

Sandus O and Lubitz B B 1961 Dielectric relaxation of aqueous glycine solutions at 3.2 centimeter wave length *J. Phys. Chem.* **65** 881–5

Saxton J 1952 Dielectric dispersion in pure liquids at very high radio-frequencies. II. Relation of experimental results to theory *Proc. R. Soc.* A **213** 344–54

Saxton J and Lane J A 1952 Dielectric dispersion in pure polar liquids at very high radio-frequencies *Proc. R. Soc.* A **213** 400–8

Schwan H P 1965 Electrical properties of bound water *Ann. New York Acad. Sci.* **125** 344–54

Schwan H P and Grant E H 1976 Complex permittivity of water at 25 °C *J. Chem. Phys.* **64** 2257–8

Smyth C P 1955 *Dielectric Behavior and Structure* (New York: McGraw-Hill) p127

Steineman A 1957 Dielektrische Eigenschaften von Eiskristallen. II. Dielektrische Untersuchungen an Eiskristallen mit eingelagerten Fremdatomen *Helv. Phys. Acta* **30** 581–610

Steineman A and Granicher H 1957 Dielektrische Eigenschaften von Eiskristallen. I. Dynamische Theorie der Dielektrizitateskonstante *Helv. Phys. Acta* **30** 553–80

Takashima S 1962 Dielectric properties of water of adsorption on protein crystals *J. Polym. Sci.* **62** 233–40

Takashima S, Casaleggio A, Giuliano F, Morando M, Arrigo P and Ridella S

1986 Study of bound water of poly-adenine using high frequency dielectric measurements *Biophys. J.* **49** 1003–8

Takashima S, Gabriel C, Sheppard R J and Grant E H 1984 Dielectric behavior of DNA solution at radio and microwave frequencies (at 20 °C) *Biophys. J.* **46** 29–34

Takashima S and Schwan H P 1965 Dielectric dispersion of crystalline powders of amino acids, peptides and proteins *J. Phys. Chem.* **69** 4176–82

Tsubomura H 1954 The nature of the hydrogen bond. I. The delocalization energy in the hydrogen bond as calculated by the atomic orbital method *Bull. Chem. Soc. Japan* **27** 445–50

Tunis M J B and Hearst J E 1968 On the hydration of DNA. I. Preferential hydration and stability of DNA in concentrated trifluoroacetate solution *Biopolymers* **6** 1325–44

Whalley E, Heath J B R and Davidson D W 1968 Ice IX: an antiferroelectric phase related to ice III *J. Chem. Phys.* **48** 2362–70

9   Non-Linear Dielectric Properties

9.1 Introduction

The discussion on the dielectric behaviour of biological polymers was limited to a linear regime in the previous chapters. Namely, dielectric properties are independent of the intensity of the input signals used for the measurement. Linear dielectric behaviour is observed when input voltages are small provided the dipole moment of the sample is not excessively large. The onset of non-linearity depends on input voltages and also on the magnitude of the dipole moment and the stability of the molecule. The mechanisms for non-linear electric polarization will be discussed in this chapter.

9.2 Non-linearity due to Dielectric Saturation

As discussed in Chapter 2, the theory of electric polarization of polar molecules was first developed by Debye (1929). Suppose a group of molecules having a dipole moment μ are distributed randomly in the absence of an electrical field, then the mean moment $\langle \mu \rangle$ will be zero. However, if a field is applied, the mean moment becomes non-zero and its magnitude is given by the following equation:

$$\langle \mu \rangle = \mu \left[\frac{1}{3} \left(\frac{\mu E}{kT} \right) - \frac{1}{45} \left(\frac{\mu E}{kT} \right)^3 + \frac{2}{945} \left(\frac{\mu E}{kT} \right)^5 - \ldots \right] \qquad (9.1)$$

where E is the field intensity. The increase in mean moment $\langle \mu \rangle$ as a function of field intensity is illustrated in figure 2.3. As shown, $\langle \mu \rangle$ increases linearly for small E but becomes non-linear for large field intensities. If E is extremely large, $\langle \mu \rangle$ approaches the value of the dipole moment for the molecule asymptotically. This means a complete orientation, called dielectric saturation. If, however, the intensity of the applied field is small, all the higher-order terms in (9.1) can be neglected. Under these circumstances, the polarizability α, which is defined as $\alpha = \langle \mu \rangle / E$, is given by the following equation:

$$\alpha = \langle\mu\rangle/E = \mu^2/3kT. \tag{9.2}$$

It is clear from this equation that α is independent of E under this condition. Equation (9.2) is, however, valid only when the truncation of higher-order terms in (9.1) is justifiable, namely when $\mu E/kT \ll 1$. Table 2.2 shows the critical field intensities for various molecules, including some biological polymers which have extraordinarily large dipole moments. Note that exceedingly large fields are required to cause non-linearity if the dipole moment is of the order of 5–6 D. However, biopolymers which have a very large dipole moment ranging from a few hundred to 10^5 D, require relatively small fields to produce non-linear dielectric behaviour. For example, only 300–500 V cm^{-1} is sufficient for the onset of non-linearity for large DNA molecules with a molecular weight of a few million. Therefore, it is not difficult to produce near-complete alignment of long DNA molecules along the field at an intensity of only a few hundred volts per centimetre.

9.3 Wien Effects

Although dielectric saturation is the commonly accepted mechanism for non-linear electric polarization, there are other causes of non-linearity. Wien (1928, 1931) observed that the applicability of Ohm's law is limited only to fields of moderate intensity. In other words, whereas the resistance R of strong and/or weak electrolytes is independent of voltage V in the linear region, i.e. low field strengths, R becomes voltage dependent at high field intensities. This means that the conductances of strong and weak electrolyte solutions increase over a wide range of field intensity. Figure 9.1 shows the field dependence of the conductance of MgSO$_4$ solution. As seen, the conductance of strong electrolytes which are almost completely ionized in aqueous solution increases markedly at high field intensities, until the increase saturates in the range over 100 kV cm^{-1}. This effect can be explained by the stripping of the counterion atmosphere by strong fields and the velocity of ion movement becoming faster than the average ionic drift velocity.

The second effect found by Wien is the field dissociation effect. This means that strong electrical fields enhance the ionization of weak electrolytes such as acetic acid and monochloroacetic acid. As shown in figure 9.2, the increase in conductance of these organic acids is a linear function of the field intensity. The dissociation of organic acids takes place in two steps, as shown below:

$$CH_3COOH \rightleftharpoons CH_3COO^-H^+ \rightleftharpoons CH_3COO^- + H^+. \tag{9.3}$$

The first step is the formation of the ion pair $CH_3COO^-H^+$ and the second step is the separation of CH_3COO^- and H^+. Electrical fields are

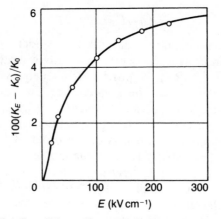

Figure 9.1 The first Wien effect. The effect of electric field on the conductance of a strong electrolyte magnesium sulphate ($MgSO_4$) solution. (After Wien 1928).

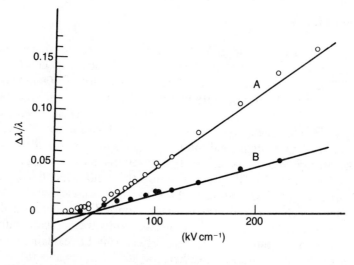

Figure 9.2 The second Wien effect. The effect of electric field on the dissociation of two weak electrolytes, acetic acid (CH_3COOH, curve A) and monochloroacetic acid ($CH_2ClCOOH$, curve B). (From Onsager 1934.) Reproduced by permission of the American Institute of Physics.

believed to shift the equilibrium of the second step.

The equation derived by Onsager (1934) defines the ratio between $K(E)$, the ionization constant in the presence of a strong field and $K(0)$, the dissociation constant in the absence of the field. This equation is shown, without any discussion of the derivation, as follows:

$$K(E)/K(0) = 1 + 2\beta q + (4\beta q)^2/2!3! + (4\beta q)^3/3!4! + \ldots \quad (9.4)$$

where q is the effective association length and is defined by

$$q = -e_1e_2/2\varepsilon kT \tag{9.5}$$

where ε is the dielectric constant of the solvent, and e_1 and e_2 are the charges of the ions. If the distance r between two ions is larger than the association length q, then the ions are considered separated and, if $r < q$, the ions are considered paired. On the other hand, 2β in (9.4) is given by the following equation:

$$2\beta = |E(e_1u_1 - e_2u_2)|/kT(u_1 + u_2) \tag{9.6}$$

where u is the mobility of ions. By definition, ukT is the diffusion coefficient and eu is the velocity of the migration of electrolytes in a field of 1 esu (300 V cm^{-1}).

The change in dissociation constant due to electrical fields can be measured in terms of the parameter $2\beta q$ as shown by (9.4). If (9.5) and (9.6) are combined, we obtain

$$2\beta q = \left(\frac{z_1u_1 + z_2u_2}{u_1 + u_2}\right) z_1z_2 \frac{|E|e^3}{2\varepsilon(kT)^2} \tag{9.7}$$

where E is electric field in esu, z_1 and z_2 are the valences of the ions and e is elementary charge. For 1:1 electrolytes, (9.7) reduces to a simple form. After conversion of E to V cm^{-1} and using 1.38×10^{-16} erg/molecule for Boltzmann's constant, we obtain

$$2\beta q = 9.636V/\varepsilon T^2. \tag{9.8}$$

Using $b = 2\beta q$, (9.4) can be written as

$$K(E)/K(0) = F(b) = 1 + b + b^2/3 + b^3/18 + b^4/180 + \ldots .. \tag{9.9}$$

The dissociation constant $K(E)$ can be defined in terms of the field enhanced concentration of free ions, C_i, as follows:

$$K(E) = \frac{C_i^2}{(C_0 - C_i)} = \frac{C_0\alpha^2}{1 - \alpha} \tag{9.10}$$

where C_0 is the total concentration of electrolyte and

$$\alpha = C_i/C_0.$$

The increase in the conductance λ is proportional to the shift of the degree of dissociation by the field, i.e.

$$\lambda/\lambda_0 = \alpha/\alpha_0 \tag{9.11}$$

where λ_0 and α_0 are the conductance and degree of dissociation for $E \rightarrow 0$. For weak electrolytes where dissociation is small, i.e. $\alpha \ll 1$, the conductance ratio λ/λ_0 is given by the following equation:

$$\lambda/\lambda_0 = (K(E)/K(0))^{1/2} = (F(b))^{1/2}. \tag{9.12}$$

If the function $(F(b))^{1/2}$ is expanded in a power series, then

$$\lambda/\lambda_0 = 1 + (1/2)b + (1/24)b^2 + \dots \tag{9.13}$$

The value of b is inversely proportional to the dielectric constant of the solvent. For example, the dielectric constants of water and benzene are 78.5 and 2.28 at 25 °C respectively. For these widely different cases, the parameter b becomes 1.384×10^{-6} V for water and 47.5×10^{-6} V for benzene. Therefore, the effect of an applied field is more pronounced when the dielectric constant of the solvent is small. Namely, $\lambda(E)/\lambda(0) = 1.37$ at $V = 5 \times 10^5$ V cm^{-1} for water and $\lambda(E)/\lambda(0) = 121$ for benzene for the same field strength. Therefore, the field induced enhancement of ionization of weak electrolytes is much more pronounced if the dielectric constant of the solvent is small. The straight lines drawn in figure 9.2 were calculated using Onsager's theory, assuming the dielectric constant of water to be 78.5. The agreement between theoretical values and observed results is exceedingly good, particularly at high field intensities.

In Onsager's theory, the quantity q, an association length, is considered fixed. In other words, each ion is surrounded by a spherical potential barrier of radius q and the application of an electrical field does not change this radius. The net rate of entry of an ion into the interior of the closed surface of this shell is given by equation (19) of Onsager's paper, i.e.

$$\int (\partial f/\partial t)dv = (u_j + u_i)kT \int \{\mathrm{grad}_n f\text{-}f\mathrm{grad}_n[(2q/r) + 2\beta x]\}\,dS \tag{9.14}$$

where f is the distribution function of an ion i in the vicinity of another ion j and grad_n denotes the normal component of the gradient at the surface S. The presence of x in this equation indicates that the field is applied in the x direction. Although the rate of entry is a function of applied field, the radius of the potential barrier q is treated as a constant.

Bass (1968) treated q as a field dependent quantity and assumed that the contour of the potential barrier surrounding an ion will be altered by the application of an electrical field. If one of the ions is fixed at the origin of polar coordinates (r, θ, φ) and the field is applied in the direction θ, then the electrical potential is given by

$$U(r, \theta) = -\frac{z_1 z_2 e^2}{\varepsilon r} - z_2 eEr \cos \theta \tag{9.15}$$

where z_1 and z_2 are the valences of these ions. The potential U has a maximum at a distance $r(\theta)$ if $\theta < \pi/2$:

$$r(\theta) = \left(\frac{z_1 e}{\varepsilon E \cos \theta}\right)^{1/2} \equiv R(\theta). \tag{9.16}$$

Clearly, for fixed E, the boundary of the barrier top depends on the angle θ. If the intensity of E is increased, the boundary (the barrier top) will penetrate the conventional association sphere when $R(0) < q$. The contour of the potential barrier will, therefore, cross over the conventional association surface. The height of the barrier on the association surface is given by

$$U(R(\theta), \theta) = -2 \left(\frac{z_1 z_2^2 e^3}{\varepsilon} E \cos \theta \right)^{1/2}. \tag{9.17}$$

According to this equation, the barrier has a minimum at $\theta = 0$, i.e. a saddle point on the axis along the applied field. Therefore, the shape of the contour of the potential barrier and its height depends upon the direction and magnitude of the applied field. If we assume r_0 as an equilibrium distance between the centres of two ions forming a pair, then the Coulombic barrier for dissociation is

$$B(0) = z_1 z_2 e^2 / \varepsilon r_0 \tag{9.18}$$

if no electric field is present. In the presence of an electrical field, the barrier height measured from the ground state becomes

$$B(E) = \frac{z_1 z_2 e^2}{\varepsilon r_0} - \alpha k T - \frac{z_1 z_2 e^2}{\varepsilon R(\theta)} - z_2 e E R(\theta) \cos \theta. \tag{9.19}$$

The second term on the RHS of this equation is envisaged as a lowering of the potential barrier due to the collision between ions and solvent molecules.

The dissociation of ion pairs is accelerated by electrical fields because of the appearance of a saddle point in the potential contour, at which the potential energy is at its lowest and the association distance at its shortest. The concept underlying Bass' theoretical formulation, therefore, is different from that used by Onsager. The ratio of $K(E)/K(0)$ obtained by Bass is

$$\frac{K(E)}{K(0)} = \frac{\exp \sqrt{8b}}{\sqrt{8b}} \left(1 - \frac{2}{3\sqrt{8b}} + \frac{4}{15(8b)} - \ldots \right). \tag{9.20}$$

In spite of the different concept introduced by Bass, the result obtained by (9.20) is similar to that of Onsager. Bass' theory, however, is based on a concept which is more flexible than Onsager's and may be close to the physical reality of non-linear ionic processes.

9.4 The Non-linear Dielectric Constant due to Conformation Changes

Non-linear electric polarization due to strong field saturation, in general, will give rise to a decrease in the dielectric constant. Inspection of (9.1) reveals that the sign of the even number terms is negative. The onset of non-linearity occurs when the magnitude of the second term becomes significant compared with that of the first term. This type of non-linear effect was first observed by Herweg (1920) with ethyl ether many years ago.

The non-linear electric polarizations which were reported by Piekara (1950, 1951), Piekara and Kielich (1958a, b) and Malecki (1965) give rise to a positive effect as well as a negative effect on the dielectric constant of polar molecules. In the theoretical treatment by Debye, it was assumed that the dipole moment of molecules is not altered by the field and that non-linearity arises from the orientation approaching the theoretical limit. The non-linear effects investigated by Piekara and Kielich are due to changes of the dipole moment and/or the polarizability of molecules by strong fields. The changes in dipole moment may be due to conformation changes of the molecule or due to the realignment of dipoles causing enhancement or partial cancellation of effective moments. The three different types of non-linear behaviour are illustrated in figure 9.3. Curve A shows a monotonic decrease due to saturation of the Debye type. Curve C is the monotonic increase and curve B shows an inversion from negative to positive effects.

In order to account for these complex non-linear effects, a correlation parameter R_s was derived by these authors. The original derivation of the correlation parameter was, however, due to Debye (1935), as shown by the equation below:

$$R_s = 3\left(1 - 4L^2 + 3L^4 + \frac{4L}{Y}(2L^2 - 1) + \frac{6L^2}{Y^2}\right) \qquad (9.21)$$

where L is the Langevin function, $L(y)$, with $Y = w/kT$, w being the interaction energy between dipoles and applied field.

Based on the assumption that the interaction between a molecule with its immediate neighbours is the predominant cause of non-linear behaviour, Piekara and Kielich derived another expression for the correlation parameter (Piekara 1950):

$$R_s = 6L/Y - (1 \mp 5L)(1 \pm L). \qquad (9.22)$$

In this theory, the Langevin function parameter $Y = w'/kT$ where w' is the interaction energy between neighbouring molecules. The correlation parameter R_s is then introduced in the expression for the second-order dielectric constant term:

$$\Delta\varepsilon = -12\pi N \left(\frac{\varepsilon + 2}{3}\right)^4 \frac{\mu^4}{45k^3 T^3} E^2 R_s \qquad (9.23)$$

where N denotes the number of molecules per cm^3 and μ is the dipole moment. The correlation parameter derived by Debye is always positive regardless of the value of y and therefore gives rise to a negative effect.

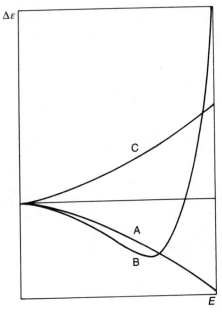

Figure 9.3 The non-linear dielectric constant in strong fields. Curves: A, due to the saturation effect discussed by Debye (1929); B, non-linearity due to configuration changes (this effect can cause a negative effect as well as positive effects); C, due to the combination of various effects, including saturation and configuration changes. (After Malecki 1962.)

This means that the sign of $\Delta\varepsilon$ is negative, i.e. a decrease in dielectric constant. In contrast, the RHS of Piekara's equation can be positive (upper sign) and negative (lower sign). The upper signs correspond to parallel alignment whereas the lower signs indicate anti-parallel alignment. For anti-parallel alignment the correlation factor R_s has a negative sign, giving rise to a positive non-linear effect.

Equation (9.23) was derived using the Lorentz internal field:

$$E = \frac{\varepsilon + 2}{3} E_0. \tag{9.24}$$

As discussed in Chapter 2, Lorentz's equation is not an appropriate representation of the local field around highly polar molecules. In particular, the interaction between neighbouring molecules plays a crucial role for determining the sign of voltage dependent electric polarization. Therefore, a correct expression for local fields is particularly important for the analysis of non-linear electric polarization. Another equation for the second-order dielectric constant term was derived by Boettcher and Bordewijk (1973) using Onsager's internal and reaction fields:

$$\frac{\Delta\varepsilon}{E^2} = -\frac{\varepsilon(n^2 + 2)^2}{(2\varepsilon^2 + n^4)(2\varepsilon + n^2)^2}\frac{4\pi\mu^4 NR_s}{45(kT)^3} \tag{9.25}$$

where n is the refractive index and R_s is the correlation parameter defined by Piekara and Kielich (equation (9.22)).

The study of the non-linear dielectric behaviour of biological macromolecules is very scant. One of the few investigations is Jones' work on myoglobin (1979), up to a field intensity of about 10^5 V cm^{-1}. As shown by figure 9.4, the sign of the voltage dependent dielectric increment is negative. Curve B of this figure illustrates the change in the dielectric constant of salt solution. The open circles with bars are the dielectric decrements of myoglobin solution after correcting the data for the non-linearity of the solvent. As shown, the voltage dependent dielectric decrement becomes quite small after the correction. Jones used a value of 0.01 for the correlation parameter R_s in (9.25) to make the theoretical value agree with the observed value. The small value of R_s seems to indicate anti-parallel alignment of myoglobin dipoles in strong fields, thus leading to partial cancelation.

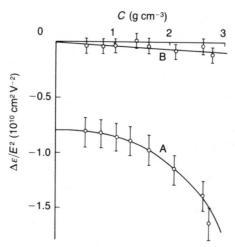

Figure 9.4 The voltage dependent dielectric increment of myoglobin solution. Curves: A, $\Delta\varepsilon/E^2$ at various concentrations of myoglobin solution without correction; B, the results after subtracting the voltage dependent dielectric decrement of salt solution. Note that the decrement after correction is extremely small. (From Jones 1979.) Reproduced by permission of Elsevier Science Publishers BV.

The non-linear dielectric properties of poly-benzyl-L-glutamate (PBLG) were investigated by Block and Hayes (1970) and by Gregson et al (1971). Only the work by Block and Hayes will be discussed in some detail. PBLG is a rod-like helical polymer in non-polar solvents and it has a large dipole moment along the helix axis. These properties make this polymer particularly suitable for the study of voltage dependent dielectric properties. Figure 9.5 illustrates the field dependence of the dielectric increment of PBLG. The highest field intensity used was $8\,\mathrm{kV\,cm^{-1}}$ which is well in the non-linear range for helical polyamino acids (see table 2.2). As shown, the sign of the non-linear effect is negative as expected but the decrease in dielectric constant is a non-linear function of E^2. The curve levels off at high field intensities without, however, sign inversion. Figure 9.6 shows the field dependence of relaxation frequency at various concentrations. Unlike the dielectric increment, the relaxation frequency is a linear function of field intensity. These observations indicate some conformation changes of PBLG in strong fields and that the non-linear behaviour of PBLG may not be a simple dielectric saturation effect.

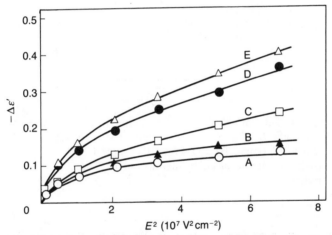

Figure 9.5 The dielectric saturation effect of PBLG as a function of E^2 at various concentrations $(10^{-4}\,\mathrm{g\,cm^{-3}})$: A, 1.59; B, 2.19; C, 3.98; D, 5.81; E, 7.96. Molecular weight $\bar{M}_\mathrm{w} = 4.6 \times 10^5$. Note the saturation of the curves with the increase in electric field intensity. (From Block and Hayes 1970.) Reproduced by permission of The Royal Society of Chemistry.

A theory was developed by Block and Hayes for field dependent dielectric behaviour using Onsager's internal and reaction fields, G and R. The calculation of these quantities was based on the use of a prolate ellipsoidal cavity instead of the spherical one used in the original work by Onsager. G_i and R_i, thus calculated, are given by the following equations.

$$G_i = \frac{2\varepsilon E}{2\varepsilon + (\varepsilon_\mathrm{c} - \varepsilon)L_i} \tag{9.26}$$

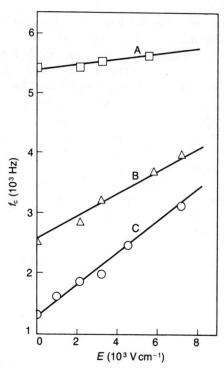

Figure 9.6 The variation of the relaxation frequency of PBLG with field strength E. The molecular weight of PBLG is 1.1×10^5 (curve A), 2.6×10^5 (curve B) and 4.6×10^5 (curve C). (From Block and Hayes 1970.) Reproduced by permission of The Royal Society of Chemistry.

where ε_c is the dielectric constant of the cavity, the shape factor L_i is given by

$$L_i = 1/e^2 - (1 - e^2)/2e^3 \ln[(1 + e)/(1 - e)] \qquad (9.27)$$

and the eccentricity

$$e = (1 - a^2/b^2)^{1/2} \qquad (9.28)$$

where a and b are the major and minor axes respectively. For long prolate spheroids, $a \gg b$ and $L_i = 1$. Therefore, the internal field reduces to

$$G_i = \frac{2\varepsilon E}{\varepsilon + 1} \qquad (9.29)$$

if ε_c is assumed to be 1. Similarly, the reaction field R_i is

$$R_i = \frac{3L_i(1 - L_i)(\varepsilon - 1)E}{4\pi[\varepsilon + (1 - \varepsilon)L_i]} \qquad (9.30)$$

where the shape factor L_i is defined by

$$L_i = \left(\frac{1}{e^2} - 1\right)\left[\frac{1}{2e}\ln\left(\frac{1 + e}{1 - e}\right) - 1\right]. \tag{9.31}$$

Using these equations, the mean dipole moment was calculated by the method described in Chapter 2. The equation for dielectric constant at low field intensities is given as

$$\Delta\varepsilon(0) = \frac{4\pi NkT\chi^2}{3g_i} \tag{9.32}$$

where $g_i = G_i/E$, N is the number of molecules and χ is given by

$$\chi = \frac{2\varepsilon}{\varepsilon + 1}\frac{\mu}{kT}. \tag{9.33}$$

The dielectric increment for high field intensities is, on the other hand, given by

$$\Delta\varepsilon(E) = \frac{4\pi NkT\chi L(\chi E)}{Eg_i} \tag{9.34}$$

where $L(\chi E)$ is the Langevin function. Thus, the ratio $\Delta\varepsilon(E)/\Delta\varepsilon(0)$ is

$$\lim_{c \to 0} \frac{\Delta\varepsilon(E)}{\Delta\varepsilon(0)} = \frac{3L(\chi E)}{\chi E} \tag{9.35}$$

where c is the solute concentration. The full expression for $L(\chi E)/\chi E$ is given by

$$L(\chi E)/\chi E = 1 - \left(\frac{2\varepsilon}{\varepsilon + 1}\right)^2\frac{\mu^2 E^2}{15k^2T^2} + \left(\frac{2\varepsilon}{\varepsilon + 1}\right)^4\frac{\mu^4 E^4}{315k^4T^4}. \tag{9.36}$$

In the derivation of these equations, it was assumed that the dipole moment of PBLG is constant and unaffected by the field. Thus, (9.36) predicts that the second-order term should be proportional to E^2 and that the sign is negative. The observed field dependent decrement is not, however, the linear function of E^2 as predicted by the theory. This indicates either that the higher-order terms are large enough to be significant or that the dipole moment changes with applied field. This suggests that the classical concept of dielectric saturation discussed by Debye may not be sufficient to explain the non-linear dielectric behaviour of helical amino acids which have a very large dipole moment and which may not be a perfect rigid rod.

9.5 Dielectric Relaxation due to Chemical Reactions

The non-linear effects discussed in the above are due to field induced changes in the ionization and molecular configurations. The field induced shift of the equilibrium constant, which played a crucial role in the Wien effects, will again be a central part of the discussion in this section. In what follows, the non-linear dielectric properties due to chemical relaxation will be discussed in detail (Bergmann *et al* 1963, DeMaeyer *et al* 1968).

As an example, let us consider a dimerization reaction of the molecular species A and B as shown below:

$$A + A \underset{k_{21}}{\overset{k_{12}}{\rightleftharpoons}} B.$$

van't Hoff's equation, which holds for ΔH and ΔV, can be extended to the change in dipole moment accompanying a chemical reaction:

$$\left(\frac{\partial \ln K}{\partial |E|}\right)_{T,P} = \frac{\Delta M}{RT} \tag{9.37}$$

where K is the equilibrium constant of the reaction and ΔM is the change in the dipole moment in the direction of the field between reactants A and the product B. The change in equilibrium constant K as a function of the strong field E is illustrated schematically in figure 9.7. The ordinate is $\ln K(E)/K(0)$. Since the curve is non-linear, the value of ΔM must be determined from the slope at each field intensity, as shown in this figure. Namely, we apply a small AC signal in addition to the strong DC pulse and measure the field induced dielectric increment, which is directly related to ΔM as will be discussed shortly. Since we are using a small signal, the measured dielectric constant does not depend on the input signal, it depends only on the intensity of the DC

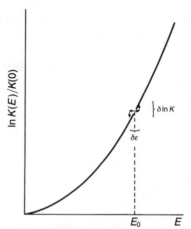

Figure 9.7 The dependence of the equilibrium constant $K(E)$ as a function of the field intensity E. The slope $\partial \ln K/\partial E$ is determined using a small measuring signal, as illustrated in this figure. (From Bergmann *et al* 1963.) Reproduced by permission of Verlag Chemie.

pulse E. In this sense, the dielectric constant is linear with regard to the small measuring signal, although it is non-linear with regard to the large perturbing field.

The reaction moment ΔM is the difference between the partial molar moments of the product, M_B, and the reactants, M_A:

$$\Delta M = M_B - 2M_A. \tag{9.38}$$

It must be pointed out that the reaction moment is not the dipole moment of the molecules but, rather, a mean moment created by the chemical reaction in the direction of applied field. The reaction moment ΔM, using Onsager's internal and reaction fields, is defined by the following equation:

$$\Delta M = N_A \frac{\varepsilon^2(n^2 + 2)^2}{2\varepsilon^2 + n^4} \frac{(\mu_B^2 - 2\mu_A^2)E}{9kT} \tag{9.39}$$

where ε is the dielectric constant of the reaction system. As shown, ΔM is proportional to E rather than to E^2 and it is also proportional to the difference between the square of the dipole moments of the product and the reactants.

The dielectric constant is related to the displacement D by the following equation:

$$\varepsilon_{S,V} = \left(\frac{\partial D}{\partial E}\right)_{S,V}. \tag{9.40}$$

Note that the differential form is used in (9.40) instead of (2.30) because of the non-linearity of the dielectric constant. Since the displacement D depends, in this case, not only on the electric field but also on the advancement of the reaction, (9.40) should be written as follows:

$$\varepsilon_{S,V} = \left(\frac{\partial D}{\partial E}\right)_{S,V,\zeta} + \left(\frac{\partial D}{\partial \zeta}\right)_{S,V,E}\left(\frac{\partial \zeta}{\partial E}\right)_{S,V}. \tag{9.41}$$

The second term on the RHS is the dielectric constant due to chemical relaxation:

$$\varepsilon_{S,V}^{ch} = \left(\frac{\partial D}{\partial \zeta}\right)_{S,V,E}\left(\frac{\partial \zeta}{\partial E}\right)_{S,V,A=0} \delta_{S,V,E} \tag{9.42}$$

where $\delta_{S,V,E}$ is defined by

$$\delta_{S,V,E} = \frac{1}{1 + j\omega\tau_{ch}}. \tag{9.43}$$

The condition $A = 0$ in (9.42) means the equilibrium state. Equation (9.42), after rearrangement, becomes (see Bergmann et al (1963) for the derivation)

$$\varepsilon^{ch} = -\frac{4\pi}{V}\frac{(\Delta M)^2}{\Delta A}\delta_{S,V,E}. \tag{9.44}$$

ΔA is defined by

$$\Delta A = \frac{-2RT}{C_0 V} \frac{2 - \alpha}{\alpha(1 - \alpha)}$$

where R is the gas constant ($= Nk$) and $\alpha = C_A/C_0$ (where C_0 is the total concentration of the polar substances). Hence, the dielectric constant due to chemical relaxation is

$$\varepsilon^{ch} = \frac{4\pi}{V} \frac{(\Delta M)^2}{2RT} \frac{C_0\alpha(1 - \alpha)}{2 - \alpha} \frac{1}{1 + j\omega\tau_{ch}}. \qquad (9.45)$$

Substituting (9.39) into (9.45), we obtain an expression which relates the dielectric constant of a reaction mixture to the reaction moment ΔM:

$$\varepsilon^{ch} = \frac{\alpha(1 - \alpha)}{2 - \alpha} \frac{C_0}{2} \frac{\varepsilon^4(n^2 + 2)^4}{(2\varepsilon^2 + n^4)^2} \frac{4\pi N}{81k^3 T^3} \frac{(\mu_B^2 - 2\mu_A^2)^2 E^2}{1 + \omega^2\tau_{ch}^2}. \qquad (9.46)$$

The relaxation time τ_{ch} is defined by

$$\tau_{ch} = \frac{1}{2k_{12}\alpha C_0 + k_{21}} \qquad (9.47)$$

and the loss tangent $(\tan \delta)^{ch}$ is given by

$$(\tan \delta)^{ch} = \frac{\varepsilon^{ch}}{\varepsilon} \frac{\omega\tau_{ch}}{1 + \omega^2\tau_{ch}^2}. \qquad (9.48)$$

The equations discussed above are the frequency domain manifestation of chemical reactions. The derivation is based on a concept that chemical reactions accompanying dipole moment changes manifest themselves in a frequency domain as a dielectric relaxation. In general, chemical reactions are measured by time domain techniques. However, if the reaction is very fast, it often becomes difficult to determine the rate constants, mainly because of instrumental limitations. In a frequency domain, however, a time constant of 10^{-7} s corresponds to 1.5 MHz, a frequency which is commonly used for dielectric measurements. Thus, the measurement is much easier in a frequency domain than in a time domain. Therefore, instead of time domain techniques, (9.47) and (9.48) can be used to determine the rate constants k_{12} and k_{21} using dielectric techniques in a frequency domain. Thus, frequency domain methods provide a unique technology for investigating fast chemical reactions or physical processes.

The block diagram of the measuring circuit is shown in figure 9.8. The capacitor C_s, the sample cell, is a part of the resonant circuit with an inductance L and a blocking capacitor C_b ($C_b \gg C_s$). A high voltage is applied to the sample via a large resistor R_b. The HF signal (0.1–100 MHz) is coupled to the resonant circuit through a capacitor C_c ($C_c \ll C_s$). The output is monitored by a HF voltmeter and its frequency is measured by a digital counter. The HF voltage across the inductance of the resonant circuit is compared with the generator output. The resonance frequency ω_0 and resonance voltage U_0 at zero field can be determined as well as the incremental values $\Delta\omega_0$ and ΔU_0 in the presence of a high DC field. The incremental loss tangent is related to measured resonance parameters by

$$(\Delta \tan \delta)_E = \frac{\gamma}{Q_0} \left(\frac{\Delta\omega_0}{\omega_0} - \frac{\Delta U_0}{U_0} \right) E \qquad (9.49)$$

where Q_0 is the circuit quality.

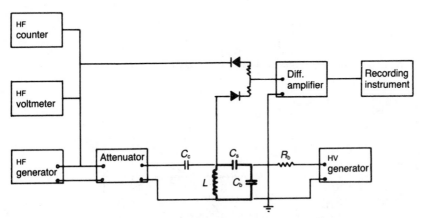

Figure 9.8 A block diagram of the measuring system. C_s is the sample cell, L is an inductor, C_b is the blocking capacitor $(C_b \gg C_a)$ and C_c is a coupling capacitor $(C_c \gg C_s)$. Reproduced with permission from DeMaeyer *et al* (1968). Copyright (1968) American Chemical Society.

A DC field of 3×10^5 V cm^{-1} is usually required to detect measurable field effects. Figure 9.9 illustrates the increase in the loss tangent, i.e. $\Delta \tan \delta$, of ε-caprolactam in carbon tetrachloride as a function of E^2. Also shown is $\Delta \tan \delta$ at various frequencies and concentrations. As seen from these diagrams, the field induced loss factor is very small, thus requiring an exceptional stability and sensitivity of the instruments used for measurement.

9.6 Chemical Relaxation with AC Field Perturbation

The technique developed by Bergmann *et al* (1963) and DeMaeyer *et al* (1968) used high-amplitude DC fields to shift the equilibrium and produce permittivity changes. In what follows, another technique which produces a dielectric increment by low-frequency high-amplitude AC

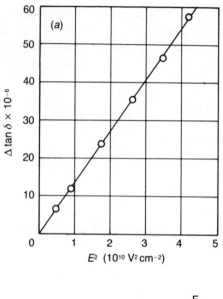

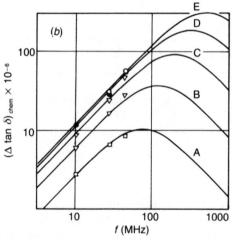

Figure 9.9 The dependence of the loss tangent $\tan \delta$ of ε-caprolactam in carbon tetrachloride. (a) The loss tangent as a function of field strength and (b) the loss tangent produced by chemical relaxation at various frequencies and concentrations. The full curves are calculated using Debye theory. Curves A–E are 0.003 M, 0.0102 M, 0.0364 M, 0.109 M and 0.338 M respectively. $E = 220$ kV cm^{-1} at $T = 22$ °C. (From Bergmann *et al* 1963.) Reproduced by permission of Verlag Chemie.

fields will be discussed (Parsoons and Hellemans 1978, Hellemans and DeMaeyer 1975, Nackaerts *et al* 1979). In this method, a low-frequency field of high amplitude and a high-frequency low-amplitude field are

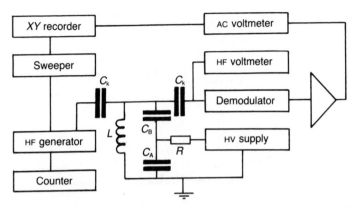

Figure 9.10 A block diagram of the apparatus used to measure the dielectric increment due to chemical relaxation using AC fields. The coil and series capacitors C_A and C_B form a parallel resonant circuit, coupled to the HF generator and the measuring instruments by a capacitor C_k. The resistor R is 1 MΩ, to ensure a constant load on the circuit. (From Parsoons and Hellemans 1978.) Reproduced from the *Biophysical Journal*, 1978, vol 24, pp 119–34 by copyright permission of the Biophysical Society.

applied simultaneously. Figure 9.10 illustrates a schematic diagram of the measuring circuit. Two identical capacitors C_A and C_B containing unknown samples and an inductor L constitute the resonant circuit. The resonance curve of the circuit without a high-voltage field is depicted schematically in figure 9.11 as an inset. When the sample is subjected to a high-voltage field, the impedance of the sample is modified and this, in turn, causes a change in the resonance curve as shown in figure 9.11. As illustrated by these two curves, impedance changes alter the resonance voltage $(\Delta V_0/V_0)$ and shift the resonance frequency $(\Delta f_0/f_0)$. Experimentally, when the frequency of low-amplitude RF fields is varied in the presence of a high-voltage field, a signal which is proportional to the difference between the resonance curves (with and without high-voltage fields) is recorded (see figure 9.11). Detecting the 3 dB points f_1 and f_2, we can calculate the resonance frequency by the following formula: $f_0 = (f_1 + f_2)/2$. The Q-value of the resonant circuit is calculated by $Q = f_0/(f_2 - f_1)$. The measured quantities ΔV_0 and Δf_0 are related to the field induced dielectric increment and the loss tangent as shown by the following equations:

$$\frac{\Delta\varepsilon}{\varepsilon} = 2\gamma(\Delta f_0/f_0) \qquad (9.50a)$$

$$\Delta\tan\phi = (\gamma/Q)[\Delta V_0/V_0 - F(\Delta f_0/f_0)] \qquad (9.50b)$$

where γ is a parameter which is determined empirically. The numerical

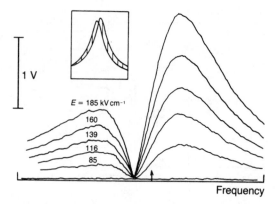

Figure 9.11 A typical resonant signal for a solution of 0.0668 mol ε-caprolactam in cyclohexane. The numbers in this figure indicate the field strengths of the low-frequency high-voltage AC fields. The resonant parameters are $f_0 = 28.171\,40$ MHz (arrow), $U_0 = 3.16$ V and $Q = 478.8$. The sweep width is 137.8 kHz. Inset: the hatched area corresponds to the difference between the resonance curve with and without high AC fields. (From Hellemans and DeMaeyer 1975.) Reproduced by permission of the American Institute of Physics.

value of the factor F is usually very close to 1. One of the results obtained using this method is shown in figure 9.12. The ordinate is the field induced dielectric increment and the abscissa is the frequency of low-amplitude RF fields. The magnitude of the low-frequency field is $138\,\mathrm{kV\,cm^{-1}}$ at 85 Hz. Curve A is the real part and curve B is the dielectric loss curve. The inset shows the proposed chemical reaction which is the source of the field induced dielectric increment. Looking at these two structures, clearly, the one on the right has a large dipole moment. This makes the value of ΔM non-zero (see equation (9.39)) and this produces a field dependent dielectric increment.

In addition to the study of the field induced dielectric increment, this system can be used for the field dissociation effect (non-Ohmic conductance) which was discussed earlier. The field dissociation effect, like the field induced dielectric increment, is frequency dependent. Therefore, the present method is quite suitable for the relaxation study of non-Ohmic conductances. Figure 9.13 shows the field induced conductance changes of tetra-butylammonium picrate (TBAP) with a field of $3\,\mathrm{kV\,cm^{-1}}$ at various frequencies of low-voltage fields.

9.7 Dielectric Relaxation and Phase Transition

In the previous section, the dielectric relaxation due to the perturbation of chemical equilibrium by strong electric fields was discussed. The

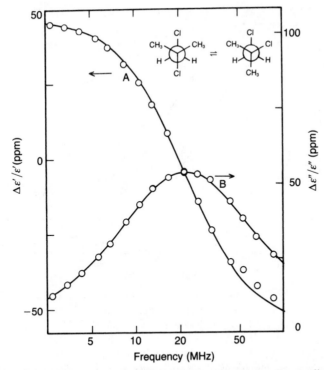

Figure 9.12 The dispersion and absorption of the non-linear dielectric effect due to chemical relaxation of 1,2-dichloro-2-methylpropane in benzene. $E = 138\,kV\,cm^{-1}$. The abscissa is the frequency of the low-amplitude RF measuring field. Inset: a possible chemical conversion of the compound. (From Parsoons and Hellemans 1978.) Reproduced from the *Biophysical Journal*, 1978, vol 24, pp 119–34 by copyright permission of the Biophysical Society.

contribution of chemical relaxation is proportional to E^2 and, therefore, diminishes quickly as the field intensity decreases. Hence, this effect cannot be observed with conventional experimental techniques unless a very strong electrical signal is used for the measurement. Schwarz (1967), however, pointed out that the effect of chemical relaxation on the dielectric behaviour can be detected if the reaction rate is either comparable or larger than the rate of dipole rotation, even in the absence of strong fields.

The development of the theory was based on the use of the following simple reaction:

$$A_1 \underset{k_{21}}{\overset{k_{12}}{\rightleftharpoons}} A_2$$

where the reactant A_1 has no dipole moment whereas the product A_2

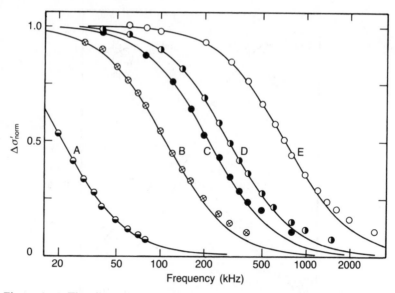

Figure 9.13 The dispersion of the field dissociation effect in a solution of TBAP at 25 °C. The effective field strength is about 3 kV cm^{-1}. The full curves are calculated by using the Debye equation. The concentrations are: 1.16×10^{-4} (curve A), 3.8×10^{-4} (curve B), 7.63×10^{-4} (curve C), 7.68×10^{-4} (curve D) and 1.12×10^{-3} (curve E) mol respectively. (From Parsoons and Hellemans 1978.) Reproduced from the *Biophysical Journal*, 1978, vol 24, pp 119–34 by copyright permission of the Biophysical Society.

has a non-zero dipole moment μ. In order to describe the orientation of the molecules A_1 and A_2, we introduce an axis which makes an angle θ with the applied field. The distribution functions γ_1 and γ_2 are defined as the concentrations of molecules A_1 and A_2 pointing into a solid angle $d\Omega = 2\pi \sin\theta d\theta$. The time derivatives of γ_1 and γ_2 are given by the following equations:

$$\frac{\partial \gamma_1}{\partial t} = -v_{12} + D_r \nabla^2 \gamma_1 \tag{9.51}$$

$$\frac{\partial \gamma_2}{\partial t} = v_{12} + D_r \nabla^2 \gamma_2 - \operatorname{div} j_E \tag{9.52}$$

where v_{12} is the velocity of reaction, D_r is the rotary diffusion coefficient and the term $\nabla^2 \gamma_i$ is given by

$$\nabla^2 \gamma_i = \frac{1}{\sin\theta} \frac{\partial}{\partial \theta} \left(\sin\theta \frac{\partial \gamma_i}{\partial \theta} \right) \tag{9.53}$$

when using the Laplace operator in spherical coordinates. The last term on the RHS of (9.52) is the angular flux of the molecule A_2 due to the torque created by a directing field. This term is given by

$$j_E = -\gamma_2 \frac{D_r}{kT} \mu E_r \sin \theta. \tag{9.54}$$

If rotational diffusion is zero, $D_r = 0$, then the chemical relaxation time dominates:

$$1/\tau_{ch} = k_{12} + k_{21}. \tag{9.55}$$

On the other hand, if there is no chemical reaction, then the relaxation process is dominated by a rotational relaxation time, $\tau_r = 1/D_r$. If both rotational and chemical relaxations are present, then the relaxation time τ_2 is written as

$$\frac{1}{\tau_2} = \frac{1}{\tau_r} + \frac{1}{\tau_{ch}}. \tag{9.56}$$

The solution of (9.52) is

$$\gamma_2 = \gamma_2^0 \left(1 + \frac{\mu E_r}{kT} \cos \theta \alpha(\omega) \right) \tag{9.57}$$

where γ_2^0 is the distribution of A_2 in the absence of electrical field E_r. The function $\alpha(\omega)$ is defined by the following equation:

$$\alpha(\omega) = \left(\frac{K_0}{1 + K_0} \frac{1}{1 + j\omega\tau_r} \right) + \left(\frac{1}{1 + K_0} \frac{1}{1 + j\omega\tau_2} \right) \tag{9.58}$$

where K_0 is the equilibrium constant for zero field. The average dipole moment ΔP of A_2 is related to the complex dielectric constant $\Delta \varepsilon^*$ by

$$\Delta P = \frac{1}{4\pi} \Delta \varepsilon^* E = g N_A \int_0^\pi \mu \cos \theta \gamma_2 2\pi \sin \theta \, d\theta. \tag{9.59}$$

The solution of this equation gives rise to a frequency dependent dielectric constant, including a chemical relaxation term:

$$\Delta \varepsilon^* = \Delta \varepsilon^0 \alpha(\omega) = \frac{\Delta \varepsilon_r^0}{1 + j\omega\tau_r} + \frac{\Delta \varepsilon_2^0}{1 + j\omega\tau_2} \tag{9.60}$$

where

$$\Delta \varepsilon^0 = g \frac{4\pi N_A \mu^2}{3kT} c_2^0 \tag{9.61}$$

$$\Delta \varepsilon_r^0 = \frac{K_0}{1 + K_0} \Delta \varepsilon^0 \tag{9.62}$$

$$\Delta \varepsilon_2^0 = \frac{1}{1 + K_0} \Delta \varepsilon^0 \tag{9.63}$$

where c_2^0 is the total concentration of A_2 in the equilibrium state and g is a directing field due to Onsager (1936).

The dielectric relaxation behaviour in the presence of a chemical reaction depends upon the relative magnitudes of the rotational and chemical relaxation times τ_r and τ_{ch}. We have to consider the following three possible cases.

Case I. $\tau_r \ll \tau_{ch}$. In this case, rotational diffusion is much faster than the rate of the chemical reaction. Using (9.56), the overall relaxation time τ_2 is practically the same as τ_r. Therefore, the second term on the RHS of (9.58) can be ignored. This means that the relaxation behaviour is dominated by rotational diffusion and consequently the chemical reaction has no effect on dipole relaxation.

Case II. $\tau_r \simeq \tau_{ch}$. Under these conditions, two relaxation terms contribute to the relaxation process with comparable weights. However, from (9.56), we readily find that the difference between τ_r and τ_2 is at most a factor of 2. In general, two dispersions cannot be differentiated clearly if the difference in relaxation times is only a factor of 2. Thus, the two terms on the RHS of (9.60) will not be distinctly separated, although the presence of two relaxation times would be discerned by a broadening of the dispersion curves.

Case III. $\tau_r \gg \tau_{ch}$. The rate of a chemical reaction is much faster than rotational diffusion. Thus, $\tau_2 = \tau_{ch} \ll \tau_r$, meaning that $\Delta\varepsilon^*$ displays two distinctly different dispersions. One is due to chemical relaxation at high frequencies and the other is due to rotational relaxation at low frequencies. This is the only case in which two dispersion regions are separable and the dielectric relaxation of chemical origin can be clearly observed. The amplitudes of $\Delta\varepsilon_r^0$ and $\Delta\varepsilon_2^0$ depend on the equilibrium constant K. If $K = 1$, then the concentrations of A_1 and A_2 are equal and the dielectric increments of both terms are exactly the same. If $K > 1$ then the increment due to chemical relaxation will be larger than that due to rotational relaxation. Note that the sum of these two increments is always constant. Thus, the presence of a chemical reaction does not affect the total amplitude of the dielectric increment but changes the relaxation process. The theory just discussed can be extended to more general cases such as dimerization reactions. Schwarz (1970) used an acid–base titration of zwitterions as an example of this type of chemical reaction.

The significance of the contribution of chemical relaxation may be explained by the following discussion. If an electric field is applied to a molecule with a dipole moment μ, the dipole axis of the molecule will change to a more energetically favourable orientation. If the dipole moment is a fixed permanent dipole, then the only way to achieve this is to reorient the entire molecule along the direction of the field E. However, a change in dipole axis can also be implemented by altering the structure of the molecule and by changing the direction of the dipole axis internally. Therefore, rotation of the entire molecule is not required in this case.

The theory discussed above is based on a tacit assumption that dipoles are the only species which are involved in chemical reactions. If the chemical process involves free ions, (9.37) must be rewritten as follows:

$$\left(\frac{\partial \ln K}{\partial |E|}\right)_{T.P} = \frac{\Delta M_E + \Sigma \Delta(Q_j \chi_{jE})}{RT} \tag{9.64}$$

where χ_{jE} is the coordinate in the field direction and Q_j is free charge. In Schwarz's theory the term $\Sigma \Delta(Q_j \chi_{jE})$ is ignored. Let us compare (9.60) with (5.43), which was derived by Scheider (1965). The dielectric increment due to chemical relaxation, as derived by Schwarz, is given by the following equation:

$$\Delta \varepsilon_{ch} = g \frac{4\pi N_A c_0}{3kT} \overline{\delta \mu^2} \left(\frac{1}{1 + j\omega \tau_2}\right) \tag{9.65}$$

whereas, the equation for dipole moment fluctuation by Scheider is

$$\Delta \varepsilon_{ch} = g \frac{4\pi N_A c_0}{3kT} \overline{\delta \mu^2} \frac{1}{1 + \tau_r/\tau_{ch}} \frac{1}{1 + j\omega \tau_2}. \tag{9.66}$$

These two equations are quite similar except for the additional factor

$$\frac{1}{1 + \tau_r/\tau_{ch}}$$

in (9.66). The origin of this term is due to the assumption used by Scheider that the free ion concentration in conducting solution is uniform and, hence, chemical equilibrium is independent of the relative directions of the dipole molecule and the applied field (Scheider 1970). In other words proton fluctuation by electrical fields is so small that the distribution of free protons remains essentially uniform. Because of this assumption, the RHS of (9.64) vanishes and this leads to a solution given by (9.66). South and Grant (1973) obtained an equation for the mean square proton fluctuation moment as shown below:

$$\varepsilon^* - \varepsilon_\infty = \frac{4\pi N}{3kT} \sum_{i=1}^{3} \left(\frac{\overline{\mu_i^2}}{1 + j\omega \tau_i} + \frac{\overline{\delta \mu_i^2}}{1 + j\omega \tau_i'}\right) \tag{9.67}$$

where the second term on the RHS is the dielectric increment due to proton fluctuation. This equation was derived with an assumption that proton fluctuation, unlike Scheider's assumption, produces a non-vanishing mean square dipole moment in the direction of applied field. The underlying concept is the same as that used by Schwarz, thus leading to the equation which is essentially the same as (9.65).

9.8 The Application of Schwarz's Theory to Polymer Reactions

As discussed, Schwarz's theory can be used for the study of fast chemical reactions without high-intensity electrical fields. Above all, the application of Schwarz's theory to the helix–coil transition of polyamino acids is of particular interest (Schwarz and Seelig 1968). The rate of the helix–coil transition for polyglutamic acid (PGA) was measured by Lumry et al (1964) using a time domain temperature jump technique. Although this method has proven quite useful for the study of fast chemical reactions, these authors found that the rate of the helix–coil transition for PGA is too fast to be determined accurately using this technique. This limitation arises partly from the difficulty of raising the temperature as a step function with very short time constants. If the relaxation time of a reaction is shorter than microseconds, the experimental difficulties become enormous. The only conclusion these authors reached is that the relaxation time of the helix–coil transition of PGA is shorter than 1 μs. Under these circumstances, the use of periodic fields and the measurement of the transition rate in the frequency domain are expected to yield more accurate data. As has been discussed, the dielectric relaxation time due to a chemical process is directly related to the rate constants of the reaction.

The formulation of this theory begins with an equation which expresses the helical content in a long polymer chain, assuming cooperative interactions between nearest neighbours:

$$\partial \bar{N}/\partial \ln K = (2/\sqrt{\sigma})[\bar{N}(1 - \bar{N})]^{3/2} \tag{9.68}$$

where $\bar{N}$ is the helical content at equilibrium. K is the equilibrium constant and σ is a nucleation parameter which has a numerical value of about 10^{-4}. This equation describes the slope of the change in helical content as a function of the equilibrium constant $\ln K$. If the equilibrium is perturbed by a sudden change in external conditions, then the helix content will change with a relaxation time τ^*, as shown by figure 9.14. We can use an electrical field to increase or decrease the helical content as indicated by the following equation:

$$(\partial \bar{N}(\theta)/\partial E)_{ch} = (\partial \bar{N}/\partial \ln K)[\partial(\ln K(\theta))/\partial E] \tag{9.69}$$

where θ is the angle between helical segments and the applied field. Since the term $\partial N/\partial(\ln K)$ is defined by (9.68) and since

$$\frac{\partial \ln K(\theta)}{\partial E} = \frac{\Delta M(\theta)}{RT} = \frac{\mu \cos \theta}{kT} \tag{9.70}$$

then

$$(\partial \bar{N}(\theta)/\partial E)_{ch} = (2/\sqrt{\sigma})[\bar{N}(1 - \bar{N})]^{3/2} \mu \cos \theta/kT. \tag{9.71}$$

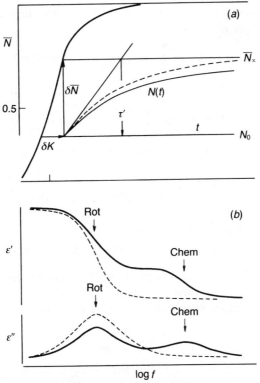

Figure 9.14 (*a*) Chemical relaxation of the helix content after a step perturbation of equilibrium. (*b*) The dielectric constant ε' and the loss factor ε'' as a function of the frequency of measurement fields f. The broken curves show the pure rotational relaxation and the full curves show the effect of fast chemical relaxation. (From Schwarz and Seelig 1968.) Reproduced by permission of John Wiley & Sons, Inc. © 1968.

Based on this equation, we can observe that the helix content increases with the electrical field if $\theta < 90°$ and decreases if $\theta > 90°$. The concentration of helical fragments $dc_h(\theta)$, which consists of n subunits and points into a solid angle $d\Omega$ is given by $dc_h(\theta) = \Sigma\, n\gamma_n(\theta)\, d\Omega$, where γ_n is the angular density function. If c_0 is the total concentration of randomly oriented helical segments, $(c_0/4\pi)\, d\Omega$ represents the concentration of segments pointing in a small solid angle $d\Omega$ for $E = 0$. The fraction of helical segments pointing in the same solid angle $d\Omega$ after the application of a field is

$$\frac{dc_h(\theta)}{(c_0/4\pi)\, d\Omega} = (4\pi/c_0) \sum_n n\gamma_n(\theta) = N(\theta). \tag{9.72}$$

The dipole moment per unit volume or the polarization contributed by helical segments is given by

$$\Delta P = N_A \sum_n \int_0^\pi n\mu \cos \theta \gamma_n(\theta) \, 2\pi \sin \theta \, d\theta \qquad (9.73)$$

where N_A is Avogadro's number. Substituting (9.72) into (9.73)

$$\Delta P = N_A \, \mu c_0 \int_0^\pi \frac{N(\theta)}{2} \cos \theta \sin \theta \, d\theta. \qquad (9.74)$$

In this equation, the contribution of rotational diffusion is neglected. The dielectric constant $\Delta \varepsilon$ is derived using the relation $\varepsilon \simeq 4\pi (\partial \Delta P / \partial E)$

$$\Delta \varepsilon_{ch}^0 = \frac{4\pi N_A \, \mu c_0}{2} \int_0^\pi (\partial N(\theta)/\partial E)_{ch} \cos \theta \sin \theta \, d\theta. \qquad (9.75)$$

Evaluating the integral using (9.71), we obtain

$$\Delta \varepsilon_{ch}^0 = (2/\sqrt{\sigma})[\bar{N}(1 - \bar{N})]^{3/2} 4\pi N_A c_0 \mu^2 / 3kT. \qquad (9.76)$$

The quantity $\Delta \varepsilon_{ch}^0$ is the full amplitude of the dielectric increment due to chemical relaxation, as shown by figure 9.14. Multiplying (9.76) by the relaxation function, we obtain the following equation:

$$\Delta \varepsilon^0 = \frac{\Delta \varepsilon_{ch}^0}{1 + j\omega\tau_{ch}}. \qquad (9.77)$$

As discussed earlier, in Chapter 5, helical PBLG exhibits an anomalous dispersion at low frequencies denoted as 'Rot' in figure 9.14. The contribution of chemical relaxation gives rise to a high-frequency dispersion denoted 'Chem' in the same figure. In order to detect the dielectric dispersion due to chemical relaxation, the concentration of dichloroacetic acid (a helix breaking solvent) is gradually increased in order to induce a helix–coil transition. The dielectric constant and loss factor are measured over the transition region. The dielectric dispersion of PBLG at various helix contents is shown in figure 9.15. This figure clearly indicates the presence of an anomalous dispersion and a peak of dielectric loss located around 10^6 Hz. Moreover, the amplitude of the dispersion is most prominent with a helical content of about 50%, as expected. However, the most important finding is the value of the relaxation time, i.e. 3×10^{-7} s. This is about one third of the value guessed by Lumry et al. The amplitude of the dielectric dispersion due to chemical relaxation is very small, as shown by figure 9.15, making accurate measurement difficult. The results obtained by Schwarz and Seelig were debated by Wada (1971) and by Marchal (1971). In spite of the controversy surrounding these experimental results, the theory developed by Schwarz and Seelig provides a powerful tool for the study of fast reactions which cannot be investigated using time domain step perturbation methods.

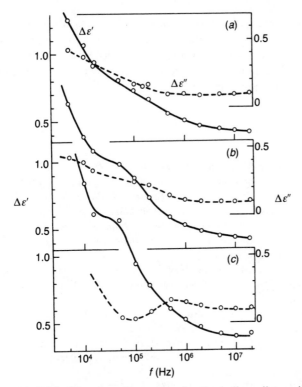

Figure 9.15 Dielectric relaxation curves for the helix–coil transition. The dielectric properties of the solvent are subtracted from that of the PBLG solution. (*a*) $T = 15.5\,°C$, $\bar{N} = 0$; (*b*) $T = 19\,°C$, $\bar{N} = 0.05$; (*c*) $T = 22\,°C$, $\bar{N} = 0.16$ (From Schwarz and Seelig 1968.) Reproduced by permission of John Wiley & Sons, Inc. © 1968.

As discussed above, if the dipole moment of the product of a chemical reaction is different from that of its reactants, then the chemical process manifests itself as a dielectric relaxation. The observation of this type of dielectric relaxation usually requires very high voltages. However, if the system is in a critical state, such as the helix–coil transition of polypeptides, the dielectric increment due to chemical relaxation can be detected if certain conditions are satisfied, even in the absence of high-intensity electrical fields. The relaxation time of chemically induced dielectric dispersion is directly related to the rate of reaction and thus this technique can be used for the study of the kinetics of fast reactions.

The mechanism for non-linear dielectric properties has been discussed in this chapter for biological polymers. However, non-linear effects are often overlooked in conventional dielectric measurements because of the requirement of high-intensity fields in which to observe them. However, in biological membranes which have a thickness of 40–80 Å, a field of $90-100 \, kV \, cm^{-1}$ exists under normal physiological conditions. Therefore, most of the membrane processes proceed under highly non-linear conditions. The non-linear impedance properties of membranes will be discussed in the next chapter.

References

Bass L 1968 Wien dissociation as a rate process *Trans. Faraday Soc.* **64** 2153–9

Bergmann K, De Maeyer L and Eigen M 1963 Dielektrische Absorption als Folge chemischer Relaxation *Ber. Bunsenges. Phys. Chem.* **67** 816–26

Block H and Hayes E F 1970 Dielectric behavior of stiff polymers in solution when subjected to high voltage gradients *Trans. Faraday Soc.* **66** 2512–25

Boettcher C J E and Bordewijk P 1973 *Theory of Electric Polarisation* 2nd edn, vol 1 (Amsterdam: Elsevier)

Debye P 1929 *Polar Molecules* (New York: Dover)

—— 1935 Der Rotationszustand von Molekulen in Flussigkeiten *Phys. Z.* **36** 100–1, 193

DeMaeyer L, Eigen M and Suarez J 1968 Dielectric dispersion and chemical relaxation *J. Am. Chem. Soc.* **90** 3157–61

Gregson M, Jones G P and Davies M 1971 High field pulse measurements of the electric anisotropy of polarizability and the electric dipole moment of PBLG *Trans. Faraday Soc.* **67** 1630

Hellemans L and DeMaeyer L 1975 Absorption and dispersion of the field induced dielectric increment in caprolactam-cyclohexane solution *J. Chem. Phys.* **63** 3490–8

Herweg J 1920 Die elektrischen Dipole in Flussigen Dielectrics *Z. Phys.* **3** 36–47

Jones P 1979 High electric field dielectric studies of aqueous myoglobin solutions *Biophys. Chem.* **9** 91–5

Lumry R, Legare R and Miller W G 1964 The dynamics of the helix-coil transition in poly-α-L-glutamic acid *Biopolymers* **2** 489–500

Malecki J 1962 Dielectric polarization of aliphatic alcohols in strong electric fields *Acta Phys. Pol.* **21** 13–43

—— 1965 Investigation of hexanol-A multimers and complexes by the method of dielectric polarization in weak and strong electric fields *J. Chem. Phys.* **43** 1351–5

Marchal E 1971 Comment on Wada's letter on dielectric evidence of chemical relaxation in the helix-coil transition of polypeptides *Chem. Phys. Lett.* **12** 9–11

Nackaerts R, Demaeyer M and Hellemans L 1979 Field dissociation effect on ion-pairs in a non-polar medium *J. Electrost.* **7** 167–86

Onsager L 1934 Deviations from Ohm's law in weak electrolytes *J. Chem. Phys.* **2** 599–615

—— 1936 Electric moments of molecules in liquids *J. Am. Chem. Soc.* **58** 1486–92

Parsoons A and Hellemans L 1978 New electric field methods in chemical relaxation spectrometry *Biophys. J.* **24** 119–34

Piekara A 1950 The phenomena of molecular orientation in polar liquids and their solutions. Part 1. Extension of Onsager's theory *Acta Phys. Pol.* **10** 37–68

——— 1951 The lowering of the freezing point in the theory of dipolar coupling *Acta Phys. Pol.* **11** 99–108

Piekara A and Kielich S 1958a A non-linear theory of the electric permittivity and refractivity of dielectric liquids in electric and magnetic fields *Acta Phys. Pol.* **18** 109–38

——— 1958b Theory of orientational effects and related phenomena in dielectric liquids *J. Chem. Phys.* **29** 1297–305

Scheider W 1965 Dielectric relaxation of molecules with fluctuating dipole moment *Biophys. J.* **5** 617–28

——— 1970 On models of dielectric relaxation due to steady state chemical processes *J. Phys. Chem.* **74** 4296–8

Schwarz G 1967 On dielectric relaxation due to chemical rate processes *J. Phys. Chem.* **71** 4021–30

——— 1970 Acid-base catalysis of dielectric relaxation of zwitterions *J. Phys. Chem.* **74** 654–8

Schwarz G and Seelig J 1968 Kinetic properties and the electric field effect of the helix-coil transition of poly-γ-benzyl-L-glutamate determined from dielectric relaxation measurements *Biopolymers* **6** 1263–77

South G P and Grant E H 1973 The contribution of proton fluctuation to dielectric relaxation in protein solutions *Biopolymers* **12** 1937–44

Wada A 1971 Dielectric evidence of chemical relaxation in the helix-coil transition of polypeptides *Chem. Phys. Lett.* **8** 211–13

Wien M 1928 Uber die Abweichungen der Elektrolyte vom Ohmischen Gesetz *Phys. Z.* **29** 751–5

——— 1931 Uber Leitfahigkeit und Dielektrizitat Konstante von Elektrolyten bei Hochfrequenz *Phys. Z.* **32** 545–7

10 Electrical Properties of Biological Membranes

10.1 Introduction

Previous chapters are devoted, almost exclusively, to the dielectric properties of biopolymers and water. It should be pointed out that the materials we dealt with in those chapters were mostly homogeneous except for in Chapter 6 where the theories of heterogeneous systems were discussed. We will discuss, in this chapter, the impedance characteristics of the membranes of nerve and muscle fibres. These topics were not included in Chapter 6. In order to facilitate an understanding of the impedance problems of biological membranes, some preliminary remarks on their structure and transport properties will be presented first.

10.2 The Structure of Biological Membranes

Biological membranes are made up of lipids and proteins, disregarding other components such as polysaccharides. Of these, the lipid content accounts for 25–70% of the total membrane material by weight. The most common lipids to be found in biological membranes are listed below.

(i) Phospholipids. These lipids consist of a polar head and long non-polar tail(s). Some of the most frequently found phospholipids are as follows.

(a) Phosphatidyl choline (lecithin). The structure of this lipid is shown below. This molecule consists of a dipolar head group $N^+ \cdots PO^-$ and non-polar tails R_1 and R_2.

$$
\begin{array}{l}
\overset{\displaystyle O}{\underset{\displaystyle |}{|}} \\
CH_2O -- P -- O -- CH_2 -- CH_2 -- N^+(CH_3)_3 \\
|\underset{\displaystyle O^-}{|} \\
CHO - CO - R_2 \\
| \\
CH_2O - CO - R_1
\end{array}
$$

(b) Phosphatidyl serine. Only the amide head group is shown below:

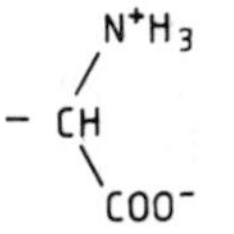

$$
\begin{array}{c}
N^+H_3 \\
/ \\
- CH \\
\backslash \\
COO^-
\end{array}
$$

(c) Phosphatidyl ethanol amine, $-CH_2-N^+H_3$.
In addition, phosphatidyl inositol, diphosphatidyl glycerol and sphingo-myelin have been found in the membranes of biological cells.

(ii) Cholesterol. The structure of cholesterol is entirely different from those of phospholipids, as shown in figure 10.1.

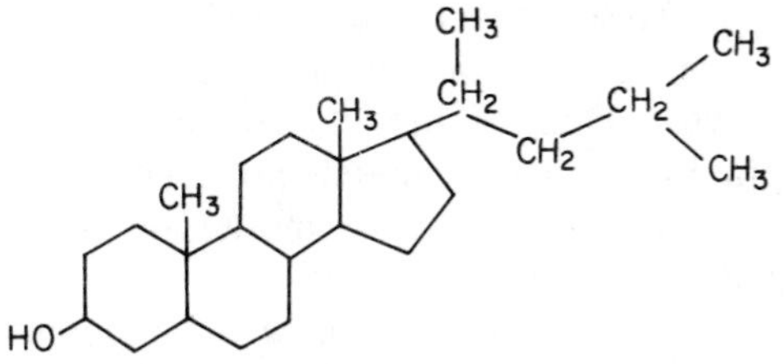

Figure 10.1 The structure of cholesterol.

(iii) Membrane proteins. There are two types of membrane protein.
(a) Extrinsic proteins. These are partially immersed in the lipid moiety. They can be extracted from the membrane using only mild detergents.
(b) Intrinsic proteins. These are tightly bound to lipids and are an integral part of the membrane. Often, both ends of these proteins protrude and make direct contact with water molecules in the suspending media and cell interior. Intrinsic proteins cannot be extracted without the use of strong detergents and without extensive disruption of the membrane. It is intuitively obvious that the surface of a membrane protein consists of hydrophobic amino acids, except for the portion which is exposed to aqueous media.

10.3 Membrane Structure and Models

As mentioned, phospholipids consist of a polar head and non-polar tails. When they are placed in a polar environment such as an aqueous medium, they form two entirely different structures. (i) Micelles. As is well known, polar heads are exposed to aqueous media and non-polar tails form the core of spherical micelles. Clearly, this is an energetically stable form because polar groups associate themselves with polar water molecules and non-polar tails make contact with non-polar groups of other lipid molecules. (ii) Another stable structure is a bilayer leaflet configuration. The bilayer of phospholipids is the basic structure of biological membranes. Lipid molecules can form flat bilayer membranes as well as spherical vesicles. The bending of a bilayer plane to form vesicles may appear to introduce instability due to mechanical strain. However, spherical vesicles are usually quite stable. In biological membranes, cholesterol molecules fit in the space between phospholipid molecules. The presence of cholesterol molecules seems to enhance the stability of bilayer membranes.

Several models have been proposed for biological membranes. The most important question is where to put protein molecules in order to make a plausible membrane model. Diagram (a) in figure 10.2 is the model proposed by Davson and Danielli (1943). As shown, proteins are placed over lipid layers. The model proposed by Robertson (1964) is similar to the Davson–Danielli model, except for the introduction of asymmetry (diagram (b)). These models are now superseded by a more recently proposed mosaic membrane model by Singer and Nicolson (1972) (diagram (c)). According to this model, protein molecules are partially and/or totally embedded in a lipid moiety. This model, although the detail is still to be elucidated for each case, seems to be sufficient as a working model for biological membrane.

The thickness of artificial lipid bilayer membranes ranges from 30 to 60 Å, depending on the length of the non-polar tails of the phospholipids and also on the conformation of these tails. The electrical capacity of lipid bilayer membranes is calculated by the following formula:

$$C_m = \varepsilon_m \varepsilon_0 A / d \tag{10.1}$$

where d and A are the thickness and area of the membrane respectively. ε_m is the dielectric constant of the membrane and ε_0 is the capacitance of free space (8.86×10^{-14} F cm^{-1}). If the thickness d is assumed to be 60 Å, and the relative permittivity is 4.0, then the membrane capacity is found to be 0.6 μF cm^{-2}. Usually, the capacity of biological membrane is known to be about 0.8–1.0 μF cm^{-2}. This finding indicates that the dielectric constant of natural membrane is somewhat larger than that of lipid membrane because of the presence of dipolar protein molecules.

10.4 The Nernst–Planck Equation

A striking observation with biological membrane is the large potential difference between external and internal surfaces. The membrane potential ranges from -20 mV to a value as large as -90 mV, depending on the type of biological cells. For example, the potential difference of

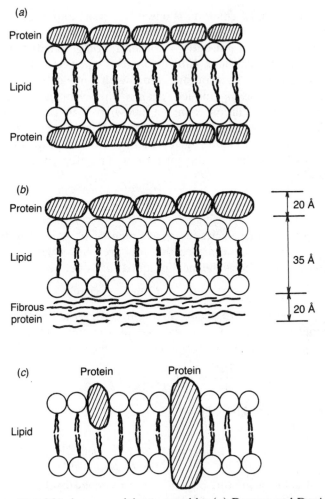

Figure 10.2 Membrane models proposed by (*a*) Davson and Danielli (1943), (*b*) Robertson (1964) and (*c*) Singer and Nicolson (1972).

nerve membrane is about -60 mV. If the membrane thickness d is assumed to be 80 Å, then the field strength in the membrane is as high as 75 kV cm^{-1}. The non-linear dielectric properties of biological materials were discussed in Chapter 9. It was found that a field of 50–100 kV cm^{-1} was sufficient to produce non-linear electric polarization for biological macromolecules. Many biological processes take place in membranes, that is, in the presence of a strong electrical field. Therefore, membrane processes including ion fluxes are often highly non-linear. It will be discussed later how these non-linear ionic processes affect the impedance characteristics of biological membrane.

The Nernst–Planck equation is one of the most important equations in membrane biophysics. In essence, this equation states that the flux of ionic species is driven not only by concentration gradients but also by an electrical potential gradient. Conversely, this equation indicates that ions may flow against concentration gradients if an electrical potential of sufficient magnitude is applied against the chemical potential. The derivation of this equation is shown below. If C is the concentration of the diffusing ion per unit volume then the flux J is defined by $J = Cv$ where v is the velocity and is given by $v = -u(\partial\mu/\partial x)$. μ is the chemical potential and u is the mobility of an ion. Therefore, flux is defined by the following equation:

$$J = -uC\frac{\partial\mu^*}{\partial x}.\tag{10.2}$$

In this equation, the electrochemical potential μ^* is used instead of the simple chemical potential because we are discussing the flux of a charged ion. The electrochemical potential μ^* is defined by

$$\mu^* = \mu_0 + RT\ln C + zF\phi\tag{10.3}$$

where μ_0 is the standard chemical potential, ϕ is the electrical potential, C is concentration, z is the valence of the ion and F is Faraday's constant. Differentiation of this equation gives

$$\frac{\partial\mu^*}{\partial x} = RT\frac{1}{C}\frac{\partial C}{\partial x} + zF\frac{\partial\phi}{\partial x}.\tag{10.4}$$

Substituting this expression into (10.2), we obtain the Nernst–Planck flux equation:

$$J = -uC\left(RT\frac{\partial\ln C}{\partial x} + zF\frac{\partial\phi}{\partial x}\right).\tag{10.5}$$

A general solution of this equation was obtained by Planck. However, the solution is complex and this makes the use of this solution for biological membrane quite cumbersome.

10.5 The Goldman–Hodgkin–Katz Equation

Another solution which was obtained by Goldman (1943), although a controversial constant field assumption was used, is simpler than Planck's solution and, with an extension by Hodgkin and Katz (1949), has often been used for the analysis of the membrane potential of nerve fibres. Goldman's solution starts by replacing the potential gradient with a simple potential difference such as

$$\frac{\partial\phi}{\partial x} = \frac{V}{d}\tag{10.6}$$

where V is the potential difference between external and internal surfaces and d is the membrane thickness. With this replacement, (10.5) reduces to a simple first-order normal differential equation (for $1:1$ electrolytes):

$$-J = uRT \frac{dC}{dx} + \frac{uVFC}{d}.$$ (10.7)

The solution of this equation for C can be transformed, using proper boundary conditions, to the following expression:

$$-J = \frac{C_0 \exp(-FV/RT) - C_d}{1 - \exp(-FV/RT)} \frac{FuV}{d}$$ (10.8)

where C_0 and C_d are the concentrations of the ion at the internal and external surfaces of the membrane respectively. The flux equations J_{Na}, J_K and J_{Cl} can be formulated separately for Na, K and Cl ions. When ion fluxes reach the stationary state, the net ion flux reduces to zero:

$$J_{Na} + J_K + J_{Cl} = 0.$$ (10.9)

Substituting the equations for J_{Na}, J_K and J_{Cl} from (10.8), we obtain the Goldman–Hodgkin–Katz equation:

$$V = \frac{RT}{F} \ln \frac{P_K(C_K)_0 + P_{Na}(C_{Na})_0 + P_{Cl}(C_{Cl})_d}{P_K(C_K)_d + P_{Na}(C_{Na})_d + P_{Cl}(C_{Cl})_0}$$ (10.10)

where P_K, P_{Na} and P_{Cl} are the permeabilities of K, Na and Cl ions respectively. Usually, the term for chloride flux is ignored in order to make a numerical calculation feasible, thus (10.10) can be rewritten as

$$V = \frac{RT}{F} \ln \frac{(C_K)_0 + b(C_{Na})_0}{(C_K)_d + b(C_{Na})_d}$$ (10.11)

where b is the ratio of Na and K permeabilities, i.e. P_{Na}/P_K. At the resting state, the sodium permeability is much smaller than the potassium permeability and the ratio b is, in general, believed to be about 0.03. Using the concentrations of Na and K in the external and internal solutions, and substituting the numerical values for R, T and F (Faraday's constant) we obtain a membrane potential of -72.8 mV. This is lower than the observed value by about 10 mV. Nevertheless, the agreement between calculated and measured values is reasonably good. On the other hand, during an excitation or action potential, the ratio b becomes much larger—a value of 20 is often used. With this value substituted into (10.11), the membrane potential at the peak of the action potential will be $+56$ mV. The result of these calculations shows that the membrane potential is dictated by the potassium permeability at the stationary state whereas the potential at the peak of excitation is dictated by the sodium permeability.

It is important to point out again that membrane impedance measurements are performed, for most biological cells, in the presence of a large bias potential. Whether these bias potentials would affect the impedance or admittance characteristics of membranes is an important question. In view of the discussions presented in the last chapter, the large field existing in biological membranes (80–100 $kV\,cm^{-1}$) should be sufficient to produce a non-linear electric polarization. In the next section, a discussion on the non-linear transport of Na and K ions will be presented.

10.6 Ion Fluxes and Membrane Depolarization

The study of ion fluxes or currents has been performed using an electronic device which is commonly called a voltage clamp circuit. In general, in order to measure ion currents, square pulses are applied across a membrane and the resultant ion currents recorded. Let us use a simple passive membrane as a model for the analysis of currents, as illustrated in figure 10.3(a). The inset in this figure shows a simple membrane model with a parallel combination of a capacitor and a resistor. The current which flows through this membrane patch can be divided into two parts: a displacement current through the capacitor, $C_m(dV_m/dt)$, and a conduction current through the resistor, V_m/R_m, where C_m and R_m are the membrane capacitance and resistance respectively. Therefore, the total membrane current is the sum of these two components:

$$I = C_m \frac{dV_m}{dt} + \frac{V_m}{R_m}. \tag{10.12}$$

The current is schematically illustrated in figure 10.3(a). One of the complications associated with the measurement of membrane currents is the presence of conductive fluids between the membrane and electrodes. This is equivalent to having a resistance in series with membrane admittance, as shown in figure 10.3(b) (see the inset). Because of the presence of this resistance, the potential across the membrane is smaller than the applied potential V_c. The time course of the increase in membrane potential can be found using (10.13), the solution of (10.12):

$$V_m = \frac{V_c R_m}{R_m + R_0} [1 - \exp(-t/\tau)] \tag{10.13}$$

where τ is the time constant which is defined by

$$\tau = C_m R_m R_0/(R_m + R_0). \tag{10.14}$$

Equation (10.13) indicates that the membrane potential builds up slowly

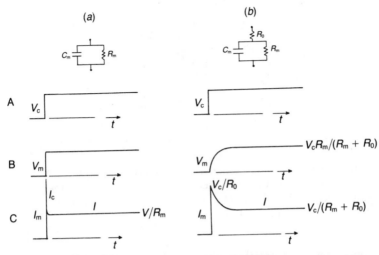

Figure 10.3 The response of a passive membrane to a step voltage V_0 applied externally. (A) Applied voltage, (B) change in membrane potential and (C) current. (a) A simple parallel RC circuit. (b) A series resistance is added to the circuit in (a).

and reaches asymptotically a limiting value $R_m V_c/(R_m + R_0)$. Likewise, the membrane current is given by the following equation:

$$I_m = \frac{V_c}{R_m + R_0}\left(1 + \frac{R_m}{R_0}\exp(-t/\tau)\right). \qquad (10.15)$$

Unlike the membrane potential, the current increases to a value V_c/R_0 at time zero and then decreases to $V_c/(R_m + R_0)$ exponentially. These are the responses of a membrane to a step voltage input if its impedance is Ohmic. A comparison of this diagram with the previous one demonstrates the effect of adding a resistor R_0 in series to membrane admittance. It will be shown later that the presence of a series resistance is one of the major sources of error for the measurement of membrane admittance, even if the value of R_0 is small.

The response of excitable membranes such as nerves to external pulses is entirely different from those illustrated in figure 10.3. When a depolarizing pulse is applied (a decreasing membrane potential) and if its magnitude exceeds 20 mV, an action potential will be generated. An action potential is an all-or-nothing signal and we cannot arrest its growth at an intermediate potential. In other words, we cannot control the membrane potential once an action potential starts developing. In order to circumvent this difficulty, Cole (1949) and Marmont (1949) developed voltage clamp techniques which are basically a negative feedback circuit, as shown by figure 10.4. In this figure, the symbol A is

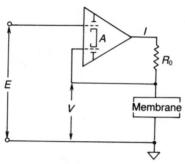

Figure 10.4 A schematic diagram of a voltage clamp circuit. A is the gain of the differential amplifier, I is current and R_0 is the series resistance. (Taken from Moore and Cole 1963.) Reproduced by permission of Academic Press, Inc.

the gain of the differential amplifier and R_0 is a series resistance due to the presence of electrolyte between the membrane and electrodes. Using the principle of negative feedback circuits, we can derive the following equation (see Moore and Cole (1963) for detail):

$$V_m = \frac{AV_c}{A+1} - \frac{R_0 I}{A+1} \tag{10.16}$$

where I is current and V_c is the applied potential. Obviously, if the gain of the differential amplifier A is very large, the second term can be ignored. Also, under this condition, the coefficient of the first term reduces to unity. Therefore, the change in membrane potential will closely follow that of applied potential. Using this circuit, we can keep the membrane potential at a desired level for the duration of the applied pulses and prevent the action potential from developing. Thus, we are able to measure ionic currents across the membrane by shifting the membrane potential from the stationary state to any desired level. Although various forms of potential change can be used, the step function is the most commonly used wave form. Using this technique, the flux of ions in nerve membrane has been investigated mostly using giant axons from squid. A typical result is displayed schematically in figure 10.5. The rather unusual current diagram shown in this figure is explained by assuming the presence of two components. One is the early inward current and the other is the late outward current. It was found that the inward current is due to Na ion flux and the outward current due to the K ion (Keynes 1951). Figure 10.5 shows the current–voltage diagram of ionic current in the nerve membrane. Curve I_p was obtained at an early stage when the inward current is a maximum and curve I_{ss} was obtained when the current had reached a steady state. Thus, curve I_p represents mostly Na current and curve I_{ss} represents slow K current.

As shown by this curve, the increase of K current is not a simple exponential function. The shape of the Na current is even more complex (curve C). Namely, the Na current reaches a peak value and then begins to diminish. This observation was interpreted as the presence of some mechanism which turns off Na current after the peak value has been reached. As shown by figure 10.5 inset (b), the I–V diagrams of nerve axon are highly non-linear. Hence, in order to explain membrane currents of nerve axon, we need a mathematical theory which is based on a non-linear scheme. This theory was developed by Hodgkin and Huxley (1952), as is discussed below.

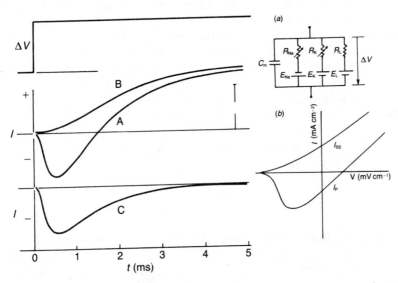

Figure 10.5 A current–time diagram of nerve axon under the voltage clamp condition. The positive current is outward and the negative sign indicates an inward direction. Curves A, B and C are the measured total current, the potassium current and the sodium current respectively. Inset (b) shows the I–V diagram. Curve I_p is the peak current and curve I_{ss} is the steady state current. Inset (a) is the equivalent circuit of the Hodgkin–Huxley equation. V_{Na}, V_K and V_L are the Nernst potentials of the Na, K and leakage channels respectively. G_{Na}, G_K and G_L are the Na, K and leakage conductances respectively. C_m is the membrane capacitance.

Each current component can be expressed individually by the following equations:

$$I_{Na} = G_{Na}(V - V_{Na}) \tag{10.17}$$

$$I_K = G_K(V - V_K) \tag{10.18}$$

where V_{Na} and V_K are 'Nernst potentials' for Na and K and are defined by

$$V_{Na} = \frac{RT}{F} \ln \frac{(C_0)_{Na}}{(C_i)_{Na}} \tag{10.19}$$

$$V_K = \frac{RT}{F} \ln \frac{(C_0)_K}{(C_i)_K} \tag{10.20}$$

where $(C_0)_{Na}$, $(C_0)_K$, $(C_i)_{Na}$ and $(C_i)_K$ are the external and internal sodium and potassium concentrations respectively. It should be noted that these expressions can be derived easily from (10.10) assuming either P_{Na} or P_K are negligibly small. These potentials arise from the uneven distribution of ions on both sides of the membrane. G_{Na} and G_K are the conductances of Na and K ions through respective 'channels'. Since the rise of ionic current is a function of time, we have to introduce time dependent parameters, i.e. m and n. Using these, ionic conductances can be defined as $\bar{G}_{Na}m$ and $\bar{G}_K n$ respectively where $\bar{G}_{Na}$ and $\bar{G}_K$ are the time independent maximum sodium and potassium conductances. The values of these new parameters m and n change as time t increases. Assuming the onset of these currents as being due to the opening of channels, the entire process can be considered as a first-order reversible kinetic process such as

$$A \underset{\beta}{\overset{\alpha}{\rightleftharpoons}} B$$

where α and β are rate constants. The variables A and B can be treated as probability functions. If we assume that m represents the probability of channels being open, then $1 - m$ represents the probability of channels being closed. Therefore, the rate of channel opening can be given by the following equation:

$$\frac{dm}{dt} = \alpha_m(1 - m) - \beta_m m \tag{10.21}$$

where α_m and β_m are the rate constants for sodium channels. This notation conforms to that used by Hodgkin and co-workers. The solution of this equation is

$$m = (m_0 - m_\infty)\exp(-t/\tau_m) + m_\infty \tag{10.22}$$

where m_0 is the initial value of m at $t = 0$. τ_m is the relaxation time and is defined by

$$\tau_m = 1/(\alpha_m + \beta_m) \tag{10.23}$$

and m_∞ is given by

$$m_\infty = \alpha_m/(\alpha_m + \beta_m). \tag{10.24}$$

Using these equations, the values of the parameters m and n can be calculated as a function of time. The rate constants α_m and β_m are voltage dependent and a set of empirical equations are available for the numerical calculation. (See Hodgkin and Huxley 1952.) In order to explain the inactivation of the sodium current, Hodgkin and Huxley introduced a parameter 'h' which, unlike m and n, decreases with time. If the parameter m, which increases with time, is multiplied by h, the product will go through a maximum value before it begins to decrease. Thus, we can simulate the inactivation of sodium current using a simple scheme. Furthermore, for the purpose of curve fitting, m and n are raised to third and fourth power, therefore, sodium and potassium currents become

$$I_{Na} = G_{Na}m^3h(V - V_{Na}) \tag{10.25}$$

$$I_K = G_K n^4(V - V_K). \tag{10.26}$$

The final form of the Hodgkin–Huxley equation is written as follows:

$$I = C_m \frac{dV}{dt} + [\bar{G}_{Na}m^3h(V - V_{Na}) + \bar{G}_K n^4(V - V_K) + G_L(V - V_L)].$$
$$\tag{10.27}$$

The last term on the RHS of this equation is added to account for the Ohmic current due to leakage. The magnitude of this term is ordinarily small unless the nerve membrane is damaged. The equivalent circuit for (10.27) is shown in figure 10.5. It is well known that the result of the numerical calculation of ionic currents under a voltage clamp condition is remarkably similar to the curve obtained experimentally. In addition, the Hodgkin–Huxley equation, which is highly non-linear, can still be solved numerically for a membrane potential V and the action potential can be simulated with striking success.

The Hodgkin–Huxley equation has been one of the major theories which have played a central role in the development of nerve physiology. Without this theory, the recent progress of neuro-physiological research would have been much slower. Although, as pointed out by many people, the Hodgkin–Huxley equation is a phenomenological theory and does not provide information with regard to the molecular mechanism, still this theory gives us an insight as to the basic scheme of nerve excitation. The central interest of excitable membrane research at present is what proteins constitute ionic channels and what forces open and close them. Recent biochemical research centres around the study of channel proteins and their chemical composition. Although these investigations are extremely tedious and time consuming, we are beginning to gain some information as to the chemical structure of channel proteins.

10.7 The Transformation of the Hodgkin–Huxley Equation

As mentioned earlier, the impedance measurement of excitable cells such as nerve and muscle must be done often in the presence of ionic currents. These currents are time variant and voltage dependent, i.e. non-linear. The question of whether or not these currents give rise to reactive as well as resistive components will be discussed in this section.

Let us begin this analysis with the Laplace transform of the Hodgkin–Huxley equation (Chandler *et al* 1962). Since the Hodgkin–Huxley equation is non-linear, we must linearize this equation before we apply the Laplace transformation. The linearized Hodgkin–Huxley equation for a small input voltage δV is shown below:

$$\delta I = C_m \frac{d\delta V}{dt} + \bar{g}_K n_\infty^4 \delta V + \bar{g}_{Na} m_\infty^3 h_\infty \delta V + \bar{g}_L \delta V$$

$$+ 4\bar{g}_K n_\infty^3 (V - V_{Na})\delta n + 3\bar{g}_{Na} m_\infty^2 h_\infty (V - V_{Na})\delta m$$

$$+ \bar{g}_{Na} m_\infty^3 (V - V_{Na})\delta h$$

$$(10.28)$$

where the parameter m_∞ is defined by (10.24). n_∞ and h_∞ are also defined by analogous equations. Using (10.23), (10.21) can be written as follows:

$$dm/dt = (m_\infty - m)/\tau_m. \qquad (10.29)$$

Using the Laplace transform variable p for d/dt we can rewrite this equation as

$$m = m_\infty/(1 + p\tau_m). \qquad (10.30)$$

The increment δm due to a small increase in voltage δV is given as

$$\delta m = \frac{1}{1 + p\tau_m} \frac{\partial m}{\partial V} \delta V. \qquad (10.31)$$

Similar equations can be derived for the parameters n and h. Using these in (10.28), and using the definition $g(p) = \delta I/\delta V$, we obtain the following expression:

$$g(p) = pC + \bar{g}_K n_\infty^4 + 4\bar{g}_K n_\infty^3 (V - V_K) \frac{\partial n/\partial V}{1 + p\tau_n} + \bar{g}_{Na} m_\infty^3 h_\infty$$

$$+ \bar{g}_{Na} m_\infty^2 h_\infty (V - V_{Na}) \frac{3\partial m/\partial V}{1 + p\tau_m}$$

$$+ \bar{g}_{Na} m_\infty^3 (V - V_{Na}) \frac{\partial h/\partial V}{1 + p\tau_h}. \qquad (10.32)$$

This equation can be simplified to the following form:

$$g(p) = pC + g_{\times} + \frac{g_n}{1 + p\tau_n} + \frac{g_m}{1 + p\tau_m} + \frac{g_h}{1 + p\tau_h} \qquad (10.33)$$

where

$$g_{\times} = \bar{g}_K n_{\times}^4 + \bar{g}_{Na} m_{\times}^3 + \bar{g}_L \qquad (10.34)$$

$$g_n = 4\bar{g}_K n_{\times}^3 (\partial n/\partial V)(V - V_K) \qquad (10.35)$$

$$g_h = \bar{g}_{Na} m_{\times}^3 h_{\times} (\partial h/\partial V)(V - V_{Na}) \qquad (10.36)$$

$$g_m = 3\bar{g}_{Na} m_{\times}^2 h_{\times} (\partial m/\partial V)(V - V_{Na}). \qquad (10.37)$$

Equation (10.33) is commonly called the linearized Hodgkin–Huxley equation (the LHH equation). For sinusoidal inputs, $p = j\omega$ and each term on the RHS represents the anomalous dispersion due to the time dependent parameters m, n and h. Equation (10.33) indicates that the admittance $g(p)$ consists of three parallel branches, as shown by figure 10.6 for the normal range of potential $V_K < V < V_{Na}$. Diagram (a) in figure 10.6 can be condensed to diagram (b) by combining chord conductances from all the conduction processes. These analyses indicate that time dependent conduction processes give rise to frequency dependent reactive components. As shown in figure 10.6, Na current produces a capacitive reactance and the K current is inductive. In this treatment, the membrane capacitance C and the chord conductance $g_{\times}$ are assumed to be fixed quantities.

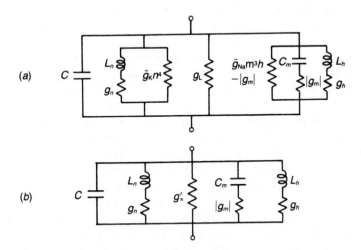

Figure 10.6 The equivalent circuit of the linearized Hodgkin–Huxley equation. Note that the Na current is predominantly capacitive and the K current is inductive. Diagram (b) is a simplified form of (a), by combining chord conductances. (From Fishman et al 1977b.)

10.8 The Capacitance of the Lipid Bilayer Membrane

Enormous amounts of information are available on the structure of lipid bilayer membranes and their transport properties. It is virtually impossible to review all this work in the limited space of this chapter. Our aim is to select a small number of papers and to review the basic problems associated with the electrical impedances of these membranes. The impedance characteristics of vesicular bilayers will not be discussed in this chapter.

Although attempts to form artificial lipid bilayer membranes go back to the late 1930s (Langmuir and Waugh 1938), serious research using lipid bilayer membranes was begun by Mueller *et al* (1962, 1963, 1964). At the beginning, lipid mixtures extracted from the brain were used. However, it became possible to use purified phospholipids and/or cholesterol derivatives to form stable membranes having a bilayer leaflet configuration.

Early measurements established a few basic features of the impedance characteristics of these membranes. Extensive measurements by Hanai *et al* (1964) found the value of the capacitance of lecithin membranes to be in the range $0.385-0.405 \ \mu F \, cm^{-2}$, as shown by table 10.1. The thickness of the membrane was calculated using the following equation:

$$C_m = \varepsilon_m \varepsilon_0 (A_1/d_1 + A_2/d_2 + \ldots + A_n/d_n) \qquad (10.38)$$

where the higher-order terms represent parts of the membrane, including the torus region near the edge of the membrane, which are thicker than the bilayer. Under normal circumstances, these terms can all be neglected and (10.38) reduces to (10.1). As shown in table 10.1, the capacitance of lipid bilayers is much smaller than those of biological membranes, i.e. approximately $0.8-1.0 \ \mu F \, cm^{-2}$. This may be partly due to the presence of hydrocarbon solvents in artificial membranes. Decane and hexadecane which are often used as the solvent have a very small dielectric constant. Although solvent molecules escape from the membrane in time, nevertheless, considerable amounts of solvent molecules remain in the membrane and increase the thickness.

White and Thompson (1973) investigated the capacitance of purified lecithin bilayer films with great precision. As mentioned before, the presence of microlenses of solvent molecules increases the thickness of the membranes and this in turn decreases the membrane capacitance. White and Thompson analyzed the error caused by the presence of solvent molecules. The membrane capacitance C_m is defined, in the presence of solvent molecules, by the following equation:

$$C_m = \frac{\varepsilon_m \varepsilon_0}{d_B} (1 - nA_1) + \frac{n \varepsilon_s \varepsilon_0 A_1}{d_1} \qquad (10.39)$$

where ε_m and d_B are the dielectric constant and thickness of the bilayer portion of the film respectively, n is the number of microlenses, and A_1 and d_1 are the average area and thickness of the microlenses respectively. ε_s is the dielectric constant of the solvent.

Table 10.1 The capacitance and thickness of lipid bilayer membrane.†

Aqueous phase	Reduced capacity, C_m (nF)	Capacity per unit area (μF cm^{-2})	Thickness of membrane‡ (Å)
10^{-1}N NaCl	6.086 ± 0.217	0.385	47.2
10^{-1}N KCl	5.900 ± 0.178	0.373	48.7
10^{-1}N CaCl$_2$	5.994 ± 0.495	0.379	47.9
10^{-3}N NaCl	6.084 ± 0.059	0.385	47.2
10^{-3}N CaCl$_2$	6.258 ± 0.059	0.395	46.6
4.18N NaCl	6.407 ± 0.230	0.405	45.5

†Taken from Hanai *et al* (1964).
‡The thickness of the non-polar part of the membrane excluding polar heads.

One of the significant conclusions reached by White and Thompson is that lipid bilayer films are not homogeneous equilibrium structures. The membrane area and/or thickness change continually yet slowly with time and the capacitance does not seem to reach an equilibrium state. A similar instability is found among biological membranes. However, the drift of electrical parameters is found mainly in membrane conductance. The capacitance of membranes usually remains constant for a long time even under *in vitro* conditions.

10.9 The Frequency Dependence of the Membrane Capacitance of a Bilayer Film

The effect of series resistance on time domain current measurements has been discussed previously. Using the circuit shown in figure 10.3(b), the effect of R_0 on membrane capacitance and conductance measurements will be discussed in this section. The admittance of a membrane including a series resistance is given by

$$Y = 1/(Z_0 + Z_m) \qquad (10.40)$$

where $Z_0 = R_0$ and Z_m is given by the following equation:

$$Z_m = 1/Y_m = \frac{1}{1/R_m + j\omega C_m}. \qquad (10.41)$$

Substituting this expression into (10.40), we obtain

$$Y = \frac{1 + j\omega C_m R_m}{(R_0 + R_m) + j\omega C_m R_m R_0}. \tag{10.42}$$

This equation can be rationalized as follows:

$$Y = \frac{[1/(R_m + R_0) + \omega^2\tau^2/R_0] + j\omega C_m R_m^2/(R_m + R_0)^2}{1 + (\omega\tau)^2} \tag{10.43}$$

where τ, the time constant, is given by

$$\tau = C_m R_m R_0/(R_m + R_0). \tag{10.44}$$

Since $Y = G_p + j\omega C_p$, we obtain, by equating the real and imaginary terms of both sides, two separate equations for C_p and G_p as shown below:

$$C_p = \frac{C_m R_m^2/(R_m + R_0)^2}{1 + (\omega\tau)^2} \tag{10.45}$$

$$G_p = \frac{1/(R_m + R_0) + \omega^2\tau^2/R_0}{1 + (\omega\tau)^2}. \tag{10.46}$$

According to these equations, the measured capacitance C_p and conductance G_p are frequency dependent quantities, even if the membrane capacitance C_m and conductance G_m are independent of frequency. If the frequency of the AC signals is very high, then (10.45) and (10.46) reduce to

$$C_p = 0 \qquad \text{and} \qquad G_p = 1/R_0. \tag{10.47}$$

On the other hand, at zero frequency, these reduce to the following expressions:

$$C_p = C_m/(1 + R_0/R_m)^2 \tag{10.48}$$

$$G_p = 1/(R_m + R_0). \tag{10.49}$$

Using these equations, we can estimate the error caused by the presence of R_0 on the magnitude of the membrane capacitance C_m. For example, if $R_0 = 10\ \Omega\ \text{cm}^2$ and $R_m = 1000\ \Omega\ \text{cm}^2$, then $C_p = 0.98 C_m$. Therefore, the error is only 2%. However, if the ratio R_0/R_m is 0.1, then the measured capacitance C_p becomes $0.82 C_m$. Thus, the error amounts to as much as 18%. The resistance of artificial lipid bilayers is extremely large and the presence of a series resistance does not cause serious errors. However, the presence of R_0 gives rise to an apparent frequency dependence in the measured capacitance C_p. For excitable membranes such as nerve and muscle fibres, the ratio R_0/R_m is about 0.01 at the stationary state. However, the membrane resistivity R_m decreases markedly during excitation. This, in turn, increases R_0/R_m significantly.

Therefore, the presence of R_0 can cause serious errors in the interpretation of membrane capacitance behaviour during the depolarization of nerves and muscles.

The first investigation of frequency dependent membrane capacitance was performed by Schwan *et al* (1966) and later by Takashima and Schwan (1974b). These two investigations indicate that the capacitance of lipid bilayer films is independent of frequency above 100 Hz. Coster and Smith (1974), however, found the capacitance of lipid films to be frequency dependent between 100 mHz and 1 kHz. As shown in figure 10.7, the capacitance and conductance show a small dispersion between 2 and 70 Hz. We can prove, using (10.45) and (10.46) that this frequency dependence is not due to a series resistance. Coster and Smith used the Maxwell–Wagner equations to explain the frequency

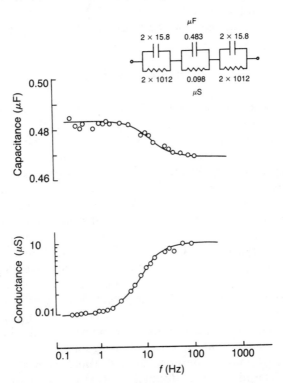

Figure 10.7 The frequency dependence of the membrane capacitance and conductance of lecithin film at ultra-low frequencies. The inset illustrates the model circuit used to obtain the full curves. The two admittances at both ends represent the polar heads of the phospholipid and the one at the centre is for the non-polar hydrocarbon tail(s). (From Coster and Smith 1974.) Reproduced by permission of Elsevier Science Publishers BV.

dependent capacitance and conductance. The model used for this calculation is shown as an inset in figure 10.7. Note that this model consists of three layers of dielectric slabs. The Maxwell–Wagner theory for two layers was discussed in Chapter 6 and the theory used by Coster and Smith is an extension of this basic equation to three layers:

$$C = \frac{\omega^2 2C_p C_H (2C_p + C_H) + C_H 4G_p^2 + 2C_p G_H^2}{(2G_p + G_H)^2 + \omega^2 (2C_p + C_H)^2} \tag{10.50}$$

$$G = \frac{2G_p G_H (2G_p + G_H) + \omega^2 (2G_p C_H^2 + 4C_p^2 G_H)}{(2G_p + G_H)^2 + \omega^2 (2C_p + C_H)^2} \tag{10.51}$$

where C_p and G_p are the capacitance and conductance of the polar region while C_H and G_H are those of the non-polar region. The smooth curves shown in figure 10.7 were calculated by substituting the numbers indicated in the inset into (10.50) and (10.51). Note that an exceedingly large value is used for the capacitance of a polar head layer.

10.10 The Membrane Admittance of Nerve Axon

The membrane admittance of squid giant axon was first investigated by Cole (1941, 1949) and Cole and Curtis (1936, 1939) using a pair of external electrodes. (See Cole (1968) for more references.) These electrodes are placed externally at two sites along or across the axon. These arrangements enable the longitudinal and transverse impedances to be measured non-invasively. However, there are numerous difficulties with external electrode methods. With the development of internal electrode techniques, attempts to measure the membrane impedance of nerve axon using external electrodes have been abandoned. However, using external electrodes, Cole and Curtis established that (i) the impedance of the giant axon of squid consists of resistive, capacitive and inductive components; (ii) the resistance of the nerve membrane changes drastically during the action potential and (iii) the capacitance remains unchanged during the excitation. These findings are still valid and well accepted as a basic description of nerve excitation. Only the last statement on constant capacitance during impulses needs scrutiny.

Early investigations of the impedance of squid axon membrane using an internal electrode were performed by Taylor (1965) and Matsumoto et al (1970). This pioneering work was followed by Takashima and Schwan's attempts (1974b) to reproduce the results of Cole and Curtis using internal electrodes. The electrode configuration used by Takashima and Schwan is illustrated in figure 10.8 (inset). The admittance of the membrane is measured using the internal electrode and an external reference electrode. Conventional low-frequency bridges, or automated impedance analyzers, or a time domain technique, which is described in Appendix F, have been used for the measurement. As discussed before, the correction for a series resistance plays a crucial role in the correct interpretation of the membrane capacitance and its frequency dependence for nerve axon. Figure 10.8 illustrates the measured capacitance (curve A) at various frequencies. The pronounced decrease of capacitance at high frequencies is due to the presence of a series resistance and does not reflect the real admittance characteristics of the nerve membrane. Curve B shows the capacitance after a correction for the series resistance. Clearly, the presence of a series resistance has a marked effect on the frequency profile of the measured capacitance. However, even after this correction, the capacitance still exhibits a small but well defined dispersion. Whether the frequency dependent capacitance is due to the anomalous dispersion of the membrane itself or due to other reasons will be discussed below.

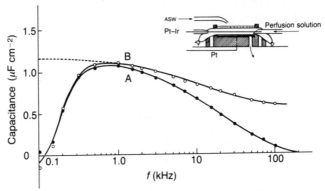

Figure 10.8 The membrane capacitance of the squid giant axon. Curve A shows measured raw data and curve B is obtained after correction for the series resistance. The inset shows the electrode arrangement. The internal axial electrode is a Pt–Ir alloy (9 : 1) with a diameter of 75 μm. The external ground electrode is pure platinum. The electrodes are coated with Pt-black. ASW is artificial sea water. The axons are internally perfused before and during measurements. (From Takashima and Schwan 1974b.) Reproduced by permission of Elsevier Science Publishers BV.

10.11 The Fringe Effect

In Takashima's experiments, a long internal electrode which spans from one end of the axon to the other was used. This is in an attempt to minimize the error due to fringe effects at the tip of the internal electrode. As shown in figure 10.9, if a field is applied using the internal and external ground electrodes, the potential profile will not be uniform near or at the tip of the internal electrode. Some current will flow radially from the tip of wire. Thus, the fringe effects due to stray field will produce a residual capacitance which turns out to be frequency dependent, as explained below. The attenuation of a potential along a nerve fibre is governed by the following equation:

$$V = V_0 \exp(-x/\lambda) \tag{10.52}$$

where V_0 is the potential at the tip of the electrode. λ is the space constant which is defined by the following equation:

$$\lambda^2 = R_m/(R_e + R_i) \tag{10.53}$$

where R_m, R_e and R_i are the resistances of the membrane, the external medium and the internal solution respectively. Hence, if λ is small, the stray field will diminish rapidly, making fringe effects small. For AC fields, (10.53) must be modified as follows (a detailed discussion of this problem will be given later):

$$1/\lambda = \sqrt{(R_e + R_i)(1/R_m + j\omega C_m)}. \tag{10.54}$$

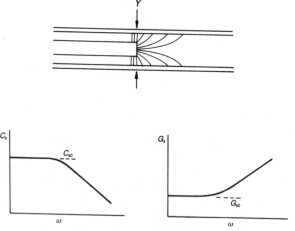

Figure 10.9 The distribution of stray field at the tip of an internal electrode. The extent of stray field depends upon the frequency of the applied field. The insets show the frequency dependence of the fringe capacitance (C_s) and conductance (G_s). $Y = G_s + j\omega C_s = [(G + j\omega C)/R]^{1/2}$. (From Schwan 1981.) Reproduced by permission of Plenum Publishing Corp.

A cursory examination of this equation reveals that the space constant decreases with increasing frequency. Therefore, the attenuation of potential will be more pronounced at higher frequencies than at lower frequencies. This means that the fringe capacitance will be frequency dependent. Thus stray field can account for, at least partially, the observed frequency dependence of measured capacitance. In order to minimize the relative contribution of the fringe effect, the length of the internal electrode was varied between 5 mm and 2.6 cm. The capacitances measured using these electrodes extrapolate to a value of $0.9\ \mu F\ cm^{-2}$ at an infinite electrode length (see figure 10.10). In principle, this value represents the real membrane capacity of squid axon without fringe effects.

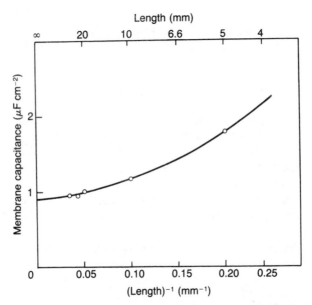

Figure 10.10 The dependence of the capacitance of squid giant axon on the length of the internal electrode. The ordinate is capacitance per unit area. An extrapolation of the curve to infinite electrode length yields, in principle, the true membrane capacitance without the fringe effect. (From Takashima and Yantorno 1977.) Reproduced by permission of the *Annals* of the New York Academy of Sciences.

Schwan (1981) estimated the magnitude of admittance due to the fringe effect using transmission line theory. Using the model shown in figure 10.9, the following equations for fringe conductance and capacitance were derived:

$$G_s^2 = \frac{G}{2R}\left\{1 + \left[1 + \left(\frac{\omega C}{G}\right)^2\right]^{1/2}\right\} \tag{10.55}$$

$$(\omega C_s)^2 = \frac{G}{2R}\left\{-1 + \left[1 + \left(\frac{\omega C}{G}\right)^2\right]^{1/2}\right\} \tag{10.56}$$

where G_s and C_s are the conductance and capacitance due to stray field respectively. C and G are the membrane capacitance and conductance per unit length respectively and R is the resistance of the axoplasm per unit length. The low- and high-frequency limiting values of G_s and C_s are

$$\omega \to 0 \qquad G_{s0}^2 = G/R \qquad \text{and} \qquad C_{s0} = C/2(RG)^{1/2} \tag{10.57}$$

$$\omega \to \infty \qquad G_{s\infty}^2 = \infty \qquad \text{and} \qquad C_{s\infty} = 0. \tag{10.58}$$

C_{s0} can be rewritten in terms of ρ, G_m and C_m (ρ is the specific resistivity of the axoplasm in Ω cm, G_m is in mS cm^{-2} and C_m is in μF cm^{-2}):

$$C_{s0} = \frac{C}{2}\left(\frac{r}{2\rho G_m}\right)^{1/2} \tag{10.59}$$

where r is the radius of the nerve fibre. Using the following values, $G_m = 1$ mS cm^{-2}, $\rho = 20$ Ω cm and $r = 200 \times 10^{-4}$ cm, C_{s0} is found to be 0.35 C. This means that approximately 35% of the total capacitance is due to the fringe effect. This result indicates that the contribution of stray field capacitance is uniformly 30–40% of the total capacitance, regardless of the magnitude of total capacity. It was observed experimentally, however, that the relative contribution of the fringe effect decreases with the increase in length of the internal electrode. The value calculated by Schwan is in agreement with the experimental results, only when the length of the electrode is very small.

10.12 The Inductive Reactance

The sharp decrease of capacitance and an increase in conductance at low frequencies in figure 10.8 are the manifestations of inductive reactance, in agreement with Cole's earlier results (1941). That this inductive reactance is due to potassium current, as had been discussed by Chandler et al (1962), was demonstrated experimentally by Takashima and Schwan (1974a) by the use of a nerve toxin, tetraethyl ammonium (TEA) which selectively blocks K current. The broken portion of curve B at low frequencies shows the measured capacitance after the K current is eliminated by TEA.

Although separate plots of capacitance and conductance against frequency are most commonly used as a representation of the dielectric data; Cole (1968) used an impedance locus. The admittance of a membrane can be transformed to a series circuit as shown by figure 10.11 (inset). This transformation can be represented by the following equation:

$$Z = \frac{1}{Y} = R_0 + \frac{R_m}{1 + (\omega R_m C_m)^2} + j\, \frac{\omega R_m^2 C_m}{1 + (\omega R_m C_m)^2}. \quad (10.60)$$

At high and low frequencies, the impedance Z reduces to R_0 and $R_0 + R_m$ respectively. On the other hand, the reactive component is non-zero only between two extreme frequencies. Figure 10.11 shows the R versus $-x$ plots obtained by Takashima and Schwan (curve A is almost identical to the one obtained by Cole and Curtis much earlier). This result clearly demonstrates that capacitive reactance dominates at high frequencies and the reactance at low frequencies is predominantly inductive. As shown by arc B, inductive reactance is completely eliminated by blocking the K current with TEA.

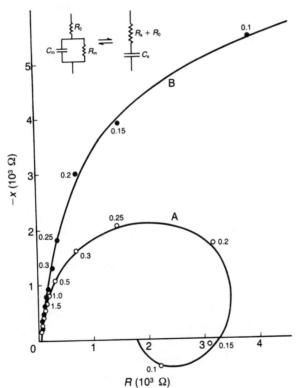

Figure 10.11 Impedance loci of squid axon membrane obtained with the internal electrode technique. Curves: A, without pharmacological agents; B, with TEA. Note the absence of inductive reactance in curve B. (From Takashima and Schwan 1974b.) Reproduced by permission of Elsevier Science Publishers BV.

The inductive reactance was demonstrated, in the above, to be due to inward K current rather than to the unique structure of the axon membrane. Using lipid bilayer films doped with an ionophore, alamethicin, Takashima (1978, unpublished) was able to simulate the inductive reactance which is similar to that observed with nerve membrane. Alamethicin is known to produce, in lipid bilayer films, a time dependent non-linear current which is similar to K current in nerves. This additional observation provides unequivocal evidence that the inductive reactance is due to delayed rectification of the time dependent K current.

Cole (1932) observed that the impedance locus of squid axon membrane is always depressed and the centre of the circular arc is located below the abscissa. If one draws a straight line, which is a tangent to the circle at the high-frequency end, the angle between the abscissa and this tangent is always smaller than $90°$. Cole called this a 'constant phase angle'. Takashima found, however, that the phase angle varied from $45°$ to nearly $90°$ as the length of the internal electrode increased. This result indicates that the small phase angle is contributed to by the fringe effect of the electrode tip. The phase angle of membrane, if the fringe effect is reduced or eliminated by the use of a long electrode, approaches $90°$. This observation indicates that membrane is nearly an ideal dielectric material.

10.13 The Variation of Capacitance with Membrane Potential

As discussed earlier, Cole and Curtis observed that the conductance of squid axon increased drastically during an action potential whereas the capacitance remained constant. Although this observation is considered a fundamental feature of nerve excitation, this conclusion must be re-examined carefully. Let us look at (10.53) once again. The space constant λ depends upon the value for membrane resistance, and the resistances of the external and internal solutions. Of these R_e and R_i are nearly constant and we have good reason to assume that the excitation of membrane has no effect on these quantities. However, the membrane resistance decreases markedly during nerve activity and this, in turn, changes the value of the space constant. Now let us turn to (10.48). Because of the constancy of R_0 ($= R_e + R_i$) and the decrease in R_m, the ratio R_0/R_m increases considerably. This means that the membrane capacitance C_m can change its magnitude, even if the measured capacitance C_p is unchanged. Figure 10.12 shows the increase in membrane capacitance (Takashima 1979) during an action potential. Measured capacitances are corrected for a series resistance, point by point, at each frequency. This result demonstrates that capacitance changes can be overlooked easily if the series resistance correction is ignored. According to this figure, the membrane capacitance of squid giant axon increases by about 25% at the peak of an action potential. Diagram (b) shows the increase in conductance. Whether the change in capacitance is due to the increase in the time variant Na current or due to the change in membrane thickness is still unknown.

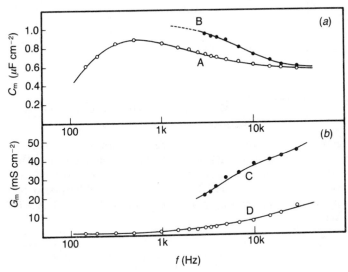

Figure 10.12 (a) The capacitance of squid giant axon in a resting state (curve A) and at the peak of an action potential (curve B). (b) The conductance of squid axon in a resting state (curve C) and at the peak of an action potential (curve D). (From Takashima 1979.) Reproduced from the *Biophysical Journal*, 1979, vol 26, pp 133–42 by copyright permission of the Biophysical Society.

The admittance change of axon membrane with depolarization can also be investigated under voltage or current clamp conditions (Cole and Baker 1941, Takashima and Yantorno 1977). The latter technique is to inject current through a large resistor and to maintain a constant membrane potential for the duration of the current pulses. Using this technique, the membrane admittance can be investigated at various levels of membrane potential with or without pharmacological agents. Figures 10.13 and 10.14 show the change in capacitance and conductance of free axon and axon with tetrodotoxin (TTX) and TEA blocking of both the Na and K currents (Takashima and Yantorno 1977). As shown, the capacitance was found to be voltage dependent only when ionic currents are present. When both Na and K currents are blocked altogether, neither the membrane conductance nor the capacitance is affected by applied pulses. These observations indicate that the non-linear admittance characteristics of nerve membrane arise primarily from

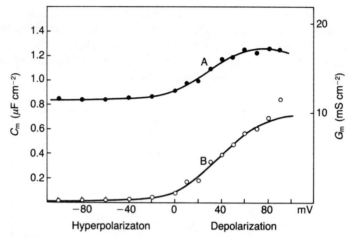

Figure 10.13 The voltage dependence of the capacitance (curve A) and conductance (curve B) of squid axon membrane without pharmacological agents. The abscissa is the applied potential. Measurements were made at 500 Hz. (From Takashima and Yantorno 1977.) Reproduced by permission of the *Annals* of the New York Academy of Sciences.

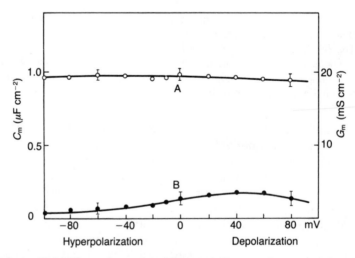

Figure 10.14 The voltage dependence of the capacitance (curve A) and conductance (curve B) of squid axon membrane. Both the Na and K currents are blocked altogether by TTX and TEA. Measurements were made at 500 Hz. (From Takashima and Yantorno 1977.) Reproduced by permission of the *Annals* of the New York Academy of Sciences.

the voltage dependence of ionic currents. Without ionic currents the nerve membrane behaves as a linear system, in spite of a large bias potential across it.

10.14 A Frequency Domain Study of Gating Current

As discussed earlier, the onset of nerve excitation is the opening of ionic channels. Hodgkin and Huxley speculated that the elementary process of channel opening is either the electrophoretic movement of charged particles or the orientation of dipolar particles in the channel. If one applies a square pulse to a dipole and measures the displacement current in the time domain, a wave form which is shown in figure 10.15 will be obtained. This is a typical wave form of the displacement current due to the orientation of dipoles. In order to observe the displacement current in nerve membrane (this is commonly called a gating current), both the Na and K currents which are much larger in magnitude must be reduced or eliminated by pharmacological agents. Since the magnitude of the gating current is very small, the signal-to-noise ratio must be increased using a signal averager. The gating current which was predicted by Hodgkin and Huxley was first observed by Armstrong and Bezanilla (1973) and Keynes and Rojas (1974) using squid giant axon.

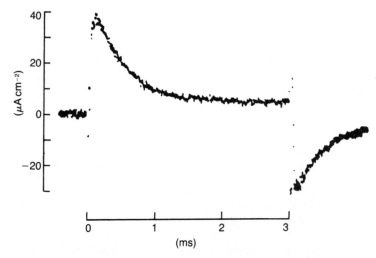

Figure 10.15 A wave form of displacement current which may be related to the opening and/or closing of sodium channels. This wave form is typical of the current due to dipolar orientation. (From Meves 1977.) Reproduced by permission of the *Annals* of the New York Academy of Sciences.

The result obtained by these authors was confirmed by other investigators (see, for example, Meves 1974). The waveform of the gating current illustrated in figure 10.15 is taken from Meves (1977). Analysis of the data led them to a conclusion that the gating current is due to the opening and closing of Na channels rather than K channels. Armstrong and Bezanilla (1975) stated that the magnitude of the gating current is equivalent to an increase of membrane capacitance by as much as 40% in the frequency domain measurement. This amounts to an increase of membrane capacitance from 1 to 1.4 μF cm^{-2}. The frequency domain measurements of gating current were performed by Fishman *et al* (1977a) and by Takashima (1978) independently. None of these measurements confirmed the predicted increase in membrane capacitance. The result obtained by Takashima is reproduced in figure 10.16. The open circles in this figure show the capacitance predicted by Armstrong and Bezanilla. The full circles are measured values at various depolarizations. The disagreement between the predicted and observed

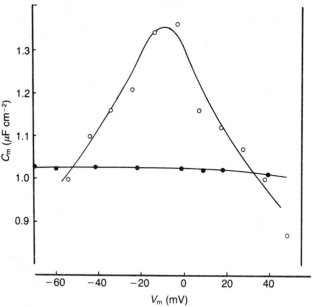

Figure 10.16 The invariance of the membrane capacitance of squid axon with applied potential (full circles). Axons were internally perfused and treated with pronase to abolish sodium inactivation. Ionic currents were blocked using TTX and TEA (From Takashima 1978.) Reproduced from the *Biophysical Journal*, 1978, vol 22, pp115–19 by copyright permission of the Biophysical Society. Open circles are the predicted capacitance increase based on the magnitude of the gating current. (From Armstrong and Bezanilla 1975.) Reproduced by permission of the *Annals* of the New York Academy of Sciences.

curves is a clear indication that there are some missing links between time and frequency domain measurements. This disagreement may be resolved, at least partially, if we assume that the magnitude of the gating current is smaller than the value estimated by Armstrong and Bezanilla and by other investigators.

10.15 A Linear Analysis of Na and K Conduction

The results obtained by Takashima and co-workers indicate that the non-linearity of the admittance of nerve axon arises from non-linear Na and K currents. This observation indicates the importance of the contribution of ionic currents to the total admittance of excitable membranes. The research performed by Fishman *et al* (1977b, 1979, 1981) aims at frequency domain analyses of impedance or admittance arising from Na and K currents. The mathematical tool used by them is the linearized Hodgkin–Huxley equation, as discussed earlier in this chapter (see equation (10.33)). Because of this, the research by Fishman *et al* is limited to a small input analysis, in order to maintain linearity.

Potassium conduction
A series gL circuit which is used for linear, time variant K conduction is shown in figure 10.17 (diagram (c)). The differential equation which

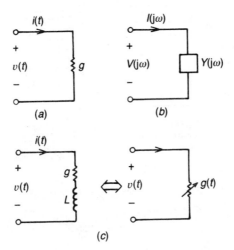

Figure 10.17 (a) The linear, time invariant conductance g. (b) An interconnected network of conductances and passive energy storage elements, capacitors and inductors. $Y(j\omega)$ is the admittance of the network. (c) A series gL circuit can exhibit the same kinetic behaviour as linear, time variant conductance. (From Fishman *et al* 1981.) Reproduced by permission of Plenum Publishing Corp.

relates the voltage $v(t)$ to current $i(t)$ is

$$v(t) = \frac{i(t)}{g} + L\frac{di(t)}{dt}. \tag{10.61}$$

If we define $\tau = gL$, this equation becomes

$$\frac{di(t)}{dt} = \frac{1}{\tau}(-i(t) + gv(t)). \tag{10.62}$$

Note the similarity between this equation and (10.29) which represents either Na or K currents in Hodgkin–Huxley's formalism. We can Fourier transform (10.62) to obtain

$$j\omega I(j\omega) = \frac{1}{\tau}(-I(j\omega) + gV(j\omega)). \tag{10.63}$$

This equation can be combined with $Y(j\omega) = I(j\omega)/V(j\omega)$ and (10.33) is obtained. This is the rationale behind linear circuit analyses. Diagram (b) illustrates the linearized HH circuit for K conduction. This circuit exhibits a resonance in impedance or an anti-resonance in admittance

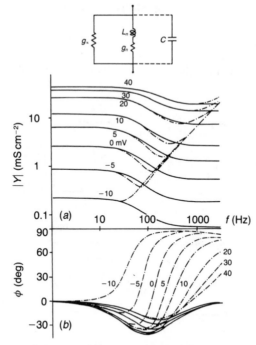

Figure 10.18 The admittance of the linear circuit (see inset). Diagram (a) shows the admittance amplitude. The full curves are for $|Y_K + Y_C|$ and the chain curves show $|Y_C|$. Diagram (b) illustrates the phase angles of $Y_K + Y_C$ (full curves) and Y_C (chain curves) at 10 °C. (From Fishman et al 1981.) Reproduced by permission of Plenum Publishing Corp.

representation. Figure 10.18 illustrates the calculated total admittance $|Y|$ $(Y_K + Y_C)$ of the circuit plotted against frequency at various membrane potentials. The anti-resonances are seen around 100 Hz. The admittance of an axon which is measured in the presence of K current is shown in figure 10.19. The curve obtained with -78 mV exhibits no anti-resonance because of the absence of K current at this potential. When the membrane is depolarized to -58 mV or to -48 mV, however, the anti-resonances become distinct. As predicted by a model calculation, the peak of anti-resonances shifts to high frequencies with depolarization.

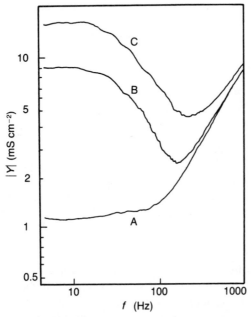

Figure 10.19 The admittance of squid axon with TTX blocking the Na current. Curves A, B and C are obtained with -78, -58 and -38 mV membrane potentials. 0.5 kF/10% Na ASW. (From Fishman *et al* 1981.) Reproduced by permission of Plenum Publishing Corp.

Sodium conduction

We can isolate Na conduction by eliminating K current using quarternary ammonium ions. The admittance of an axon without K conduction is displayed in figure 10.20. The anti-resonance which we observed with K conduction is replaced by an anomalous resonance at membrane potentials of -10 to -30 mV.

The phase angle at these membrane potentials is greater than $90°$ indicating that Na conductance is negative between -10 and -30 mV, as demonstrated by time domain measurements. These data, however,

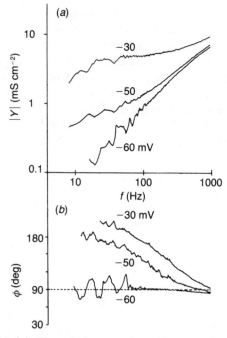

Figure 10.20 (*a*) The admittance of squid axon without K current, $|Y_{Na} + Y_C + Y_L|$. (*b*) The phase angles of the same axon at the same membrane potentials. 290 mM Cs at 10 °C. (From Fishman *et al* 1979.) Reproduced by permission of Elsevier Science Publishers BV.

represent the sum of admittances of the capacitance, negative Na conduction and, presumably, the small positive leakage conduction g_L. Let us first assume that $g_L = 0$ and calculate, using the LHH equation, the admittance of Na conductance and capacitance theoretically and compare the results with experimental data. The result of this calculation is shown in figure 10.21. This figure demonstrates that Na conductance itself exhibits an anomalous resonance because of its negative value. Adding the admittance due to membrane capacitance, we obtain a set of curves which are similar to those observed experimentally. However, the calculated phase angle is 180° at all potentials while the observed phase angle at the resting potential is 90°. The admittance was recalculated, changing the value of the leakage conductance g_L from 0 to 0.3, 0.35 and 0.4 mS cm^{-2}. Figure 10.21 shows the results of these recalculations. First of all, the calculated total admittance $|Y_{Na} + Y_C + Y_L|$ compares very well with that observed. Moreover, with $g_L = 0.35$ mS cm^{-2}, the calculated phase angle approaches 90°. From this, we can draw a conclusion that positive leakage current

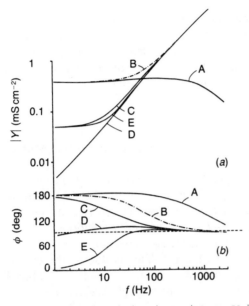

Figure 10.21 (*a*) The admittance $|Y_{Na}|$ and $|Y_{Na} + Y_C|$ calculated using the LHH equation at a membrane potential of -60 mV with $g_L = 0$ (curves A and B respectively). Curves C, D and E are calculated with $g_L = 0.3$, 0.35 and 0.4 mS cm^{-2} ($|Y_C + Y_L + Y_{Na}|$). (*b*) The phase angles calculated under the same conditions. $\bar{g}_K = 0$, $V = 0$, LHH at 10 °C. (From Fishman *et al* 1979.) Reproduced by permission of Elsevier Science Publishers BV.

neutralizes negative Na conduction, thus the total admittance is dominated by membrane capacitance, hence, a phase angle of 90°.

As has been discussed, the admittance or impedance characteristics of nerve membrane are dominated by ionic currents in a non-linear regime. Even in the linear domain, the presence of ionic currents should modulate the admittance characteristics of nerve membrane as Fishman and co-workers' research clearly demonstrates. Their work attempts a linear analysis of ionic conduction in the frequency domain. In spite of the use of simple linear models, the presence of various components such as capacitance and even leakage conductance makes the analyses quite complex, thus requiring considerable skill and insight into the correct interpretation of experimental data.

The giant axon of squid is the most exhaustively investigated nerve preparation because of its large size and remarkable stability. Other preparations such as frog sciatic nerve, lobster and crayfish axons are also available for electrophysiological research. However, the preparation of these nerve specimens is much more difficult and no data are available on their admittance or impedance characteristics. However, in view of the fact that the ionic currents of these nerve preparations are quite similar to those of squid axon, we have good reason to believe that the admittance characteristics of these nerve membranes are not drastically different from those of squid axon.

10.16 The Linear Passive Impedance of Muscle Membrane

The membrane capacitance of biological cells including nerves is almost uniformly $1\,\mu F\,cm^{-2}$. This is due to the fact that the thickness of biological membranes is uniformly 60–80 Å and their dielectric constant is no more than 4–7. One of the exceptions of this rule is muscle membrane. The total membrane capacitance of skeletal muscle is known to be as high as 6–$8\,\mu F\,cm^{-2}$. This is due to the extensive infolding of surface membrane to form transverse tubules (T tubules). Therefore, if we divide the total membrane capacitance by the total membrane area including T tubules, the membrane capacitance per unit area may reduce to a normal value of $1\,\mu F\,cm^{-2}$.

Phenomenologically, the presence of T tubules manifests itself as a small change in the wave form of the action potential (Gage and Eisenberg 1969b). However, the ionic current diagram of skeletal muscle membrane is not significantly different from that of nerve membrane under voltage clamp conditions (Hille and Cambell 1976). Hence, the effect of T tubules on the voltage clamp current is difficult to detect using time domain analyses. The presence of T tubules manifests itself most prominently in the admittance or impedance characteristics of muscle fibres. Because of this, the impedance of skeletal muscle membrane has been investigated by a number of researchers using external and internal electrodes.

The impedance characteristics of skeletal muscles were first studied, using external electrodes, by Guttman (1939) and also by Schwan (1954). There are advantages and disadvantages of the external electrode technique. Above all, this technique is suitable for the measurement of surface admittance, which gives rise to an exceedingly large anisotropic dielectric constant. However, the presence of large interfacial polarization tends to interfere with the analysis of the tubular admittance. The use of internal electrodes enables one to study the membrane admittance unequivocally without interference from interfacial polarizations. The measurement of the impedance of skeletal muscle membrane was performed by Tasaki and Hagiwara (1957) using a pair of glass microelectrodes. One of the electrodes was used for AC current injection and the other for voltage recording. Using this technique, the capacitance of the sartorius muscle of toad was found to range between

5 and 9 μF cm^{-2} and was frequency independent in the range 45 Hz–2 kHz.

Some years later, the impedance of frog sartorius muscle was investigated by Falk and Fatt (1964) using internal electrodes. The electrode arrangement used by Falk and Fatt is shown in figure 10.22 (inset). The impedance of muscle is calculated as the ratio of voltage to current ($Z = V/I$). However, the voltage which is produced by an injected current attenuates along the muscle fibre before it reaches the voltage electrode. Hence, the potential detected by a voltage electrode at $x = x_1$ is bound to be lower than the value at $x = 0$. The correction for the voltage attenuation between current and voltage electrodes is a crucial step in this technique. Before we discuss the results by Falk and Fatt, a general treatment of signal transmission in conducting fibres using the cable equation will be presented.

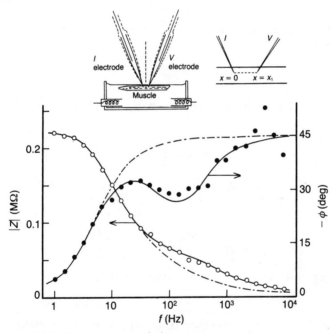

Figure 10.22 The impedance $|Z|$ and the phase angle ϕ of muscle fibre against frequency. The inset shows the electrode arrangement. I and V are the current and voltage electrodes. The second inset on the right shows the x coordinates of the electrodes. (From Falk and Fatt 1964.) Reproduced by permission of The Royal Society.

10.17 The Transmission of Signals along Conducting Fibres

The model used for the transmission of a voltage signal along a conducting fibre is illustrated in figure 10.23. If a current I_0 is injected at $x = 0$, the attenuation of the induced membrane potential $(V_m = V_i - V_e)$ is given by

$$\frac{\partial V_m}{\partial x} = -r_i i_i + r_e i_e \tag{10.64}$$

where r_i and r_e are the resistances of the internal and external solutions in $\Omega\ cm^{-1}$. i_i and i_e are currents in these media. If we define the total current I as the sum of i_i and i_e, (10.64) becomes

$$\frac{\partial V_m}{\partial x} = -i_i(r_i + r_e) + r_e I. \tag{10.65}$$

Differentiating this equation with respect to x, we obtain, with the assumption that $\partial I/\partial x = 0$, the following equation:

$$\frac{\partial^2 V_m}{\partial x^2} = -\frac{\partial i_i}{\partial x}(r_i + r_e). \tag{10.66}$$

We define membrane current i_m as the rate of i_i leaking out to an external medium, i.e.

$$-i_m = \frac{\partial i_i}{\partial x}. \tag{10.67}$$

The two branches of membrane admittance divide the current into two parts:

$$i_m = C_m \frac{\partial V_m}{\partial t} + \frac{V_m}{r_m} \tag{10.68}$$

where r_m is the membrane resistance in $\Omega\ cm$ and C_m is the membrane capacitance in $F\ cm^{-1}$. Combining (10.66), (10.67) and (10.68) we obtain

$$C_m \frac{\partial V_m}{\partial t} + \frac{V_m}{r_m} + \frac{1}{r_e + r_i}\frac{\partial^2 V_m}{\partial x^2}. \tag{10.69}$$

Multiplying both sides by r_m and using definitions such as

$$\tau = C_m r_m \quad \text{and} \quad \lambda^2 = r_m/(r_e + r_i) \tag{10.70}$$

we obtain the cable equation (see Hodgkin and Rushton 1946):

$$-\lambda^2 \frac{\partial^2 V_m}{\partial x^2} + \tau \frac{\partial V_m}{\partial t} + V_m = 0 \tag{10.71}$$

where τ and λ are time and space constants respectively.

Time domain analysis of the transmission of electrical signals uses a step current pulse and measures the induced voltage as a function of time and distance. For these analyses, most of the information is extracted from transient responses. For this reason, a complete solution of the cable equation is required if a time domain analysis is desired.

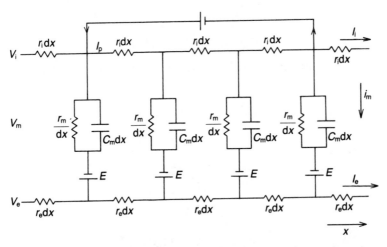

Figure 10.23 The core conductor model which is used to analyze the transmission of subthreshold signals. The units of the membrane parameters are: r_i = resistance of internal fluid ($\Omega\ cm^{-1}$), r_e = resistance of external fluid ($\Omega\ cm^{-1}$), r_m = membrane resistance ($\Omega\ cm$), C_m = membrane capacitance ($F\ cm^{-1}$), i_i and i_a are internal and external currents, and i_m is membrane current. (From Plonsey 1969.)

The general solution of (10.71) is shown below:

$$V_m(x, t) = \frac{r_i I_0 \lambda}{4}\left\{ e^{-x/\lambda}\left[1 - \operatorname{erf}\left(\frac{x}{2\lambda}\sqrt{\frac{\tau}{t}} - \sqrt{\frac{t}{\tau}} \right) \right] \right.$$

$$\left. - e^{x/\lambda}\left[1 - \operatorname{erf}\left(\frac{x}{2\lambda}\sqrt{\frac{\tau}{t}} + \sqrt{\frac{t}{\tau}} \right) \right] \right\} \qquad (10.72)$$

where the error function (erf z) is given by the following equation:

$$\operatorname{erf} z = \frac{2}{\sqrt{\pi}} \int_0^z \exp(-w^2)\,dw. \qquad (10.73)$$

The numerical value of the error function ranges between -1 and $+1$. For a transient solution, the voltage $V(x, t)$ depends on time and coordinates. A complete graphic representation of (10.72) is illustrated in figure 10.24. It should be noted that the increase in voltage with time follows the error function and its decay with distance is exponential.

At steady state, the induced voltage is no longer dependent on time. Thus, the $\partial V_m/\partial t$ term in (10.71) can be eliminated:

$$- \lambda^2 \frac{d^2 V_m}{dx^2} + V_m = 0. \qquad (10.74)$$

The steady state solution is given by

$$V_m = Ae^{-x/\lambda} + Be^{x/\lambda}. \qquad (10.75)$$

Of these, only the first term on the RHS is required for a positive x coordinate.

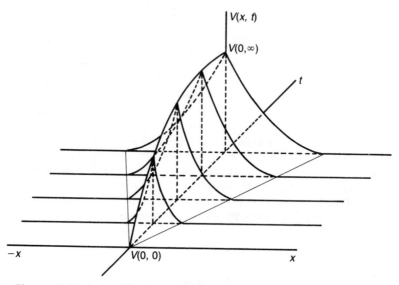

Figure 10.24 A graphic representation of the potential profile V along the distance and time axes. V is plotted in the z direction, the distance along the fibre is in the x direction and time is in the y direction.

10.18 The Calculation of the Space Constant for AC Inputs and Muscle Impedance

Equation (10.75) indicates that if we inject a current at $x = 0$ and record the voltage after the stationary state has been reached, the measured voltage attenuates exponentially along the fibre if injected current is not large enough to elicit an action potential. In other words, the cable equation dictates the transmission of subthreshold signals. The subthreshold transmission should be differentiated from the suprathreshold transmission of the action potential. The attenuation of voltage along the fibre is the major source of error for the measurement of impedance of muscle or nerve fibres using the electrode arrangement illustrated in figure 10.22.

The impedance Z measured at a certain distance away from $x = 0$ is given by the following equation:

$$Z = Z_0 \exp(-x/\lambda) \qquad (10.76)$$

where λ is the space constant which is defined by (10.70). Z_0 is the impedance measured at $x = 0$ and is defined, neglecting r_e, as follows:

$$Z_0 = \tfrac{1}{2}(r_i/Y_m)^{1/2} \qquad (10.77)$$

where

$$Y_m = 1/r_m + j\omega C_m.$$

For AC inputs, (10.70) is modified to (10.54). The space constant is now a complex quantity. The real part causes the attenuation of signals and the imaginary part causes oscillation. Therefore, the wave form of the transmitted signal can be envisaged as a damped oscillation. Let us first separate the real and imaginary parts of Z_0. Note the changes of the symbols r_i, to r, $1/r_m$ to g and C_m to c:

$$Z_0 = \frac{\sqrt{r}}{2} \sqrt{\frac{g - j\omega c}{g^2 + \omega^2 c^2}}. \qquad (10.78)$$

We start the procedure by equating the numerator to $M + jN$. By squaring, we obtain

$$g - j\omega c = M^2 + 2jMN - N^2. \qquad (10.79)$$

Equating the real and imaginary terms on both sides

$$g = M^2 - N^2 \qquad (10.80)$$

$$-c\omega = 2MN. \qquad (10.81)$$

Solving these equations for M and N, we obtain

$$M = \left(\frac{g}{2}[1 + \sqrt{1 + (\omega c/g)^2}]\right)^{1/2} \qquad (10.82)$$

$$N = \frac{-\omega c}{2}\left(\frac{2}{g}\right)^{1/2}[1 + \sqrt{1 + (\omega c/g)^2}]^{-1/2}. \qquad (10.83)$$

Introducing new symbols m and n where

$$m = (1/2)^{1/2}[1 + \sqrt{1 + (\omega c/g)^2}]^{1/2} \qquad (10.84)$$

and

$$n = \frac{1}{\sqrt{2}}\left(\frac{\omega c}{g}\right)[1 + \sqrt{1 + (\omega c/g)^2}]^{-1/2} \qquad (10.85)$$

the impedance Z_0 is now written as

$$Z_0 = \frac{(r/g)^{1/2}}{2[(\omega c/g)^2 + 1]^{1/2}}(m - jn). \qquad (10.86)$$

This is the impedance at $x = 0$ where current is injected and the induced potential is unattenuated. Using a similar method, the space constant λ can also be separated into real and imaginary parts. Using this and Z_0 in (10.76), we finally obtain

$$Z = \frac{(r/g)^{1/2}}{2[(\omega c/g)^2 + 1]^{1/2}} \{[m \cos n(rg)^{1/2}x - n \sin n(rg)^{1/2}x]$$
$$-j[n \cos n(rg)^{1/2}x + m \sin n(rg)^{1/2}x]\} \exp[-m(rg)^{1/2}x]. \qquad (10.87)$$

Note that this equation reduces to (10.86) when $x = 0$.

The amplitude of the impedance and its phase angle are plotted against frequency. The same data can be presented as a complex locus. Figure 10.22 and figure 10.25 illustrate these plots. The unique feature of these plots is the presence of a large impedance at low frequencies and another one, which is relatively small, at high frequencies. This clearly indicates the presence of two major sources of impedance in muscle fibre. Likewise, the complex plane locus shows the presence of two arcs, a large one at low frequencies and a small one at high frequencies. The low-frequency impedance was attributed to transverse tubular membranes and the high-frequency impedance to surface membrane. The full and chain curves are calculated using the models shown by the insets in this figure. Obviously, the complex model with two time constants (model (b)) gives a much better agreement with the data than the simple one time constant model (model (a)).

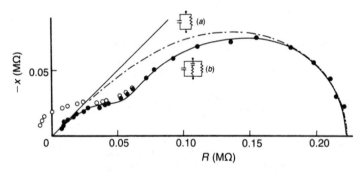

Figure 10.25 The impedance locus of skeletal muscle. Note the presence of two loci: the large arc at low frequencies and the small one at high frequencies. The full curve was calculated with the two time constant model (see insets) and the chain curve was calculated using the one time constant model. The open and full circles were obtained by separate measurements. (From Falk and Fatt 1964.) Reproduced by permission of The Royal Society.

Using the two time constant model, the capacitances and resistances of surface and T tubule membranes, C_m, C_e, and r_m, r_e are determined.

In addition, the time constants $(C_e + C_m)r_m$ and $C_e r_e$ are calculated. Table 10.2 shows the numerical values of these parameters. Although the values fluctuate from one sample to another, nevertheless, the capacitance C_e is almost twice as large as that of the surface membrane. As shown in this table, the two time constants are vastly different from one another, indicating substantial differences in the electrical properties of surface and tubular membranes.

Freygang *et al* (1967) found that the two time constant circuit proposed by Falk and Fatt was basically correct. However, the phase angle measured was found to deviate systematically from the curve calculated by Falk and Fatt's circuit, indicating that the real tubular system is more complex. In addition, these authors found that hypertonicity increased the area of the T tubules, resulting in an increase in C_e, but the value of r_e remained unchanged. These observations suggest that C_e reflects the capacitance of a T tubular membrane but r_e may be located outside the tubules.

Eisenberg and Gage (1967) selectively disrupted T tubules by treating frog sartorius muscle with glycerol prior to the measurement of impedance. These authors found that glycerol treatment caused a reduction of the total capacitance from 6.7 to 2.25 μF cm^{-2} while the specific membrane resistance and internal resistance were found unaffected. The resistance of muscle fibres is mainly due to chloride conduction which is concentrated on the surface membrane. This may explain the invariance of membrane resistance after glycerol treatment. Based on these observations, the large capacitance which amounts to about 4 μF cm^{-2} is attributed to T tubular membrane and the remaining 2.2 μF cm^{-2} is due to surface membrane. However, the total area of T tubules was estimated to be 4.5 times larger than that of surface membrane (Eisenberg *et al* 1972, Valdiosera *et al* 1974) while the ratio of tubular capacitance and surface membrane capacitance is only 2. This means that either the estimate of surface membrane capacitance is too large or the capacitance of the tubular membrane is underestimated. Gage and Eisenberg (1969a) estimated the fraction of tubules left in glycerol treated muscle fibres to be about 1.6%. This translates to approximately 6% of the area of surface membrane. The correction of surface membrane capacitance for this additional area leads to a new value of 2.1 μF cm^{-2}, which is still much higher than that of nerve membrane. In addition to glycerol treatment, low ionic strength was found, by Vaughan *et al* (1972), to eliminate the capacitance of T tubules, reducing the total capacitance to 2.2–2.3 μF cm^{-2}. No satisfactory explanation for this observation has yet been found. It is interesting to note that elimination of T tubule capacitance always reduces the total capacitance to some 2 μF cm^{-2} instead of 1 μF cm^{-2}. It is unlikely, however, that the lipid composition of muscle surface membrane is different from those of other biological cells.

Table 10.2 The capacitance (C_c and C_m) and resistance (r_c and r_m) of tubular and surface membranes.[†]

Experiment	$(C_m+C_c)r_m$ (ms)	$C_c r_c$ (ms)	C_c (μF cm^{-2})	C_m (μF cm^{-2})	r_c (Ω cm^2)	r_m (Ω cm^{-2})	(C_c+C_m) (μF cm^{-2})
1	23.5	1.38	3.29	2.46	418	4100	5.75
2	22.2	—	0.87	2.61	—	6380	3.48
3	12.6	0.81	5.22	2.09	155	1725	7.31
4	16.6	0.78	2.80	3.20	277	2770	6.00

[†]Taken from Falk and Fatt (1964).

10.19 The Vaseline Gap Method Using Single Fibres

As has been discussed, the intracellular microelectrode technique is the major tool used for the study of the impedance of skeletal muscle fibres. However, the microelectrode technique requires cumbersome corrections for various parasitic capacitances, and a possible non-linearity of glass microelectrodes. Takashima (1985) used a Vaseline gap method. (The detail of this method is found in Hille and Cambell's paper (1976).) As illustrated in figure 10.26, a single fibre isolated from frog semitendinous muscle is placed across a gap (A) which is electrically isolated from pools B and C using Vaseline. Two electrodes are placed in pool A and pool B or C and the membrane admittance is measured. Since, under ideal conditions, no current will flow outside the fibre, we only measure the admittance across the muscle membrane.

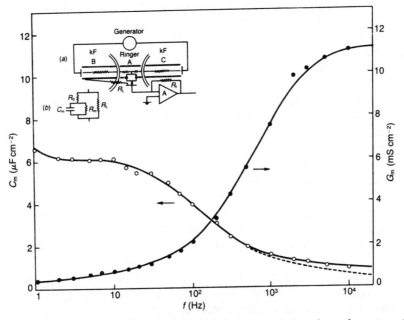

Figure 10.26 The frequency profile of the capacitance and conductance of frog semitendinous muscle measured using the Vaseline gap method (see inset (a)). Ag–AgCl electrodes are placed in pools A and B or C. Inset (b) shows the equivalent circuit including the series resistance R_s and the leakage resistance R_L. The broken portion of curve A is obtained after correction for the cable attenuation of input signals between electrodes. (From Takashima 1985.) Reproduced by permission of Springer-Verlag.

The most difficult problem of this technique, however, is the leakage currents through the Vaseline gap and, therefore, careful analyses are required. The equivalent circuit of a gap system with a membrane is shown in the inset of this figure. Analysis of this circuit leads to the following two equations. Whereas impedance is a suitable representation for the microelectrode method, an admittance representation is more convenient for the gap technique:

$$Y(\text{Re}) = \frac{1}{R_L} + \left(\frac{1}{R_0 + R_m} + \frac{1}{R_0}(\omega\tau)^2\right)\Big/L(\omega) \qquad (10.88)$$

$$Y(\text{Im}) = \left(\frac{C_m R_m}{R_0 + R_m} - \frac{R_0 R_m C_m}{(R_0 + R_m)^2}\right)\Big/L(\omega) \qquad (10.89)$$

where

$$L(\omega) = 1 + (\omega\tau)^2 \qquad \text{and} \qquad \tau = R_0 R_m C_m/(R_0 + R_m) \quad (10.90)$$

where R_L, R_0 and R_m are the leakage resistance, the series resistance and the resistance of a membrane respectively. C_m is total capacitance. Inspection of these equations readily reveals that the leakage resistance R_L affects only the real part of admittance, i.e. conductance whereas the imaginary part, i.e. capacitance, is not affected by leakage current. This means that we cannot determine membrane conductance accurately using this method unless the leakage current is zero. However, with careful sealing (Hille and Cambell's estimate of R_L is about 10 MΩ), we can obtain reasonable conductance values consistently between 0.5 and 1.2 mS cm^{-2}. The length of muscle fibre is about 0.5–1.0 cm. Therefore, the resistance of sarcoplasm becomes quite large. Hence, the correction of data for the series resistance due to sarcoplasm is an essential step to obtaining a correct membrane capacitance. Capacitance and conductance versus frequency plots are shown in figure 10.26. The total capacitance is found to be about 6–6.5 μF cm^{-2} and that of the surface membrane is about 1.5 μF cm^{-2}. The latter value is somewhat smaller than the one found by Eisenberg and Gage (1967). The admittance elements along the T tubule connect to the external solution via the lumen with a resistance R_t (see the inset in figure 10.26). Hence, the total capacitance C and conductance G including surface and tubular membranes are, based on this circuit, given by

$$C = \frac{C_m}{1 + (\omega C_m R_m)^2} + \frac{C_T}{1 + (\omega T)^2}\left(\frac{R_T}{R_T + R_t}\right)^2 \qquad (10.91)$$

and

$$G = \frac{\omega^2 R_m C_m}{1 + (\omega C_m R_m)^2} + \frac{\omega^2(R_t C_T R_T^2) + (R_T + R_t)}{1 + (\omega T)^2}\left(\frac{1}{R_t + R_T}\right)^2 \qquad (10.92)$$

where C_m and R_m are the capacitance and resistance of surface membrane and C_T and R_T are the capacitance and resistance of tubules respectively. T is the time constant for tubules as shown below:

$$T = \frac{R_t R_T C_T}{R_t + R_T}.$$ (10.93)

At low frequencies, the total capacitance reduces to

$$C = C_m + C_T \left(\frac{R_T}{R_t + R_T}\right)^2.$$ (10.94)

Therefore, the observed low-frequency limiting capacitance must be corrected for the term in parentheses. This can be estimated using (10.92) and figure 10.27. At low and high frequencies, the measured conductance G reduces to

$$G_0 = 1/(R_t + R_T) \qquad \text{for } \omega \to 0$$ (10.95)

$$G_\infty = 1/R_t \qquad \text{for } \omega \to \infty.$$ (10.96)

In figure 10.27, two intersections of the arc with the abscissa represent $1/G_0 \ (= R_t + R_T)$ and $1/G_\infty \ (= R_t)$ respectively. Numerically, they are $8.5 \times 10^5 \ \Omega$ and $1.1 \times 10^5 \ \Omega$. Hence, the value of R_T is $7.4 \times 10^5 \ \Omega$. Using these we can calculate the ratio of $R_T/(R_t + R_T)$ and obtain a value of 0.87. Therefore, the capacitance of the T tubule is $(6.4-1.5) \times (0.87)^2 = 6.45 \ \mu\text{F cm}^{-2}$. Thus the total capacity is $7.97 \ \mu\text{F cm}^{-2}$.

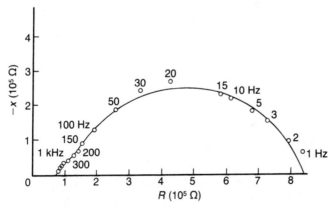

Figure 10.27 The impedance locus of the same sample as figure 10.26. The data represent one time constant behaviour. Samples show either one time constant or two time constant behaviour. (From Takashima 1985.) Reproduced by permission of Springer-Verlag.

The impedance plot shown in figure 10.27 exhibits only a small hump at high frequencies. However, it should be pointed out that the impedance or admittance characteristics of muscle fibres vary from one sample to another. Some fibres show a distinct two time constant behaviour while other fibres indicate simple one time constant characteristics This variability is due to the characteristic frequency of T tubules rather than to that of surface membrane. For some fibres, the dispersion of C_T is found at very low frequencies and is distinctly separate from that of surface membrane, hence, two time constant behaviour. With other fibres, C_T dispersion is found at higher frequencies, close to that of surface membrane, hence, one time constant behaviour. The equivalent circuit of muscle fibres including surface membrane and tubular systems has been discussed by many investigators. The circuit used by Falk and Fatt is shown in figure 10.25. The circuits (a) and (b) shown in figure 10.28 were proposed by Eisenberg (1967) for crab muscle. The numbers in parentheses are circuit parameters. Diagram (a) is redundant and can be reduced to diagram (b).

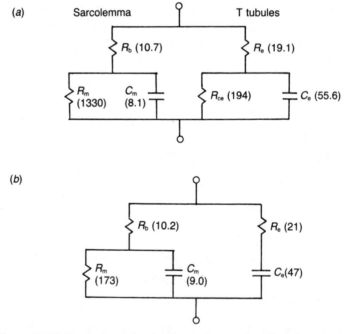

Figure 10.28 Equivalent circuits for crab muscle. Circuits (a) and (b) have the same resistance and reactance values for the parameters indicated. Note that diagram (b) has the minimum number of circuit elements. (From Eisenberg 1967.) Reproduced from *The Journal of General Physiology*, 1967, vol 50, pp 1785–806 by copyright permission of the Rockefeller University Press.

10.20 The Voltage Dependence of the Capacitance of Muscle Fibre

The voltage dependent displacement current of muscle fibre was first investigated by Schneider and Chandler (1973). The electrode arrangement used by these investigators is shown in figure 10.29(a). The ionic currents were all eliminated by the use of pharmacological agents. The membrane current density at $x = l$ is related to the potential difference $\Delta V(= V_2 - V_1)$ by

$$i_m = 2\Delta V/3R_i l^2 \qquad (10.97)$$

where R_i is the internal resistance per unit length. Figure 10.29(b) shows two voltage recordings. Diagram (b) A shows a voltage change V_1 from -79 to $41\,\text{mV}$ and a corresponding current. Diagram (b)B shows ΔV from -126 to $41\,\text{mV}$ and the current. Figure 10.29(c)

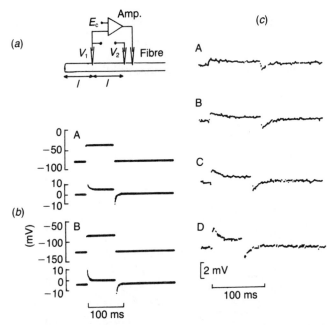

Figure 10.29 (a) A diagram of the three-microelectrode technique. (b) A The traces of V_1 and ΔV (= $V_2 - V_1$) for a pulse from -79 to $41\,\text{mV}$. (b) B Another trace from -126 to $41\,\text{mV}$. (c) The difference between ΔV for various depolarizing pulses: A, $(-79\,\text{mV}, \quad 25\,\text{mV}) - (-107\,\text{mV}, \quad 25\,\text{mV})$; B, $(-79\,\text{mV}, 33\,\text{mV}) - (-117\,\text{mV}, 33\,\text{mV})$; C, $(-79\,\text{mV}, 41\,\text{mV}) - (-126\,\text{mV}, 41\,\text{mV})$; D, $(-79\,\text{mV}, 56\,\text{mV}) - (-144\,\text{mV}, 56\,\text{mV})$. (From Schneider and Chandler 1976.) Reprinted by permission from *Nature*, vol 242, p244. Copyright © 1976 Macmillan Journals Limited.

illustrates the differences between displacement currents obtained with various depolarizing voltage steps. Analysis of these diagrams led these authors to conclude that these time dependent non-linear membrane currents are due to displacements of charges bound within the membrane. They further suggested that these charge movements may be a process which is similar to the orientation of dipoles. Schneider and Chandler (1976) carried out a more detailed research on the displacement current in muscle fibre using a similar electrode arrangement.

Adrian and Almers (1974, 1976a, b), Almers *et al* (1975) and Adrian and Peres (1979) published a series of papers on the voltage dependent capacitance of muscle fibres. The electrode arrangement used is shown in figure 10.30. The capacitance can be calculated using the following equation:

$$C_{\text{eff}} = \frac{k}{V_1(\infty)} \int_0^\infty \left(V_2(t) - \frac{V_2(\infty)}{V_1(\infty)} V_1(t) \right) dt \tag{10.98}$$

where k is defined by

$$k = a/3l^2 R_1 \tag{10.99}$$

where R_1 is the resistivity of myoplasm and 'a' is the radius of the fibre.

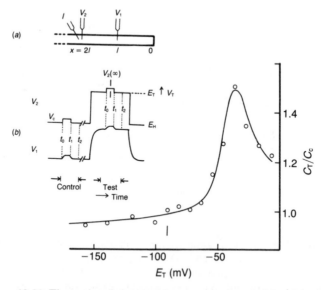

Figure 10.30 The ratio of the membrane capacitance (C_T/C_c) measured at E_T and at the holding potential E_H (−90 mV). The abscissa is the potential E_T. Inset (a) shows the three-electrode arrangement. Electrode 3 delivers current. The end of the fibre is at $x = 0$. Inset (b) shows the pulse sequence. (From Adrian and Almers 1976a, b.) Reproduced by permission of Cambridge University Press.

$V(t)$ and $V(\infty)$ are transient and steady state voltages. In order to measure C_{eff} as a function of membrane potential, small step voltages are applied from a holding potential E_H and then the same potential is superimposed on a depolarizing potential. The pulse sequence is depicted in figure 10.30 (inset). Using this method, Adrian and co-workers determined the ratio of the capacitances of depolarized membrane and, at holding potential, C_T/C_c. Figure 10.30 illustrates this ratio at various membrane potentials. A marked increase in the capacitance with depolarization should be noted. The increase reaches a peak value of 1.4 at around -40 mV. The complexity of transient capacitance measurements with muscle fibre arises from the uneven current distribution in T tubules. If we inject a current pulse to shift the potential of surface membrane, the induced potential spreads radially inward along the T tubules. However, due to the finite space constant, the potential would attenuate as it goes down along the tubules, causing an uneven potential distribution. Only when the space constant is infinite, will the potential distribution be uniform and we are able to measure the capacitance of the entire T system. Otherwise, the measured capacitance represents only a fraction of the T system's capacitance. The spreading of potential along the T system is a slow process and this, in turn, causes a delay in capacitance increase. The displacement current found by Huang (1982) clearly shows two components, indicating a temporal separation of two charge movements.

Almers (1978) suggested, after careful analysis, that neither a sodium nor potassium gating can account for the magnitude of the observed capacitance increase. The number of Na and K channels in muscle membrane is simply not enough to produce the observed charge movement. Almers interprets the charge movements in terms of a voltage transmission from the tubular system to the sarcoplasmic reticulum via a junction called a 'triad'. This type of ion movement appears to cause charge build-up across membranes and produces a voltage dependent capacitance. This interpretation seems to conform to that by Schneider and Chandler.

The experiments discussed above were all performed using step pulses in the time domain. The only frequency domain experiment reported so far was performed by Takashima (1985) by superimposing a small sinusoidal measuring signal on step bias potentials. All the experiments performed by this author used isolated single fibre and the Vaseline gap method, as already described. In most of his measurements, Na and K currents were eliminated by TTX and TEA in bathing solution. Some of the results obtained are shown in figure 10.31. As illustrated, the capacitance of the fibre increases by about 15% when bias potentials are preceded by a large negative prepulse (-100 mV). This increase is smaller than that observed by Adrian and co-workers. However, the difference is due to the frequency limitation of Takashima's experiments and is, therefore, limited to the detection of the voltage dependence of surface membrane capacitance only. The difference between the results by Adrian and co-workers and those by Takashima indicates that the major portion of the voltage dependent capacitance of muscle fibre arises from T-tubular membrane.

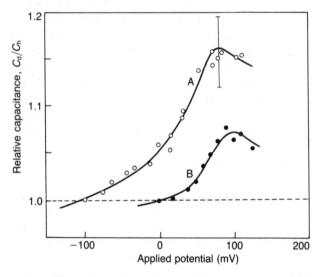

Figure 10.31 The voltage dependence of the membrane capacitance of skeletal muscle fibre. The ordinate is relative capacitance C_d/C_h, the ratio between the capacitances with depolarizing pulses and those with a hyperpolarizing prepulse. Curve A was obtained with a -100 mV hyperpolarizing prepulse and curve B was obtained without a prepulse. (From Takashima 1985.) Reproduced by permission of Springer-Verlag.

10.21 A New Technique for the Measurement of the Membrane Capacitance of Small Cells

Various techniques for the impedance measurement of biological membranes have been discussed in this chapter. The preparations such as nerves and muscles, although some of them are quite small, are fibrous cells and various internal electrode techniques are available. If the diameter of the fibre is too small, still Vaseline and/or other gap methods can be used to measure membrane impedance or admittance directly. However, these techniques cannot be used for small spherical biological cells, such as erythrocytes or lymphocytes. For these samples, what is called a 'suspension technique' has been used for many years (see Schwan 1957, Cole 1968). Cells are suspended in appropriate solvents and the dielectric constant and conductivity are measured in a wide frequency range. The membrane capacitance can be determined using inhomogeneous dielectric theories, which were discussed in Chapter 6. This method has been successful for small cells for which no other measuring techniques are available. However, the suspension technique fails to provide detailed information on the impedance characteristics of these cells. Above all, it is virtually impossible to extract the information on membrane conductance. The conductance of passive membranes is simply too small to be detected by this method (Lassen *et al* 1978, Hunter 1977, Tosteson *et al* 1973). Under these circumstances, some new technology is called for.

The recent development of a new method, the patch clamp technique, enables one to determine the membrane conductance of small biological cells. This technique, which was developed by Sakmann and Neher (1983), uses a specially prepared small glass pipette with a diameter of 0.5–1 μm. The tip of the pipette is pressed against the external wall of the cell. If properly done, the pipette makes very tight contact with the cell membrane, with a leakage resistance higher than 10 GΩ. Therefore, the current between the pipette and the external ground electrode will flow only through the membrane, be it a conduction current or a displacement current. With this electrode arrangement, gross membrane conductance and also single channel current can be measured. This technique can also be used for the frequency domain measurement of membrane capacitance using sinusoidal inputs instead of square pulses (Takashima *et al* 1988).

Suggested Reading

Adelman W J 1971 *Biophysics and Physiology of Excitable Membranes* (New York: Van Nostrand–Reinhold)

Cole K S 1968 *Membranes, Ions and Impulses* (Berkeley, CA: University of California Press)

Lakshminarahanaiah N 1969 *Transport Phenomena in Membrane* (New York: Academic)

Plonsey R 1969 *Bioelectric Phenomena* (New York: McGraw-Hill)

Schanne O F 1978 *Impedance Measurements in Biological Cells* (New York: Wiley)

Schwan H P 1963 in *Physical Techniques in Biological Research* vol VI, ed. W Nastuk (New York: Academic) ch 6

Tasaki I 1982 *Physiology and Electrochemistry of Nerve Fibers* (New York: Academic)

Warren R C 1987 *Physics and the Architecture of Cell Membranes* (Bristol: Hilger)

References

Adrian R H and Almers W 1974 Membrane capacity measurements on frog skeletal muscle in media of low ion content *J. Physiol.* **237** 573–605
—— 1976a The voltage dependence of membrane capacity *J. Physiol.* **254** 317–38
—— 1976b Charge movement in the membrane of striated muscle *J. Physiol.* **254** 339–60
Adrian R H and Peres A 1979 Charge movement and membrane capacity in frog muscle *J. Physiol.* **289** 83–97
Almers W 1978 Gating currents and charge movements in excitable membranes *Rev. Physiol. Biochem. Pharmacol.* **82** 96–190
Almers W, Adrian R H and Levinson S R 1975 Some dielectric properties of muscle membrane and their possible importance for excitation-contraction coupling. *Ann. New York Acad. Sci.* **264** 278–92
Armstrong C M and Bezanilla F 1973 Charge movement associated with the opening and closing of the activation gates of the Na channels *J. Gen. Physiol.* **63** 533–52
—— 1975 Currents associated with the ionic gating structure in nerve membrane *Ann. New York Acad. Sci.* **264** 265–77
Chandler W K, FitzHugh R and Cole K S 1962 Theoretical stability properties of a space clamped axon *Biophys. J.* **2** 105–27
Cole K S 1932 Electric phase angle of cell membranes *J. Gen. Physiol.* **15** 641–9
—— 1941 Rectification and inductance of the squid axon during activity *J. Gen. Physiol.* **25** 29–51
—— 1949 Dynamic electrical characteristics of the squid axon membrane *Arch. Sci. Physiol.* **3** 253–8
Cole K S and Baker R F 1941 Transverse impedance of the squid axon during current flow *J. Gen. Physiol.* **24** 535–49
Cole K S and Curtis H J 1936 *Cold Spring Harbor Symp. Quantitative Biology* vol 4, pp73–89
—— 1939 Electric impedance of the squid axon during activity *J. Gen. Physiol.* **22** 649–70
Coster H G L and Smith J R 1974 The molecular organization of bimolecular lipid membrane *Biochim. Biophys. Acta* **373** 151–64
Davson H and Danielli J F 1943 *Permeability of Natural Membranes* (New York: MacMillan)
Eisenberg R S 1967 Equivalent circuit of single crab muscle fibers as determined by impedance measurements with intra-cellular electrodes *J. Gen. Physiol.* **50** 1785–806
Eisenberg R S and Gage P W 1967 Frog skeletal muscle fibers: changes in electrical properties after disruption of the transverse tubular system *Science* **158** 1700–1
Eisenberg R S, Vaughan P C and Howell J N 1972 A theoretical analysis of the capacitance of muscle fibers using a distributed model of the tubular system *J. Gen. Physiol.* **59** 360–73
Falk G and Fatt P 1964 Linear electrical properties of striated muscle fibers observed with intracellular electrodes *Proc. R. Soc.* B **160** 69–123

Fishman H M, Moore L E and Poussart D 1977a Asymmetry currents and admittance in squid axon *Biophys. J.* **19** 177–83

—— 1981 *The Biophysical Approach to Excitable Systems* ed. W J Adelman and D E Goldman (New York: Plenum) pp65–95

Fishman H M, Poussart D and Moore L E 1979 Complex admittance of Na$^+$ conduction in squid axon *J. Membrane Biol.* **50** 43–63

Fishman H M, Poussart D, Moore L E and Siebenga E 1977b K$^+$ conduction description from the low frequency impedance and admittance of squid axon *J. Membrane Biol.* **32** 255–90

Freygang W H, Rapoport S I and Peachey L D 1967 Some relations between changes in the linear electrical properties of striated muscle fibers and changes in ultrastructure *J. Gen. Physiol.* **50** 2437–58

Gage P W and Eisenberg R S 1969a Capacitance of the surface and transverse tubular membrane of frog sartorius muscle fibers *J. Gen. Physiol.* **53** 265–78

—— 1969b Action potentials, afterpotentials, and excitation-contraction coupling in frog sartorius fibers without transverse tubules *J. Gen. Physiol.* **53** 298–310

Goldman D E 1943 Potential, impedance and rectification in membranes *J. Gen. Physiol.* **27** 37–60

Guttman R 1939 The electrical impedance of muscle during the action of narcotics and other agents *J. Gen. Physiol.* **22** 567–91

Hanai T, Haydon D A and Taylor J 1964 An investigation by electrical methods of lecithin-in-hydrocarbon films in aqueous solution *Proc. R. Soc.* A **281** 377–91

Hille B and Cambell D T 1976 An improved Vaseline gap voltage clamp for skeletal muscle fibers *J. Gen. Physiol.* **67** 265–93

Hodgkin A L and Huxley A F 1952 A quantitative description of membrane current and its application to conductance and excitation in nerve *J. Physiol.* **117** 500–44

Hodgkin A L and Katz B 1949 The effect of sodium ions on the electrical activity of the giant axon of squid *J. Physiol.* **108** 37–77

Hodgkin A L and Rushton W A H 1946 The electrical constants of a crustacean nerve fibre *Proc. R. Soc.* B **133** 97–132

Huang C L-H 1982 Pharmacological separation of charge movement components in frog skeletal muscle *J. Physiol.* **334** 375–87

Hunter J J 1977 Human erythrocyte anion permeabilities measured under conditions of net charge transfer *J. Physiol.* **268** 35–49

Keynes R D 1951 The ionic movements during nervous activity *J. Physiol.* **114** 119–50

Keynes R D and Rojas E 1974 Kinetics and steady-state properties of the charged system controlling sodium conductance in the squid giant axon *J. Physiol.* **239** 393–434

Langmuir I and Waugh D F 1938 The adsorption of proteins at oil-water interfaces and artificial protein-lipoid protein *J. Gen. Physiol.* **21** 745

Lassen U V, Pape L and Vestergaard-Bogind B 1978 Chloride conductance of the amphiuma red cell membrane *J. Membrane Biol.* **39** 27–38

Marmont G 1949 Studies on the axon membrane: I. A new method *J. Cell. Comp. Physiol.* **34** 351–82

Matsumoto N, Inoue I and Kishimoto U 1970 The electric impedance of the squid axon membrane measured between internal and external electrodes *Japan. J. Physiol.* **20** 516–26

Meves H 1974 The effect of holding potential on the asymmetry currents in squid giant axons *J. Physiol.* B **243** 847–67

—— 1977 Activation, inactivation, and chemical blockage of the gating current in squid giant axons *Ann. New York Acad. Sci.* **303** 322–38

Moore J W and Cole K S 1963 in *Physical Techniques in Biological Research* vol VI, ed. W Nastuk (New York: Academic) p263

Mueller P, Rudin D O, Tien H T and Wescott W C 1962 Reconstruction of cell membrane structure in vitro and its transformation into an excitable system *Nature* **194** 979

—— 1963 Methods of formation of single bimolecular lipid membranes in aqueous solution *J. Phys. Chem.* **67** 534

—— 1964 Formation and properties of bimolecular lipid membranes. Recent progress *Surf. Sci.* **1** 379

Robertson J D 1964 in *Cellular Membranes in Development* ed. M Locke (New York: Academic)

Sakmann B and Neher E 1983 *Single Channel Recording* (New York: Plenum)

Schneider M F and Chandler W K 1973 Voltage dependent charge movement in skeletal muscle: a possible step in excitation-contraction coupling *Nature* **242** 244–6

—— 1976 Effects of membrane potential on the capacitance of skeletal muscle fibres *J. Gen. Physiol.* **67** 125–63

Schwan H P 1954 Die Elektrischen Eigenschaften von Muskelgewebe bei Niederfrequenz *Z. Naturf.* b **9** 245–51

—— 1957 in *Advances in Medical and Biological Physics* vol 5, ed. S H Lawrence and C A Tobias (New York: Academic) p147

—— 1981 in *The Biophysical Approach to Excitable Systems* ed. W J Adelman and D E Goldman (New York: Plenum) p3

Schwan H P, Huang C C and Thompson T E 1966 *Abstract Biophys. Soc. Ann. Mtg., Boston, Mass.*

Singer S J and Nicolson G L 1972 The fluid mosaic model of the structure of cell membranes *Science* **175** 720–31

Takashima S 1978 Frequency domain analysis of asymmetry current *Biophys. J.* **22** 115–19

—— 1979 Admittance change of squid axon during action potentials: change in capacitance component due to sodium current *Biophys. J.* **26** 133–42

—— 1985 Passive electrical properties and voltage dependent membrane capacitance of single skeletal muscle fibers *Pflugers Archiv* **403** 197–204

Takashima S, Asami K and Takahashi Y 1988 Frequency domain studies of impedance characteristics of biological cells using micropipet technique *Biophys. J.* **54** 995

Takashima S and Schwan H P 1974a in *Liquid Crystals and Ordered Fluids* vol 2, ed. J F Johnson and R S Porter (New York: Plenum) pp199–209

—— 1974b Passive electrical properties of squid axon membrane *J. Membrane Biol.* **17** 51–68

Takashima S and Yantorno R E 1977 Investigation of voltage dependent membrane capacity of squid axon *Ann. New York Acad. Sci.* **303** 306–21

Tasaki I and Hagiwara S 1957 Capacity of muscle fiber membrane *Am. J. Physiol.* **188** 423–9

Taylor R E 1965 Impedance of the squid axon membrane *J. Cell. Comp. Physiol.* **66** 21–6

Tosteson D C, Gunn R B and Wieth J O 1973 in *Erythrocytes, Thrombocytes, Leukocytes* ed. E Gerlach, E Moser, E Deutsch and W Wilmanns (Stuttgart: Thieme)

Valdiosera R, Clausen C and Eisenberg R S 1974 Measurement of the impedance of frog skeletal muscle fibres *Biophys. J.* **14** 249–334

Vaughan P C, Howell J N and Eisenberg R S 1972 The capacitance of skeletal muscle fibers in solutions of low ionic strength *J. Gen. Physiol.* **59** 347–59

White S H and Thompson T E 1973 Capacitance, area and thickness variations in thin lipid films *Biochim. Biophys. Acta* **323** 7–22

Appendix A

A brief discussion of Hückel molecular orbital (HMO) theory is given below. HMO is the simplest of the molecular orbital theories. A discussion of more advanced molecular orbital theories is found in Streitwieser (1961) and Salem (1966).

Hückel Molecular Orbital Theory

The molecular orbitals of conjugated organic molecules are defined as a linear combination of atomic orbitals as shown by equation (A.1), where C_1, C_2, ... are normalized coefficients which satisfy equation (A.2):

$$\Psi = C_1\phi_1 + C_2\phi_2 + \ldots + C_n\phi_n \qquad (A.1)$$

$$C_1^2 + C_2^2 + \ldots + C_n^2 = 1. \qquad (A.2)$$

These wave functions are the eigenfunctions of a one-electron Hamiltonian operator H:

$$H = -\frac{h^2}{8\pi^2 m}\nabla^2 + \frac{e^2}{r} \qquad (A.3)$$

where h is Planck's constant, m is the reduced mass, r is the distance and e is electronic charge. The energy associated with the molecular orbital can be given by

$$E = \int \Psi H \Psi \, d\tau \left(\int \Psi\Psi \, d\tau\right)^{-1} \geq E_0 \qquad (A.4)$$

where E and E_0 are the calculated and real energies. The variation principle states that the calculated energy E is always greater than or equal to the real energy E_0. Therefore, the coefficients of equation (A.1) must be so chosen as to minimize E:

$$\frac{\partial E}{\partial C_j} = 0 \qquad C_j = C_1, C_2, \ldots, C_j. \qquad (A.5)$$

Substituting (A.1) into (A.4) and rearranging, we obtain

$$E = \sum_r \sum_s C_r C_s \int \phi_r H \phi_s \, d\tau \left(\sum_r \sum_s C_r C_s \int \phi_r \phi_s \, d\tau\right)^{-1}. \qquad (A.6)$$

For brevity, we introduce the following notation:

$$H_{rs} = \int \phi_r H \phi_s \, d\tau \tag{A.7}$$

$$S_{rs} = \int \phi_r \phi_s \, d\tau. \tag{A.8}$$

Substituting in (A.6) we obtain

$$E = \sum_r \sum_s C_r C_s H_{rs} \left(\sum_r \sum_s C_r C_s S_{rs} \right)^{-1} \tag{A.9}$$

or

$$\sum \sum C_r C_s H_{rs} - E \sum \sum C_r C_s S_{rs} = 0. \tag{A.10}$$

Differentiating (A.10) with coefficients C_j and equating the derivatives to zero (see equation (A.5)), we obtain

$$C_1(H_{11} - ES_{11}) + C_2(H_{12} - ES_{12}) + \ldots + C_n(H_{1n} - ES_{1n}) = 0$$

$$C_1(H_{21} - ES_{21}) + C_2(H_{22} - ES_{22}) + \ldots + C_n(H_{2n} - ES_{2n}) = 0$$

$$\vdots \qquad\qquad \vdots \qquad\qquad \vdots \qquad\qquad \vdots$$

$$C_1(H_{n1} - ES_{n1}) + C_2(H_{n2} - ES_{n2}) + \ldots + C_n(H_{nn} - ES_{nn}) = 0. \tag{A.11}$$

In these equations, H_{rr} and H_{rs} are called Coulomb and resonance integrals respectively. Also, S_{rr} and S_{rs} are called overlap integrals. At this stage, we introduce various simplifications. Namely, if atoms r and s are not bonded, we assume that the resonance integral H_{rs} is numerically small and can be ignored, i.e. $H_{rs} = 0$. Also the overlap integral S_{rs} is either 1 or 0 depending upon whether $r = s$ or $r \neq s$. With these simplifications and introducing new symbols $\alpha = H_{rr}$ and $\beta = H_{rs}$, we can rewrite (A.11) as follows:

$$C_1(\alpha - E) + C_2\beta + \ldots + C_n\beta = 0$$

$$C_1\beta + C_2(\alpha - E) + \ldots + C_n\beta = 0$$

$$\vdots \qquad \vdots \qquad\qquad \vdots \qquad \vdots$$

$$C_1\beta + C_2\beta + \ldots + C_n(\alpha - E) = 0. \tag{A.12}$$

This equation has non-trivial roots only if the following determinant is zero:

$$\begin{vmatrix} \alpha - E & \beta & \cdots & \beta \\ \beta & \alpha - E & \cdots & \beta \\ \vdots & \vdots & & \vdots \\ \beta & \beta & \cdots & \alpha - E \end{vmatrix} = 0. \tag{A.13}$$

Using the method discussed above, let us write a determinant for the benzene molecule (see figure A.1)

$$
\begin{array}{c|cccccc|}
 & 1 & 2 & 3 & 4 & 5 & 6 \\
\hline
1 & \alpha - E & \beta & 0 & 0 & 0 & \beta \\
2 & \beta & \alpha - E & \beta & 0 & 0 & 0 \\
3 & 0 & \beta & \alpha - E & \beta & 0 & 0 \\
4 & 0 & 0 & \beta & \alpha - E & \beta & 0 \\
5 & 0 & 0 & 0 & \beta & \alpha - E & \beta \\
6 & \beta & 0 & 0 & 0 & \beta & \alpha - E \\
\end{array} = 0.
$$

(A.14)

We can divide the determinant by a constant (β in this case) without changing its value. Introducing another symbol $x = (\alpha - E)/\beta$, we can rewrite the above determinant as follows:

$$
\begin{vmatrix}
x & 1 & 0 & 0 & 0 & 1 \\
1 & x & 1 & 0 & 0 & 0 \\
0 & 1 & x & 1 & 0 & 0 \\
0 & 0 & 1 & x & 1 & 0 \\
0 & 0 & 0 & 1 & x & 1 \\
1 & 0 & 0 & 0 & 1 & x \\
\end{vmatrix} = 0.
$$

(A.15)

Solving this equation, we obtain the intrinsic roots in the form of

$$E = \alpha - x\beta.$$

(A.16)

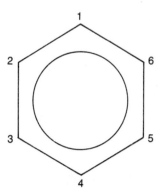

Figure A.1 The structure of a benzene molecule. The carbon atoms are numbered for the calculation of molecular orbitals.

The energy E consists of two terms, i.e. α and $x\beta$. While α represents the energy of non-bonded atoms, the second term represents the stability of molecular orbitals. The integral β is a negative quantity.

Therefore, if the value of x is negative, the second term gives rise to a decrease in energy, i.e. a stability of bonds. On the other hand, if the value of x is positive, the second term increases the total energy and the molecular orbital is unstable, i.e. antibonding.

It is customary to use spectroscopic data to evaluate resonance integrals empirically. For example, if we plot the frequency of the first $\pi-\pi$ transition of a polyene series against the HMO energy differences in units of β, we obtain a straight line. From the slope of this plot, the value of β is determined to be -3.08 eV (-71.1 kcal mol^{-1}). However the value of the β-unit depends upon whether the molecules are aliphatic or aromatic. In general, the β-unit for aromatic compounds such as benzene is smaller than that of aliphatic compounds and the value of β is estimated to be -0.69 eV (-16 kcal mol^{-1}). Once the latent roots are found, they are substituted into (A.12), and the coefficients C_1, C_2, ..., C_n are calculated by solving simultaneous equations as Appendix B explains.

References

Salem L 1966 *The Molecular Orbital Theory of Conjugated Systems* (Menlo Pk: W A Benjamin)

Streitwieser A Jr 1961 *Molecular Orbital Theory for Organic Chemists* (New York: Wiley)

Appendix B

Calculation of the Molecular Orbitals of Butadiene

The structure of butadiene is shown in figure B.1. By inspection of this structure, we can set up the secular equation at once as shown below:

$$\begin{vmatrix} x & 1 & 0 & 0 \\ 1 & x & 1 & 0 \\ 0 & 1 & x & 1 \\ 0 & 0 & 1 & x \end{vmatrix} = 0. \tag{B.1}$$

Figure B.1 The structure of butadiene. Spheres represent carbon atoms.

In butadiene, the resonance integrals β_{13}, β_{24}, and β_{14} are zero because these integrals are formed between non-bonded atoms. Solving this equation, we obtain a set of latent roots:

$$x = -1.618, -0.618, 0.618, 1.618.$$

Substituting these roots into (A.16), we obtain

$$E_1 = \alpha + 1.618\beta$$
$$E_2 = \alpha + 0.618\beta$$
$$E_3 = \alpha - 0.618\beta$$
$$E_4 = \alpha - 1.618\beta. \tag{B.2}$$

These energy levels are illustrated schematically in figure B.2. In order to calculate the coefficients of molecular orbitals, we substitute for the latent roots in (B.1). For example, if $x = -1.618$

$$-1.618 \, C_1 + C_2 = 0$$
$$C_1 - 1.618 \, C_2 + C_3 = 0$$
$$C_2 - 1.618 \, C_3 + C_4 = 0$$
$$C_3 - 1.618 \, C_4 = 0. \tag{B.3}$$

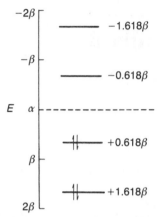

Figure B.2 The energy levels of the molecular orbitals of butadiene. (After Streitwieser 1961.)

Solving these equations simultaneously, we obtain values for C_1, C_2, C_3 and C_4 with the normalization condition

$$C_1^2 + C_2^2 + C_3^2 + C_4^2 = 1. \tag{B.4}$$

Repeating the same procedure for all of the four latent roots, we finally obtain the complete solution:

$$E_1 = \alpha + 1.618\beta \qquad \psi_1 = 0.371\phi_1 + 0.600\phi_2 + 0.600\phi_3 + 0.371\phi_4$$

$$E_2 = \alpha + 0.618\beta \qquad \psi_2 = 0.600\phi_1 + 0.371\phi_2 - 0.371\phi_3 - 0.600\phi_4$$

$$E_3 = \alpha - 0.618\beta \qquad \psi_3 = 0.600\phi_1 - 0.371\phi_2 - 0.371\phi_3 + 0.600\phi_4$$

$$E_4 = \alpha - 1.618\beta \qquad \psi_4 = 0.371\phi_1 - 0.600\phi_2 + 0.600\phi_3 - 0.371\phi_4.$$

$$\tag{B.5}$$

Using these molecular orbitals, we can calculate the dipole moment of butadiene which, because of its symmetry, is zero. For the calculation of polarizability, excited- as well as ground-state molecular orbitals are needed. Whereas the dipole moments of symmetric molecules vanish, the polarizabilities of these molecules are non-zero.

References

Streitwieser A Jr 1961 *Molecular Orbital Theory for Organic Chemists* (New York: Wiley)

Appendix C

Normally the delta function is defined as

$$\delta(t - t_0) = 0 \qquad \text{for } t \neq t_0 \qquad \text{(C.1)}$$

and

$$\int_{-\infty}^{\infty} \delta(t - t_0)\, dt = 1. \qquad \text{(C.2)}$$

The first equation indicates that the δ-function has undefined magnitude for $t = t_0$ and is zero otherwise. The second equation indicates that the area under the function is unity. Combining these two properties, the δ-function appears as an impulse with an infinite height.

Appendix D

Determination of the Distribution Function

As already discussed, the use of empirical parameters in the Cole–Cole equation (3.71) and the Fuoss–Kirkwood equation (3.88) is equivalent to using the distribution functions shown by equations (3.76) and (3.91). These functions can be derived using the following method (Fuoss and Kirkwood 1941). First of all, equation (3.53) can be rewritten as follows:

$$\frac{\varepsilon - \varepsilon_\infty}{\varepsilon_0 - \varepsilon_\infty} = Q(\omega) \tag{D.1}$$

where

$$Q(\omega) = \int_0^\infty \frac{G(\omega)}{1 + j\omega\tau} d\tau. \tag{D.2}$$

We can separate, as before, the real and imaginary parts of (D.1), i.e.

$$Q(\omega) = J(\omega) - jH(\omega) \tag{D.3}$$

where

$$J(\omega) = \int_0^\infty \frac{G(\tau)}{1 + (\omega\tau)^2} d\tau \tag{D.4}$$

and

$$H(\omega) = \int_0^\infty \frac{G(\tau)\omega\tau}{1 + (\omega\tau)^2} d\tau. \tag{D.5}$$

We can rewrite (D.4) and (D.5) introducing two new variables, i.e. $x = -\log \omega/\omega_m$ and $s = \log \tau/\tau_m$ (where ω_m is the frequency at which $H(x)$ is maximum and $\tau_m = 1/\omega_m$):

$$J(x) = \int_{-\infty}^\infty \frac{F(s)}{1 + e^{2(s-x)}} ds \tag{D.6}$$

$$H(x) = \int_{-\infty}^\infty \frac{F(s)e^{(s-x)}}{1 + e^{2(s-x)}} ds \tag{D.7}$$

where $F(s) = \tau G(\tau)$. The function $F(s)$ must satisfy the normalization condition, namely

$$\int_{-\infty}^{\infty} F(s)\,ds = 1. \tag{D.8}$$

Equation (D.7) can be simplified by introducing

$$\text{sech}(s - x) = \frac{2}{e^{(s-x)} + e^{-(s-x)}}. \tag{D.9}$$

Substituting this expression into (D.7), we obtain the following equation:

$$H(x) = \int_{-\infty}^{\infty} \text{sech}(s - x)F(s)\,ds \tag{D.10}$$

which is of the form

$$H(x) = \int_{-\infty}^{\infty} K(s - x)F(s)\,ds. \tag{D.11}$$

The function $F(s)$ can be determined if we know the function $H(x)$. The solution of (D.11) has been obtained as

$$F(s) = \frac{1}{\pi}\left[H\left(s + j\frac{\pi}{2}\right) + H\left(s - j\frac{\pi}{2}\right)\right]. \tag{D.12}$$

Let us apply this equation to Fuoss–Kirkwood theory to derive the distribution function (3.91). Equation (3.88) can be rewritten as follows:

$$H(x) = H(O)\,\text{sech}(\beta x). \tag{D.13}$$

Substituting $x = s + j(\pi/2)$ and $x = s - j(\pi/2)$ into this equation and using (D.12), we obtain

$$F(s) = H(O)\left(\frac{1}{\cosh(\beta s + j\beta\pi/2)} + \frac{1}{\cos(\beta s - j\beta\pi/2)}\right)\frac{1}{\pi}. \tag{D.14}$$

Using the relation

$$\cosh(\beta s + j\beta\pi/2) = \cosh(\beta s)\cos(\beta\pi/2) + j\sinh(\beta s)\cos(\beta\pi/2) \tag{D.15}$$

equation (D.15) is found to be

$$F(s) = \frac{2}{\pi}H(O)\frac{\cos(\beta\pi/2)\cosh(\beta s)}{\sinh^2(\beta s) + \cos^2(\beta\pi/2)}. \tag{D.16}$$

The function $H(O)$ can be determined using the normalization condition (equation (D.8))

$$\frac{2}{\pi}H(O)\int_{-\infty}^{\infty} \frac{\cos(\beta\pi/2)\cosh(\beta s)}{\sinh^2(\beta s) + \cos^2(\beta\pi/2)}\,ds = 1. \tag{D.17}$$

The value of this integral is π/β. Thus, on substituting it into (D.17), we obtain

$$H(O) = \beta/2 \tag{D.18}$$

which when substituted into (D.16) gives

$$F(s) = \frac{\beta}{\pi} \frac{\cos(\beta\pi/2)\cosh(\beta s)}{\sinh^2(\beta s) + \cos^2(\beta\pi/2)}. \tag{D.19}$$

This is identical to (3.91).

The same method can be applied to the Cole–Cole equation. Defining ε''_m as the maximum value of ε'' at which the frequency ω is ω_m then

$$\frac{\varepsilon''}{\varepsilon''_m} = \frac{1 + \cos(n\pi/2)}{\cosh(nx) + \cos(n\pi/2)}. \tag{D.20}$$

In this equation, ε'' and ε''_m are equivalent to $H(x)$ and $H(O)$, therefore

$$H(x) = H(O) \frac{1 + \cos(n\pi/2)}{\cosh(nx) + \cos(n\pi/2)}. \tag{D.21}$$

Substitution of $x = s + j\pi/2$ and $x = s - j\pi/2$ into this equation leads to the following expression:

$$F(s) = \frac{2}{\pi} H(O) \frac{\cos(n\pi/2)(1 + \cos(n\pi/2))}{\cosh(ns) + \cos(n\pi)}. \tag{D.22}$$

As before, $H(O)$ can be determined using the normalization condition, i.e.

$$H(O) = \frac{1}{4} \frac{\sin n\pi}{\cos(n\pi/2)(1 + \cos(n\pi/2))}. \tag{D.23}$$

Substituting this into (D.22) we obtain the distribution function for the Cole–Cole equation:

$$F(s) = \frac{1}{2\pi} \frac{\sin n\pi}{\cosh(ns) + \cos(n\pi)}. \tag{D.24}$$

As has been emphasized, the physical significance of these functions is not understood. However, it has been proven experimentally that the distribution of relaxation times for many dielectric materials can be explained phenomenologically using these functions in dispersion equations.

References

Fuoss R M and Kirkwood J G 1941 Electrical properties of solids. VIII. Dipole moments in polyvinyl chloride-diphenyl systems *J. Am. Chem. Soc.* **63** 385–94

Appendix E

The Four-electrode Technique

Most of the dielectric measurements discussed so far are performed using two-electrode techniques. In this method, two electrodes of various geometries are used for the measurements of capacitance and resistance of dielectric samples. If the sample is non-conductive, as are most organic molecules, the presence of electrode surfaces does not produce an additional capacitive component. However, if the sample is conductive, as are most aqueous biological samples, there will be an accumulation of charge on the surfaces of the electrodes and this entails the formation of electrical double layers. Under this condition, the equivalent circuit of conductive solutions placed between two electrodes is illustrated schematically by figure E.1. C_p and R_p are the capacitance and the resistance of the double layer respectively. Often this is called electrode polarization. The presence of electrode polarization in series with sample admittance causes a marked increase in the measured total capacitance and resistance at low frequencies. The magnitude of the capacitance due to electrode polarization is so large that the correction for it is one of the major steps toward successful measurement with conductive samples. However, the correction is very cumbersome and, so far, no simple technique has been available.

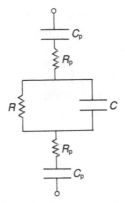

Figure E.1 The equivalent circuit for a conductive dielectric sample with electrode polarization impedance, C_p and R_p.

The use of the four-electrode technique was first proposed by Schwan (1963) as a potentially useful method for dielectric measurement with conductive biological materials. However, the first successful measurement was reported by Hanss (1973) with DNA solution, as discussed earlier. In the following, the system developed by Hayakawa *et al* (1975) will be discussed in some detail. In the four-electrode technique, two pairs of metal electrodes are used. One pair is used to inject AC current, I, into the sample and the other pair is used as potential electrodes. The potential electrodes are followed by differential amplifiers having a very high input impedance. This enables the detection of induced potential difference V without interference due to electrode polarization. The ratio V/I is the impedance of the sample Z and its real and imaginary parts are given by the following equations:

$$Z' = -\frac{1}{C_0}\left(\frac{\varepsilon_0}{\sigma}\right)^2 \omega\left[\left(\frac{\sigma}{\omega\varepsilon_0} - \varepsilon''\right) + \frac{R_p}{\omega}\left(\frac{\sigma}{\varepsilon_0}\right)^2 C_0\right] \qquad (E.1)$$

$$Z'' = \frac{1}{C_0}\left(\frac{\varepsilon_0}{\sigma}\right)^2 \omega\left[\varepsilon' + \frac{1}{\omega^2 C_p}\left(\frac{\sigma}{\varepsilon_0}\right)^2 C_0\right] \qquad (E.2)$$

where C_0 is the vacuum capacitance of the sample cell and ε_0 is the permittivity of free space. The second term in brackets in the above equations is due to electrode polarization and, therefore, these terms reduce to zero, in theory, for four-electrode systems. The block diagram of this system is shown in figure E.2. Using E_1 and E_4, the sum of two AC currents with frequencies ω and ω_0, measuring and reference frequencies are injected into the sample. The use of a reference signal stems from the necessity to correct for the conductance drift of the specimen. The impedance difference at measuring and reference frequencies is given by

$$\Delta Z'(\omega) = -\frac{1}{C_0}\left(\frac{\varepsilon_0}{\sigma}\right)^2 \Delta(\omega\varepsilon''). \qquad (E.3)$$

The amplitudes of these currents are kept constant and equal. The induced potential V across the potential electrodes E_2 and E_3 is detected by a differential amplifier. The potential which is primarily due to the DC conductance of the specimen is eliminated by a summing amplifier with a proper feedback resistor R', equal to $\Delta Z'(\omega_0)$, the specimen impedance at ω_0. The output of the summing amplifier V_{sig} is defined by the following equation:

$$V_{sig} = \hat{I}[(Z'(\omega) - R')\cos\omega t + Z''(\omega)\sin\omega t$$
$$+ (Z'(\omega_0) - R')\cos\omega_0 t + Z''(\omega_0)\sin\omega_0 t]. \qquad (E.4)$$

If R' is exactly equal to $Z'(\omega_0)$, the impedance $Z'(\omega)$ would be

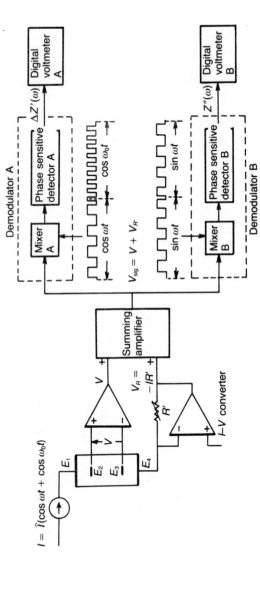

Figure E.2 A block diagram for the four-electrode system. See the text for detail (Hayakawa *et al* 1975). Reproduced by permission of the *Japanese Journal of Applied Physics*.

obtained by demodulating V_{sig} with respect to $\cos \omega t$. Another technique which is added in this system is to implement further matching of R' with $Z'(\omega_0)$. The reference signal to mixer A is a rectangular wave with two alternating frequencies between ω and ω_0 in every half-switching cycle. The rectangular wave is in phase with $\cos \omega t$ during one half-switching cycle, and then is in phase with $\cos \omega_0 t$ during the other half-cycle. For mixer B, the reference signal is a rectangular wave which is in phase with $\sin \omega t$ and $-\sin \omega t$ in each half-cycle. The output signals of mixers A and B are $Z'(\omega) \cos \Omega t$ and $Z''(\omega) \cos \Omega t$ where Ω is the switching frequency. From these we obtain the signals $\Delta Z'(\omega)$ and $Z''(\omega)$ by use of the phase sensitive detectors A and B. The values of ε' and ε'' for the sample are calculated from these quantities. The measuring cell used for the four-electrode system is illustrated in figure E.3.

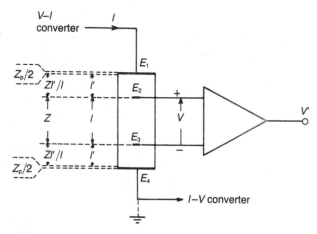

Figure E.3 A sample cell with current (E_1 and E_4) and voltage (E_2 and E_3) electrodes. A differential amplifier is shown. (Hayakawa *et al* 1975). Reproduced by permission of the *Japanese Journal of Applied Physics*.

A recent paper by Cole (1977) discusses another four-electrode system. Applying this system to DNA solution, Cole measured the dielectric constant at 3 Hz, without the effect of electrode polarization.

References

Cole R H 1977 Dielectric theory and properties of DNA in solution *Ann. New York Acad. Sci.* **303** 59–73

Hanss M 1966 Mesure de la dispersion dielectrique en tres basse frequence de solutions de DNA *Biopolymers* **4** 1035–41

Hayakawa R, Kanda H, Sakamoto M and Wada Y 1975 New apparatus for measuring the complex dielectric constant of a highly conductive material *Japan. J. Appl. Phys.* **14** 2039–52

Schwan H P 1963 in *Physical Techniques in Biological Research* vol 6, ed. W L Nastuk (New York: Academic) p323

Appendix F

Use of Pseudo-random Noise for Dielectric Measurement

The measurement of dielectric constant and conductivity can be performed in a time domain or in a frequency domain. For frequency domain measurements, ordinarily, impedance or admittance bridges are used. Some instruments can be operated manually with great precision. However, present day measurement systems can be operated using an on-line computer although their resolution may be somewhat lower than manual systems. May it be manual or automated, we still have to select frequencies sequentially and repeat the measurements at each frequency. On the other hand time domain methods use various types of electrical excitation and calculate the transfer function in order to obtain desired dielectric parameters.

The excitation signal must satisfy the following conditions: (i) wide band—all frequencies of interest are simultaneously present with sufficient energy density; (ii) low peak-to-average power ratio and (iii) deterministic without statistical fluctuations. The stimulus has a precisely known magnitude and phase characteristics. From these, the excitation signal $S(n)$ can be considered periodic of period N with the total number of sample points $T = N\Delta t$. Periodic signals can be expressed by a Fourier series:

$$S(n) = \sum_{k}(A_S(k)\cos 2\pi kn + B_S(k)\sin 2\pi kn) \qquad (F.1)$$

where k is the index of discrete components with frequencies $1/T$, $2/T$, ..., k/T. The Fourier transformed signal $S(k)$ can be expressed alternately by a complex form, as a product of magnitude and phase components:

$$S(k) = \text{Mag}_S(k)\,\text{Arg}_S(k) \qquad (F.2)$$

where

$$\text{Mag}_S(k) = [A_S^2(k) + B_S^2(k)]^{1/2} \qquad (F.3)$$

and

$$\text{Arg}_S(k) = \tan^{-1}(B_S(k)/A_S(k)). \qquad (F.4)$$

Likewise, the transfer function of the system $H(k)$ and the steady state system response $R(k)$ can be defined as

$$H(k) = \text{Mag}_H(k)\,\text{Arg}_H(k) \qquad (F.5)$$

$$R(k) = \text{Mag}_R(k)\,\text{Arg}_R(k). \qquad (F.6)$$

From linear system theory, the unknown function $H(k)$ is measured directly as

$$\text{Mag}_H(k) = \text{Mag}_R(k)/\text{Mag}_S(k) \qquad (F.7)$$

$$\text{Arg}_H(k) = \text{Arg}_R(k) - \text{Arg}_S(k). \qquad (F.8)$$

The implicit assumption is that both the stimulus and the response are defined on precisely the same discrete time axis n and over the same period of length N. Obviously from (F.7) and (F.8), it would be convenient if $\text{Mag}_S(k) = \text{constant}$ and $\text{Arg}_S(k) = 0$. Although impulses are a natural choice as a stimulus, this must be rejected since their peak/average ratio is too high. Instead, a pseudo-random binary sequence (PRBS) is selected by Husimi and Wada (1976) and by Poussart and Ganguly (1977) in independent studies. Although there are some differences between the techniques developed by these two groups, the principle underlying the techniques is basically the same. The current discussion is, however, primarily based on the work by Poussart and Ganguly.

The key feature of a PRBS is that it can be made to approximate some of the statistical properties of a truly random process, in particular, its magnitude is the same as that of a discrete impulse while its time wave form has a 1:1 peak/average ratio. PRBS can be generated with $N = 1023$ by a technique of a ten-stage digital shift register with appropriate feedback. A digital discrete Fourier transform (DFT) implemented by the FFT algorithm is based on a base 2 sequence, i.e. 2^p. However, the period of PRBS can only be of the length $2^p - 1$. This discrepancy is solved by using time expansion–compression in order to precisely match the total period of a PRBS sequence ($p = 10$, 1023 clock pulses covering T seconds) with the duration of the time window of 1024 sampling pulses, also covering T seconds.

Figure F.1 illustrates the magnitude functions of various stimulus signals. Part (a) is white random noise. The large statistical fluctuations which would require averaging are clear. Part (b) is the impulse. Part (c) is a PRBS without synchronization as discussed above. The 1023 bit sequence, augmented by one additional bit is loaded into the 1024 point DFT. Note that even a minute mismatch introduces considerable deviation from the flat spectrum at high frequencies. Part (d) is a PRBS with synchronization with a DFT period. Note the near-perfect magnitude baseline, flat and smooth. If the input PRBS signal is properly Fourier transformed and the coefficients $A_S(k)$ and $B_S(k)$ calculated, we can determine the functions $\text{Mag}_S(k)$ and $\text{Arg}_S(k)$ using (F.3) and (F.4).

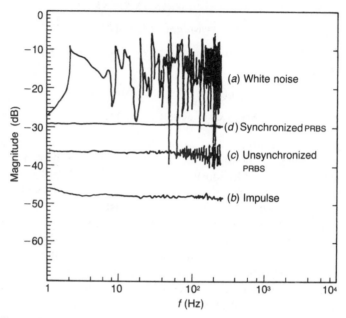

Figure F.1 An illustration of various wideband stimulus signals. The curves are shifted along the ordinate for the sake of clarity. (From Poussart and Ganguly 1977.) © 1977 IEEE.

The system discussed above has been applied to the investigation of the impedance of squid giant axon. A small PRBS stimulus (2 mV peak-to-peak) was superimposed on the voltage–clamp-voltage step function. After the initial transient had elapsed, the steady state response is gated to the DFT sampling circuit. The full dynamic range of the instrument is preserved by an automatic baseline compensation. With response functions Mag_R and Arg_R thus determined, the transfer functions Mag_H and Arg_H can be calculated using (F.7) and (F.8). The final data are displayed as the amplitude of admittance $|Y|$ and phase angle $\angle Y$. The data can also be displayed as the capacitance and conductance of the sample, as done by Husimi and Wada. The results obtained with squid axon are shown in figure F.2 as an example.

References

Husimi Y and Wada A 1976 Time domain measurement of dielectric dispersion

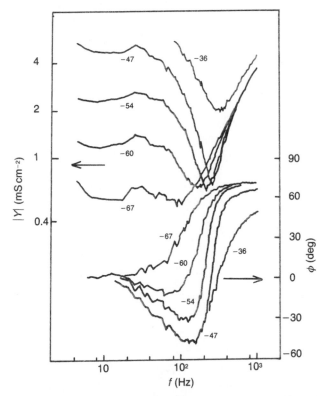

Figure F.2 The complex admittance measured with squid giant axon membrane for different mean clamping potentials in mv. The acquisition time for each pair of magnitude and phase functions is 250 ms. (From Poussart and Ganguly 1977.) © 1977 IEEE.

as a response to pseudorandom noise *Rev. Sci. Instrum.* **47** 213–19
Poussart D and Ganguly U S 1977 Rapid measurement of system kinetics. An instrument for real-time transfer function analysis *Proc. IEEE* **65** 741–7

Author Index

Subject Index